THEORY OF THE EARTH

THEORY OF THE EARTH

Thomas Nail

Stanford University Press
Stanford, California

STANFORD UNIVERSITY PRESS
Stanford, California

©2021 by the Board of Trustees of the Leland Stanford Junior University.
All rights reserved.

No part of this book may be reproduced or transmitted in any form or by any means, electronic or mechanical, including photocopying and recording, or in any information storage or retrieval system without the prior written permission of Stanford University Press.

Printed in the United States of America on acid-free, archival-quality paper

Library of Congress Cataloging-in-Publication Data
Names: Nail, Thomas, author.
Title: Theory of the Earth / Thomas Nail.
Description: Stanford, California : Stanford University Press, 2021. | Includes bibliographical references and index.
Identifiers: LCCN 2020037817 (print) | LCCN 2020037818 (ebook) | ISBN 9781503614956 (cloth) | ISBN 9781503627550 (paperback) | ISBN 9781503627567 (epub)
Subjects: LCSH: Geology—Philosophy. | Geology, Stratigraphic—Anthropocene. | Earth (Planet)—History—Philosophy.
Classification: LCC QE6 .N35 2021 (print) | LCC QE6 (ebook) | DDC 551.01—dc23
LC record available at https://lccn.loc.gov/2020037817
LC ebook record available at https://lccn.loc.gov/2020037818

Cover design: Rob Ehle

Cover image: Xiaojiao Wang | Shutterstock

Typeset by Kevin Barrett Kane in 10.2/14.4 Minion Pro

Contents

List of Illustrations vii

Introduction   1

Part I: Geokinetics: The Kinetic Earth

1   The Flow of Matter   19

2   The Fold of Elements   32

3   The Planetary Field   42

Part II: History of the Earth

A. Mineral Earth

4   Centripetal Minerality   61

5   Hadean Earth   74

B. Atmospheric Earth

6   Centrifugal Atmospherics   91

7   Archean Earth I: Pneumatology   105

8   Archean Earth II: Biogenesis   116

C. Vegetal Earth

9   Tensional Vegetality   135

10   Proterozoic Earth   153

D. Animal Earth

11 Elastic Animality   177
12 Phanerozoic Earth I: Kinomorphology   191
13 Phanerozoic Earth II: Terrestrialization   208

Part III: The Kinocene: A Dying Earth

14 Kinocene Earth   227
15 Kinocene Ethics   242
    Conclusion: The Future   271

Notes 277
Index 319

Illustrations

TABLE I.1	Theory, form, content, and patterns of motion	13
TABLE I.2	Patterns of motion	14
FIGURE 2.1	Lorenz attractor	36
FIGURE 2.2	Cycle and period	38
FIGURE 2.3	Conjunction, thing, object, and image	40
FIGURE 3.1	Field of circulation	48
FIGURE 3.2	Knots	49
FIGURE 10.1	Mushrooms are fruiting bodies made of hyphae	157
FIGURE 10.2	Dandelion vortex	172
FIGURE 11.1	Subcutaneous tissue from a young rabbit, highly magnified	185
FIGURE 12.1	Dendritic structure of material evolution	195
FIGURE 12.2	Choanoflagellates	197
FIGURE 12.3	Sponge vortex	200
FIGURE 12.4	Jellyfish vortex	202
FIGURE 13.1	Animal dendrite body	212
FIGURE 15.1	Declining phytomass totals through history in gigatons of carbon	246
FIGURE 15.2	Human energy expenditure through history in yottajoules	248
FIGURE 15.3	Terrestrial energy expenditure through history in yottajoules	252

THEORY OF THE EARTH

Introduction

WE NEED A NEW THEORY OF THE EARTH. Most people are accustomed to treating the earth as a relatively stable place that they live on and move on. Today, however, this stable ground is becoming increasingly unstable—for some of us more than others.[1]

Due to the widespread use of global transportation technologies, for example, there are now more people and things on the move than ever before in history. Vast amounts of materials are in constant circulation as billions of humans ship plants, animals, and technologies around the world. More than half the world's plant and animal species have now been forced into migration due to climate change.[2] The earth is becoming so mobile that even its glaciers are speeding up. Karl Marx was not thinking of receding glaciers or greenhouse gases when he said "all that is solid melts into air," but that is what is happening.

Geological time used to refer to slow, gradual processes, but today we are watching the land sink into the sea and forests transform into deserts in our lifetimes. We can even see the creation of entirely new geological strata made of plastic, chicken bones, and other waste that could remain in the fossil record and affect geological formations for thousands, even millions, of years to come.[3]

Some human groups are now changing the entire earth so dramatically and permanently that geologists have begun calling our age the

Anthropocene.[4] It no longer makes sense to think of humans as transient occupants moving on a relatively stable earth. Humans are geological, atmospheric, and hydrological agents entangled in all the earth's processes, which are now increasingly in flux. The arrival of the Anthropocene, more than any human historical event, is finally awakening us to the realization that we have never lived on a stable earth. There is significant literature now on climate change and the role of humans as geological agents.[5] Nevertheless, I argue that the most radical import of the Anthropocene is the unpredictable agency and mobility of the earth itself.

In other words, defining the Anthropocene by human historical markers such as agriculture, the industrial revolution, and nuclear bombs should not cause us to lose sight of the most important lesson of our time. *Nature and humans have never been separate systems.* The Anthropocene is not only about humans and what they have done to the earth. It is about the earth and what it is doing to itself through humans.

However, the participation of the earth in climate change in no way negates the need for ethical action on the part of humans.[6] Climate change is a significant problem that demands radical social change. Some historical actors and social systems are particularly responsible for ecological destruction, while others are disproportionally affected by its consequences.[7] But these problems will not be solved using our old paradigm of humans *as separate from* nature. New epochal problems require new philosophical and historical orientations, which is why this book tries to provide a new theory and ethics of the earth for the present.

We tend to think of the world in terms of stasis rather than process. In our zeal to halt our runaway energy consumption, we act as if the goal were to conserve, accumulate, and stabilize energy use. And yet humans, as part of nature, have evolved alongside other life forms in a way that maximizes our collective energy use, flow, and movement.

But it has gotten to the point now that we won't even let our trash degrade. We make things from plastics that last for tens of thousands of years and then bury them underground. Vast islands of plastic are floating in our oceans like quasi-immortal beasts. The net effect of all this is that the planet's *own* energy consumption is *slowing down*, with disastrous consequences.

We continue to think of the earth in terms of stability and conservation, against our best interests. This book is motivated by the advent of increasing planetary mobility, which pushes us to think about the earth and its history in a whole new way. We need a different history and ethics that will help us to go *with* the flow of planetary energy processes, not *against* them.

TWO PROBLEMS

The Scottish "father of modern geology," James Hutton (1726–1797), published his groundbreaking work, *Theory of the Earth,* more than two hundred and thirty years ago, in 1788. Hutton wrote at a time when humans knew little about geological processes or the age of the earth. The 18th century was a time when geology was still a wide-open field.

Like all new sciences, geology was mostly theoretical at first. Over time, it was separated from philosophy and made into a physical science. As more time passed, other sciences, such as chemistry, physics, biology, and cosmology, had philosophy turn its attention back to them, but geology has still not become a subject of philosophy again.[8] To my knowledge, there is no definitive book-length work on the philosophy of geology in existence today.[9] However, given our present historical situation, I think it is high time for philosophy to rethink the history of the earth.

I wrote this book because I think there are at least two significant problems with our theories and treatments of the earth in the Western tradition.[10] These problems are at the heart of the current ecological crisis, and whether or not we overcome them will play a significant role in the survival of future planetary forms of life.

Stasis

The first problem is that of stasis. Historically, we have tended to view the earth as the stable object par excellence.[11] Many prehistoric mythologies described the earth as the primordial womb or egg from which all things were born and to which they cyclically return.[12] In the ancient Near East and the classical world, most people thought of the earth as the stable center of the universe, a static sphere upon which the whole cosmos turned.[13] For Copernicus, the earth itself was still a relatively unchanging sphere, even if it rotated around the sun. Even Hutton defined the earth as a profoundly slow, uniform, and relatively stable cycle of balanced change.[14]

The theory of plate tectonics, in the 1960s, was the first major geological revolution to question the stability of the earth itself. However, even then, the near consensus of "uniformitarianism" still described tectonic movements as slow, uniform, and relatively homogeneous. Even when we have acknowledged that the earth moves, we have rarely and only recently begun to acknowledge that the movements of the earth are profoundly and unpredictably affected by, and integrated with, nonlinear and non-geological cosmic, biological, and chemical processes.[15]

We have treated and, in various ways, continue to treat the earth as a kind of unmoved mover.[16] We either act as if our scientific knowledge about the earth is a separate thing, unconditioned by the earth itself, or we think that the earth that existed before us and will exist after us is somehow radically unrelated to us.[17] Most geologists still believe that there are uniform and mechanical laws of geology.[18] Most of us in the West are unconscious uniformitarians. We still act like the earth is largely stable but punctuated by exceptional environmental disasters.[19]

Meanwhile, we have new technologies, including high-precision geochronology and satellite observation, along with detailed data on the earth's temperature, precipitation, river flow, glacier behavior, groundwater reserves, sea level, and seismic activity. We can now directly see that many of the earth's processes are neither as slow nor as constant as we thought.[20] All of our significant predictions about climatic change failed to anticipate how rapid and nonuniform the changes have been so far and how integrated the earth's systems have proven to be.[21] Climate scientists still have no working models to explain sudden "tipping points" in the earth's history, where temperatures suddenly rise 10 to 15 degrees in less than ten years.[22]

Treating the earth as stable, uniformly predictable, linear, or mechanistic allows us to continue to act as if we can pollute it and extract as much as we want from it without significant or uncontrollable consequences. If the earth is just a bunch of mechanical stuff, we can treat it however we want and then mitigate the problems with geo-engineered solutions.[23] So far, however, no such technical fixes exist that are feasible, nor are any likely to appear.[24] As George Bataille once remarked, "All that we recognize as truth is necessarily linked to the error represented by the 'stationary earth.'"[25]

History

The second, related problem is that we have treated the earth as an ahistorical substance lacking genuine novelty.[26] For most of recorded Western history, humans have thought of the earth primarily as a passive object or as the product of natural, divine, or mechanical laws. The natural sciences frequently explain the movements of the earth according to causes other than the earth itself (laws, forces, principles of uniformity, etc.). Geological histories are thus typically histories written about the earth, not histories written as practices of the earth itself.

The anthropocentric assumption is that only when nature becomes aware of itself in the human being can we say that it becomes genuinely historical and meaningful.[27] Western historians have long believed that only humans can have a history, because only humans are self-conscious and genuinely novel agents.[28] People too often think that the earth's systems simply form the backdrop or stage upon which real history, i.e. human history, occurs.[29]

This anthropocentric narrative is evident from the almost complete historical erasure of earth processes as active contributors to some of the most significant events in Western history. The Holocene glacial retreat, the medieval warm period, and the "little ice age" all played significant active roles in shaping human history. Yet historians frequently leave these events out of the books.[30] Earth processes like volcanoes, fires, hurricanes, earthquakes, and tsunamis also continue to shape history in crucial and active ways.[31]

Even when we acknowledge the activity of the earth, we tend to do so while thinking of the earth as a living and vital subject like ourselves.[32] Unfortunately, this is still a biocentric image of the earth. This image misunderstands inorganic matter as being like organic matter when the historical situation is precisely the opposite. The earth is mostly *not* alive. The earth is part of much larger non-living cosmic cycles and patterns that are not fully captured with the idea of the planet as an organic individual (Gaia).[33] The earth is neither in stasis nor in homeostasis; it is neither mechanistic nor vitalistic; it is neither an object nor a subject. Instead, I argue, it is a turbulent process operating far from equilibrium.[34]

I do not think, as some do, that we have arrived at the "end of nature," in which nothing exists unmixed with human activity.[35] The origins of this idea were well-intentioned but wrong and are now potentially dangerous. The idea was that if we emphasized how significant and widespread human intervention in nature was, that would help us see that nature is a human ethical issue we should take seriously.[36] However, the flawed assumption of this position is that reality can be otherwise than it is only through human activity.

Unfortunately, the focus on human structures[37] and human-nature hybrids[38] has tended to obscure the profoundly nonhuman indeterminacy of the earth and the cosmos.[39] Not everything is or has been a human hybrid. Human-nature hybrids are only a very tiny portion of nature.

There is today a marked reluctance (whether implicit or explicit) on the part of humanists and social scientists to interrogate the prehuman material conditions of human beings.[40] Critical and social theories always seem to begin and end with human histories rather than with the deep historical prehuman earth as the turbulent and mobile condition that is immanent to humans themselves.[41]

On the one hand, I think that the geosciences need to recognize the historical and social conditions of their claims about the earth. On the other hand, the social sciences and humanities, in turn, need to recognize the geological conditions of their concepts and social structures.[42] Moreover, both need an immanent critique of the earth as their shared material kinetic condition.[43] For all the recent interest in things and objects in the theoretical humanities, there has been ironically little attention given to the earth.[44]

The danger of starting all our histories with classical Greece or early human evolution is that it gives us an inflated sense of our importance. For example, if humans do not take the earth's deep and turbulent history seriously, we are more likely to think that we can dominate or geo-construct it at will.[45] If we want to overcome the nature-culture duality, we need to start taking the cultural history of the earth seriously.[46] Starting our histories with European modernity or even human history only reasserts an implicit division between nature and humans, whatever we might say to the contrary.

I worry that if we think the earth has no genuine historical agency, we may foolishly think that it can have no real effect on human history. Natural scientists often treat earth systems as passive mechanical processes following universal laws, punctuated by random changes. However, we ignore the truly indeterminate movement of the earth at our peril. The deep history of the earth is not a secondary or derivative history merely told by humans about something that they are not. The earth is the immanent material condition of human historicity itself.[47] Humans are the earth and therefore bear its history. In my view, our ability to see this ought to be the real point of the Anthropocene.

The aim and novelty of my work here in *Theory of the Earth* is to overcome these two problems, the problems of stasis and of history, by inverting their static and ahistorical assumptions. What new philosophy and geology might await us if only we took seriously the earth's genuinely unpredictable power of movement? What would it mean to reconsider human ethics and politics as terrestrial and geological formations?

A HISTORICAL ONTOLOGY OF THE EARTH

The Anthropocene marks a new period in geological history. It forms the limits of a previous epoch and provides the outline of a new one, defined in part by the increasing mobility and instability of the earth.[48] However, the advent of the present is never limited to the present alone. Now that our present has emerged, it is possible, in a way that it was not before, to inquire into the conditions of its emergence and discover something new about the nature and history of the earth's constitutive mobility.[49]

Most of our existing theories assume that the earth is homeostatic, uniform, stable, or capable of being stabilized by life. However, it seems to me that the recent increase in planetary mobility, sudden climatic change, and emergent feedback patterns in earth systems ought to draw our attention to this instability.[50] More importantly, it should draw our attention to a previously hidden dimension of the earth's fundamental instability, only now coming into view: the earth is suddenly proving to be more mobile and eccentric than we thought possible. It's time to start taking this seriously. It's time for, among other things, a different conceptual framework.

The approach of this book is not to write a philosophy *about* the earth, as a distinct substance separate from philosophical practice or humans. Humans and their philosophies are not outside of or separate from the earth's systems. *Theory of the Earth* is also not a "natural philosophy," "cultural history," or "geophilosophy" that studies human thoughts about nature or the earth's relationship to human thought or culture.[51] The focus of this book is instead on the earth *itself as a theoretical practice*.[52] Recent works have done a good job of showing the importance to humans of geological and material processes. *Theory of the Earth* goes one step farther, theorizing these deep geological and material processes themselves.

This book is also not a philosophy of geological science as a human institution.[53] *Theory of the Earth* makes extensive use of contemporary earth and natural sciences but does not critically engage them all using the full repertoire provided by science and technology studies. There are already plenty of books that do this, including my own.[54] My purpose and usage here are entirely different. I cite scientific studies in this book not because I naively accept them as universal truths about the objective facts of nature nor because of so-called science envy.

Instead, *Theory of the Earth* treats the earth sciences as real historical ontological dimensions of our present. Rather than trying to prove that knowledge and nature are endlessly open to human revision and reconstruction, my goal here is to demonstrate the performative reality that the earth itself has produced *as* our scientific knowledge of it. Knowledge is not something we have *about* the earth, as if the earth were something separate from us. Knowledge is something that a region of the earth performatively *does* to itself and with itself.[55] This book is a study of the deep historical and material conditions of this earthly knowledge performance.

Before there were humans, the earth moved independently of what humans thought about it.[56] However, this deep historical earth and its cosmic flows are not radically unrelated to humans. The present is the key to the past because some of the past coexists immanently within the present, within us. We cannot go back and change the earth's deep history, but insofar as it is literally in our bones, we are immanently related to it.[57]

We are not cut off from access to the earth, nor stuck inside our heads. Our heads are not entirely our own—they too belong to the earth. We have access to the earth and the cosmos because we *are* them, albeit only a small region of them.[58] We can, therefore, know something about this deep history precisely because it is the material condition of our very existence. Our bodies and cultures are material memories or traces of the deep history of the cosmos and the earth. This is the earth I am primarily interested in. By this, however, I do not mean that there is only one true objective earth that humans can know absolutely, or even progressively, through science. The earth is neither a single objective reality nor a mere construction of human scientific knowledge.

The methodology of this book is what I call "historical" or "material" ontology. It is historical and material in the sense that our practical inquiry always begins from somewhere historically particular: the present-day earth. From the specificity of the present,[59] the world *is* a specific way, a way that includes us as a region of that same world.[60] Sensation, knowledge, and the historical present are not separate from the world just because we are humans.

This method is ontological in the sense that our situated descriptions are real aspects or dimensions of reality. My method is neither about the earth in itself, independent of or unrelated to us (naive realism), nor about the earth as it is strictly for us (constructivism). There is no division; we are a region of the real earth itself. Its deep history persists into the present as our immanent deep history. The earth really and performatively constructs itself.

In other words, my question is not "what is the earth like in itself?" or "what is human language, mind, economics, or power like such that it is possible to think of the earth?" My question is, "what are the material and historical conditions of the earth, up to and including us?" Multiple human structures shape contemporary reality. These structures are, in turn, conditioned by other real, terrestrial processes that have been around since long before humans walked the earth.[61] This is what I am interested in: the deep conditions of the present.[62] What is especially interesting is that these conditions have turned out to be more profoundly eccentric than we ever imagined.

This work aims to locate the historical conditions of this present-day eccentric mobility.[63] It is not a universal history but a single situated account, among others, from the vantage of the present. I do not offer any final word or universal theory of the earth.[64] Reality does not mean totality. Human history is open because the movement of the earth is open, not the other way around.

The history of the earth is like a double image. In the well-known images of the old/young woman and the duck/rabbit, both figures are really there in each case. Both descriptions are true and different at the same time. The earth, however, is not just two images but a vast multiplicity of images, and the perceiver of those images is only a region of the image itself that actively changes the image by looking at it.[65]

The natural and earth sciences tend to act as if there were one fixed objective world and a single set of universal natural laws about that one nature.[66] However, there are as many natures as there are paths leading from past to present. All the paths are real, just as each figure in the double images is real. If humans are part of the earth, then so is this book. What are the cosmic and terrestrial conditions for this book and for the body writing it?

Theory of the Earth is both a theory of the earth before humans and, at the same time, a theory of the immanent material conditions of the human itself as a region of the earth's deep history.[67] They are the same ongoing history. The historical ontology of the earth is thus not situated because we are humans but, rather, we are humans because we are a historically situated region of the earth's present.

Theory of the Earth is, therefore, not a theory in the traditional sense of an abstract and universal mental representation of the world. Instead, it is a "theory" in the etymological sense of the Greek word *theōría*, as a "movement, sending, or process." Theory is, therefore, a performative process that describes the structure of the immanent movements that constitute it.[68]

MOVING TOWARD A KINETIC THEORY OF THE EARTH

Theory of the Earth reconsiders the immanent history of the earth from the perspective of the increasingly unstable mobility that defines the Anthropocene. It thus provides a uniquely movement-oriented or "kinetic" theory of the earth. This methodology has two significant consequences.

First, by focusing on the movement of the earth, we are able to avoid problematic theories of the earth as an "active," "generative," "vital," "living," subject or as a "passive," "law-driven," "mechanical," "dead" object. I find it unhelpful to divide, oppose, and choose one side of these binaries against the other.

Matter in motion is the immanent historical condition for both subjective and objective dimensions of the earth. There are different patterns or regimes of motion, but movement has no historical opposite. There is, strictly speaking, nothing in the universe that is not in motion.[69] Even space and time themselves are products of motion—not the other way around.[70]

Motion is neither determined nor random. Patterns of nature are emergent features of a universe in motion. There are no laws of nature before there is a universe in which those laws are emergent features. In short, motion allows us to overcome the dualisms we have projected onto the earth.

Second, focusing on movement allows us to see the material continuity between beings that have historically been thought of as categorically and ontologically divided. Movement, for example, flows between cosmos, planet, life, humans, animals, plants, rocks, microbes, and so on, down to the smallest vibrations of matter. The movement of matter plays a constitutive role at every level. Rather than project our own life and subjectivity back onto the earth (Gaia), this book begins its history prior to life and the earth to show how they emerged as a material process.

As mentioned previously, the assumptions of stasis and stability are at the core of the Western project.[71] They are at the top of the great chain of being. Our most straightforward definitions of motion, as a transition from point A to B, assume a static background and internally static, self-identical, points "A" and "B." Even when we consider closed cycles, loops, and orbits, we assume a change that merely oscillates between A and B without any fundamental instability in the line itself.[72]

The material basis for this abstract idea of a static background and identical points is the earth. One of the main reasons we have assumed planetary stability in the first place is because most of human history has taken place during a geological epoch of relative climatic stability, the Holocene. In other words, our idea of motion is historically and geologically particular—but we have taken it to be universal.[73] This is the great epochal error of our time.

However, if the earth is a non-uniform and turbulent mover, as I argue, then the movement from A to B is much more like a continual transformation of the whole line AB itself.[74] The earth is not uniform. Its movement is turbulent, unstable, and entangled with the cosmos in ways that we are only now discovering. This has radical and undertheorized consequences for our understanding of the earth and of motion.

A THEORY, HISTORY, AND ETHICS OF THE EARTH

I have organized this book into three major parts covering the theory, history, and ethics of the earth.

Part I: Geokinetics

I propose a new movement-oriented theoretical framework of the earth as an alternative to the traditional ones defined by stability. Instead of thinking about the earth as an object, subject, substance, or essence in isolation from the cosmos, I introduce a process theory of the earth. I call this a "geokinetic" theory because it treats the hydrosphere, lithosphere, atmosphere, and biosphere as fully integrated earth processes that flow, cycle, and circulate through one another.

The kinetic theory of the earth begins from the contemporary observation that the earth is much more fluid and unpredictable than we ever thought possible. The earth flows. We are now aware of the deep historical coproduction, or "sympoiesis," of all kinds of material flows that we used to study separately. Flows of rock, flows of water, flows of air, flows of life, and even vast cosmic flows of matter are profoundly interdependent processes. What if we retold the history of the earth from this perspective?

In my previous books, I began my historical ontologies with early human prehistory in order to study the longue durée of the emergence of politics, ontology, art, and science in the Near Eastern and Western traditions. In all of these works, I attempted to show the hidden and constitutive primacy of movement and matter. Although I started with human history, the goal was to show the transversal historical patterns of motion that moved through human and nonhuman processes alike.[75]

But where did these patterns of motion come from in the first place, if they were not the sole invention of human beings? *Theory of the Earth* is an answer to this question. What I call "geokinetics" is the study of the deep historical and material conditions for the emergence of, among other things, human politics, ontology, art, and science. In my movement-centered philosophy, I named my study of these areas, "kinopolitics," "kinology," "kinesthetics," and "kinemetrics" to emphasize the primacy of movement. A central thesis of this book and of "geokinetics" is that humans and their culture are continuous with cosmic and terrestrial processes of kinetic dissipation (see Table I.1).[76]

Human culture is only a regional and specific expression of what nature has already been doing in a general sense for a very long time. I realize that this is a big claim, and I do not expect most readers to agree with it immediately. But if it is accurate, it has enormous consequences.

Part II: History of the earth

Another consequence of my movement-oriented perspective is that it makes possible a new history of the earth. Against mechanical and vitalistic theories, I argue that the history of the earth is about the indeterminate dissipation of energy through four patterns of motion.

Introduction 13

TABLE I.1 Theory, form, content, and patterns of motion

THEORY	FORM	CONTENT	HISTORICAL PATTERNS
Kinopolitics	Relation	Border	territorial, political, juridical, economic
Kinology	Modality	Surface	space, eternity, force, time
Kinesthetics	Quality	Image	function, form, relation, difference
Kinemetrics	Quantity	Object	ordinal, cardinal, intensive, quantum
Geokinetics	Nature	Earth	mineral, atmospheric, vegetal, animal

Each of these patterns is associated with the rise and prevalence of a different planetary structure in the earth's history. Minerals emerged through a centripetal motion, the atmosphere through a centrifugal motion, plants through a tensional motion, and animals through an elastic motion. In each historical eon, a new regime rises to predominance, while all the older ones persist and mix with it. Now, in the 21st century, we find our contemporary earth at the intersection of all four major historical regimes. These are the limits not of what the earth can do, but rather of what the earth has done so far (see Table I.2).

The earth is not just a rock. In fact, a rock is not just a rock. The profound uncertainty of the earth's systems today prompts us to completely reconsider our previous categories, substances, teleologies, and hierarchies. We need new definitions and histories for these new hybrid processes of mingled minerals, atmospheres, plants, and animals. We need a process theory of the earth based on patterns, not substances.

The conditions of the present are not locatable in the present alone nor in human history alone. Deep history, in all its uneven flux, is the key to understanding our planetary present. The past does not go away but persists and coexists, to varying degrees, in the present. In other words, there are humans only because there are rocks.[77]

A new kinetic history of the earth will help us to see more fully the present earth that we *are* and how to live better on it. This history is critical if we are to move away from our current tendencies toward mechanism,

TABLE I.2 Patterns of motion

	CENTRIPETAL	CENTRIFUGAL	TENSIONAL	ELASTIC
Politics	territorial	political	juridical	economic
Ontology	space	eternity	force	time
Art	function	form	relation	difference
Science	ordinal	cardinal	intensive	quantum
Nature	mineral	atmospheric	vegetal	animal

vitalism, uniformitarianism, geo-constructivism, and homeo*stasis*. This book is an immanent critique of our moving earth.

Part III: The Kinocene

The third consequence of the kinetic theory of the earth is that it will provide us with a new perspective on contemporary life—what I am calling the "Kinocene." There are as many Anthropocenes as there are ways to think about the present. That is a good thing.[78] So without wishing to negate the others, I would like to propose the addition of one more. The Kinocene is an age defined by the earth's post-Holocene return to itself as an increasingly mobile, turbulent, and dynamically entangled process.

This transition is historically gendered, raced, economic, and asymmetrical, and in our examination of this transition, we must also think about the real possibility of human extinction, something we want to avoid. But it is also crucial to recognize that the Kinocene would be nothing without the contributions of the earth itself. These include fossil fuels, metallurgic compounds, positive climate feedback processes, hydrologic conditions, and the plants and animals that also transform the climate.[79]

Of all the names for our geological epoch, we should not forget the earth itself as a constitutive part of this transition.[80] The twin narratives of humans as earth-destroyers and as earth-savers are two parts of the same anthropocentric dilemma.[81]

By thinking only about our own movements of energy expenditure and conservation on a "relatively static earth," we have failed to see ourselves as part of the larger cosmic and terrestrial drama of increasing flow rate and mobility. By damaging the earth's dissipative processes (especially the biosphere), humans have slowed down the kinetic movement of energy throughout the planet. Fossil fuel capitalism has increased human energy consumption, but only at the cost of decreasing planetary energy consumption by much more.

I rewrite natural and human history from the broader perspective of movement. This offers a new ethical orientation to our "Kinocene" present and to the cosmos. My thesis is that, if humans want to survive, then the most geohistorically likely way forward is to contribute to the earth's massive process of energy expenditure, including land fertility, biodiversity, and climate stability. This shift requires us to reject our current biocentric emphasis on conservation in favor of expenditure and flux.

Today, unprecedented increases in the earth's unpredictable mobility prompt us to reconsider all our planetary paradigms. These changes challenge us to reconsider the nature of nature as well as the deep history of the earth. Perhaps most importantly, the earth's turbulent mobility forces us to rethink our ethical relationship to one another, the planet, and the cosmos at large. The Kinocene is calling us to become what we are: the earth.[82]

PART I
GEOKINETICS: THE KINETIC EARTH

1   The Flow of Matter

THE EARTH FLOWS because the matter of the cosmos flows through it. In this chapter, I take the flow of matter as the starting point for a new theory of the earth. This is because I think the earth is much more like a process than it is like a stable object (Spaceship Earth) or autonomous subject (Gaia).

The earth is a material process continuous with the expansion of the universe that produced, and continues to produce, the earth. The earth is not a vacuum-sealed object cut off from the outside. Nor is it an unchanging or uniformly changing substance following autonomous processes. Geology flows from cosmology.

Flows of matter continually compose, cycle through, and flow out of the earth. The earth is only the regional circulation of a much larger kinetic and entropic process. Historically speaking, philosophy, politics, and much of geology have not taken this ongoing flow of cosmic matter seriously.

This has led to an inverted understanding of the earth and our relationship to it. We have posited ourselves and the earth as profound reversals of the general movement of the universe, which flows, cycles, and dissipates entropically.[1] The crowning achievement of reversal is anthropocentrism.[2] We look at the universe and think how wasteful it is. Against its waste, we think life, and human life in particular, is so special because it fights against the waste of cosmic entropy. We have cast ourselves as the heroes in a universal drama of life against death.[3]

In this book, however, I offer a different perspective. The earth, I argue, is not so much a "planet" as it is a process of terrestrialization. It is the cosmos continually made earth. Every product presupposes a kinetic process, and this is where I propose to start with the earth. Part I of this book aims to rectify the missing theory of motion behind our thinking about the earth. It begins with the cosmos as the immanent material condition of the earth.

This accomplishes two significant moves. First, it abandons any notion of the earth as an absolute ground, of itself, of history, of humans, or thought. Geophilosophy and geoscience have both granted unjustified primacy and autonomy to the earth. This is why the instability of the Anthropocene has caught them so off guard. The earth is not behaving like the good ground it is supposed to be. The Anthropocene is less an age of humans than of the inhuman. And second, it provides a new conceptual vocabulary with which to talk about how indeterminate fluctuations of matter can produce and sustain emergent kinetic patterns continuous with the larger cosmos and with one another. If the matter that makes the earth is unstable, then so is the earth.

In Part I of this book I start with the idea that the earth is made of material flows. From there, I lay out a kinetic theory of the earth that I call "geokinetics." Geokinetics has three aspects: the flow of matter, the fold of elements, and the circulation of planetary fields. These are the three terms in my conceptual framework. I hope that they will position us well to put forward a process theory of the earth. This framework, although it may sound abstract in some places, will help us, in Part II, to reinterpret the deep history of the earth, including the emergence of minerals, the atmosphere, plants, and animals. Ultimately, in Part III, these concepts will also form the theoretical basis for an ethics of living well in the Kinocene.

FLOW

The earth flows. This is the first and central thesis from which the entire conceptual framework of geokinetics follows. What must the nature of the earth be to make it capable of being in the unstable motion we see so prominently today?

Etymologically, the earth is literally dirt. But where did this dirt come from to make it capable of continual terrestrialization? What are the historical, material, and kinetic conditions such that the earth came to be dirt? This dirt is no native, but a cosmic migrant. It is already part of a much larger flow of matter that we need to take seriously.

The earth is matter thrown into motion. Without the constant thermodynamic flow of energy from the universe, there is no accumulation, dissipation, or recombination of matter into a stable earth. The thermodynamic transfer of energy into and out of non-equilibrium states is what has allowed the earth to emerge, persist, and distinguish itself from other aspects of the cosmos. But the flows of matter that composed the dirt of the earth also have their own conditions that take us further back, to an even more mobile flow of matter.

Before the Big Bang, 13.8 billion years ago, when the universe was younger than 10–43 seconds old, there were only indeterminate quantum fluctuations. These fluctuations occurred at a size smaller than the smallest measurable length (1 Planck length) and at indeterminately high temperatures. There was no void; no metaphysical singularity; no stasis. There was not even movement in the traditional sense of something moving from point A to point B in space, either. There was no space or time.

Yet there was matter as energy, and there was motion as a continual transformation of the whole. There was no stable background of spacetime, no ground, and no foundation upon which the Planck Epoch could have emerged. Matter was neither this nor that, neither here nor there, neither continuous nor discontinuous, but pure indeterminate flux.

Before the Planck Epoch, the universe was neither random, determinate, nor probabilistic. The cosmos was neither one nor many because its energy was as indeterminate as its position and momentum. The laws of physics had not yet emerged, and even the conservation of energy could not be guaranteed.

In cosmological time, we can call this the "Indeterminate Epoch." Through completely relational and nonrandom processes of its own, energy began to iterate itself into a single Planck-sized pattern called the Planck Epoch. This was not a singularity, "cosmic egg," or "primeval atom," as the 20th-century astronomer Georges Lemaître thought.[4] Instead, it was the first emergent form of the universe: a fluctuating but metastable region of spacetime with the smallest size and highest temperature theoretically measurable.

This flux and flow, however, were still too small and hot for the four fundamental forces, including gravity and electromagnetism, to be divided from one another. So they remained continuous aspects of the same flow. This is what cosmologist call the "Grand Unification Epoch."

Around 10^{-32} seconds after the Big Bang, a rapid inflation of spacetime occurred in which an enormous amount of spacetime unfolded from this cosmic flux. The universe expanded to 10^{78} times its previous volume, or the equivalent of going from 1 nanometer to 10.6 lightyears long, in a fraction of a second. Again, this movement was not a spatiotemporal movement of something across or against a fixed background of spacetime—it was an expansion of spacetime itself. In other words, the flow of the universe was not a movement from here to there but, rather, the creation of the here and there. There was no extensive movement of *something*, but the immanent kinetic unfolding of the universe into and out of itself.

Then came the Inflationary Epoch, the production of spacetime itself—the material condition of all discrete beings. Since light moves through spacetime and not the other way around, inflation flowed faster than the speed of light.

One of the most important, although not yet experimentally demonstrated, ideas in theoretical physics today is that spacetime is an emergent feature of a moving universe. In other words, spacetime is not a substance or force but a metastable process. It is like the "bubbles" or "foam" stirred up by a more primary turbulent process of quantum matter in motion.[5] What physicists call "quantum gravity theory" is the attempt to provide a quantum theory of spacetime and thus unify the main frameworks of theoretical physics: quantum physics and general relativity.[6]

This is a dramatic and perhaps abstract-sounding way to begin to rethink something that is, to us, the most concrete: the earth. However, the indeterminacy of the universe is a crucial first step in rethinking the earth as a process. The indeterminate flow of the universe is the immanent material condition of the earth. It shapes the way we think about what earth processes are, that is, material processes with a genuine capacity for novelty and motion. Starting from quantum cosmology, then, we should expect, rather than be surprised by, the mobility and instability of the earth.

Einstein's cosmology of a static spherical cosmos assumed spacetime but did not explain its emergence. In this way, it was much like older geological models that assumed a stable primordial earth but did not explain it. Similarly, early cosmologies of the Big Bang assumed a uniformly expanding universe, just as geologists believed in a uniformly moving earth.

These old geologies and cosmologies also have parallels in the philosophical ontology of time that treats time as universal and given.[7] Growing up in the Holocene, humans have had a long terrestrial bias for stability.[8] Is it possible that our philosophical, religious, and scientific pretensions to universality and stability have arisen from the extremely particular geokinetic situation that is the Holocene?

The cosmological theory of sudden chaotic inflation,[9] the indeterminate fluctuations of quantum gravity,[10] and discoveries of the earth's sudden and unpredictable climate history[11] have overthrown these old cosmologies, geologies, and philosophies. I argue here that a much more fitting philosophical perspective for our time is that of motion.[12]

Why should we continue to model our philosophies on foundations, our cosmologies on background spacetimes, and our geologies on Holocene uniformities, when the evidence is pointing us in the opposite direction? Our earth is not a ground, but a tiny metastable region of an unstable and dying universe.

We have got the problem entirely upside down. We shouldn't shake off our anthropocentrism in favor of bio- or geocentrism.[13] There is no privileged Archimedean point and foundation for knowledge. Because we have taken the earth to be a stable and uniform place, we have imagined that our universe must be similarly stable and that it must all come from some static unmoved mover, God, void, or homogeneous theological singularity. The outrageous lengths to which humans go in order to explain movement by something else never ceases to amaze.[14]

Rarely do geologists or philosophers extend their thinking to the origins of the universe more broadly. When they do, their reflections are often out of date or limited to our solar system. But the increasing shift in physics from substances (spacetime) to kinetic processes (quantum gravity) should also prompt us to reconsider substance-based approaches to the earth sciences and philosophy.

The oft-cited theory that before the Big Bang, there was nothing, and then afterward, there was something, is typical of substance-based and dualistic metaphysics.[15] In the beginning, there was stasis, and then there was movement. This story hardly moves beyond Aristotle's unmoved mover. The only real alternative, in my view, is an indeterminate process cosmology without beginning or end, stasis or movement, being or nonbeing.[16]

Our metastable earth is the product of indeterminate cosmic flows. It is their regular flux and flow that continually supports and reproduces every inch of our spacetime. The geological transformation of the earth is part of the same continual alteration of the universe. We live on an earth in which the matter of the cosmos is deeply entangled. For example, indeterminate quantum ripples in early spacetime eventually developed into later galaxy clusters and planets like ours. Many of the earth's most important physical and chemical changes are related to processes outside the earth and even outside our solar system.[17]

The enormously crucial takeaway from these indeterminate quantum fluctuations is that they change our whole conception of what the universe is. We live in one big fluctuating metastable process, without beginning or end.[18] The earth is not an exception to the rest of our unstable and fluctuating universe. From this starting point of indeterminate flows, a whole new theoretical framework for thinking about the earth can emerge.

MATTER

The earth flows and moves; matter is what is in motion. However, matter is not a substance. "Earth" is "dirt, soil, or material," but soil is not static stuff. Soil is exceptionally mobile and heterogeneous. Soil is the ground of the earth. However, it is the most visibly unstable and shifting kind of ground because it is made by erosion and decomposition.

Soil is a kinetic process composed of traces from deep cosmic and terrestrial time: minerals, liquids, organic materials, gases, and organisms. Soil is composed of the earth and recomposes the earth by regulating the atmosphere, supporting life, and storing water. Soil is the earth composting itself, and dirt is the co-creation of all the earth's processes, continually transforming one another. It is billions of years of cosmic and terrestrial history in motion.

How arrogant that we ever spoke of the earth as a static or passive ground, an "immovable ark,"[19] or foundation for anything, when the whole thing is fluctuating compost. How could we ever have talked of a geology or geography of "the" earth when a single handful of dirt offers such a simple material demonstration of its deep and mobile history?[20]

In a single handful, we hold our atmosphere, our minerals, life, and death. Soil is a bit of our terrestrial and cosmic deep past. That the most apparently stable of all objects (the earth) is phonetically identical with the most mobile (earth) is a beautiful lesson. That which the ancients most maligned as being at the bottom of the great chain of being (earth) is also the concrete foundation upon which they stood to model the spherical cosmos. The earth was both the condition for the geocentric cosmos and the filthy material soil that covered their feet.

But the earth is not a sphere, and neither is the universe. The earth is more like an eddy in a river through which flows of matter continuously stream. It is replenished and depleted in a vortical cosmic dance. "The world is a vampire," as The Smashing Pumpkins sang: a vampire living from the death of the sun. The universe must die to keep living.

Before Aristotle, there was no word to designate "matter in general." This is because matter was not a substance, stuff, or kind of thing, but a process frequently associated with weaving in archaic Greek culture.[21] Aristotle appropriated the older Greek word *hyle*, meaning "firewood, wood, or forest" and redefined it to mean "substance or stuff in general."

In this way, the term *hyle* went from meaning something earthly and mobile—growing, decomposing, and composing—to being a passive substrate molded by ideal forms and laws of nature. Earth and matter have suffered the same fate in the Euro-Western tradition. The return to one also suggests a return to the other. A new theory of the earth thus calls for a new theory of process materialism in which matter is again an earthly flow.

We cannot correctly understand the earth as a being because it is a becoming. Trying to determine the being of matter is a similarly flawed approach. Matter neither is nor is not, but flows, folds, and circulates. Matter is not a being but a becoming. The quantum theory of matter, for example, at least in some interpretations,[22] is neither an empirical nor a metaphysical definition. It is an historical-ontological description of matter as indeterminate.

In the West, matter is the historical name of what is in motion. Matter and motion stand together at the bottom of the great chain of being in the dirt. They are less like static beings than anything above them on the chain, and that is because they are not beings at all. They are dirty becomings, whose "being" continually changes as they change and who therefore have an "indeterminate" being.

This is also why the moving earth is not identical to itself. It is a geokinetic material process, not a substance or subject. Geokinetics is, therefore, a kind of "kinetic materialism." In this book, I treat the earth not as a discrete, deterministic, or probabilistic being but rather as matter-in-motion, that is, as an open relational process without fixed characteristics or essences.

Matter is what is in motion, but matter is not reducible to motion. Motion in itself, without a determinable matter, is a pure and immobile abstraction.

Matter and motion have suffered a shared fate in Western history.[23] Western culture has almost always subordinated them to some higher category. In classical Greece, matter and motion were subordinate to eternal forms and unmoved movers. In the medieval world, they were subservient to the vital forces or *vis inertia*, which directed their motions and formed their matters. Finally, in the modern world, they were subordinate to mechanism, rationalism, and natural laws.[24]

And just as matter and motion share a subordinate position, they also share a possible twin liberation. If motion is ontologically primary, as I argue, then so is the matter that moves. If matter is fundamental, then so is its motion. Without matter, the concept of movement remains a "false" or idealist category.[25] Without movement, matter remains static, discontinuous, and dead.

This book uses the idea of kinetic materialism to rethink the earth as a process.[26] If the earth is a material process, then it cannot be adequately defined as an object or a subject. We need a new, process-based, framework. All "universal" ideas of matter come from material and historical beings in motion on the earth. Even the belief that matter and the earth are relatively passive bodies is possible because of the relatively stable Holocene period.

Kinetic materialism is, therefore, neither a Copernican revolution, in which it is we who move around the stars, nor a Ptolemaic counterrevolution, in which we are at rest while the stars move. It is a Hubblean revolution, where everything is in motion. This is not a reductionistic theory of matter, because there is no single idea of matter that can define it once and for all. Matter is an open process of motion whose being changes as it moves.

For matter to become earth, it must be able to flow. Thus, I propose that material flows are a good starting point for rethinking the earth.

Pedesis

The earth is, first and foremost, cosmic matter on the move—migrant dirt. Geologists call the soil of the earth the "pedosphere," from the Greek root *-ped*, meaning "foot, track, earth, or ground." The pedosphere thus deserves special attention because it contains, mixes, and composts all the other spheres: lithosphere, biosphere, hydrosphere, and atmosphere. The pedosphere is not just one spatial region of the earth among or between others.

The pedosphere is the concrete place where all the other spheres break down, commingle, mix, and directly coproduce one another. The pedosphere is a kind of woven, braided knotwork where the other spheres touch, translate, and transform one another. The pedosphere stores and releases carbon, regulates water distribution and filtration, and makes minerals available to the surface. Plants and animals, including humans, are made of pedosphere.

In one sense, the pedosphere is the place where particular creatures walk: the ground. In this sense, it is a frequent site of direct physical composition and decomposition. It is the place where the tracks and traces of terrestrial entanglement are impressed on the surface. It is where waste becomes food for mobile life. It is where the sky leaves a memory of its movement through precipitation and where rock leaves a souvenir of its tectonic path. It is the place where the earth walks, moves, and wanders. The pedosphere is the foot of the walker and the ground that walks and is remade by walking.

In another sense, the pedosphere invokes a much larger movement of the earth itself as a planet, from the Greek word *planáō*, "to wander." The pedosphere is not just terrestrially but cosmically mobile. The earth walks a path through the solar system. Its orbit is not strictly circular, elliptical, or regular. The earth is an eccentric wanderer. It shakes itself with quakes; it moves off its rotation. Its orbit around the sun varies in eccentricity between more circular and more elliptical in response to the irregular orbits of Jupiter and Saturn. All the planets are wanderers, swerving together in vast cycles and epicycles, each lasting hundreds of thousands of years.[27] Each cycle is a unique iteration.

The earth even leaves a trail or trace of heat as it moves through the solar system and distributes the sun's radiation. And as our solar system moves through the Milky Way, it also moves up and down through the gravities of other star systems and asteroid belts, further changing planetary orbits and bombarding our planets with asteroids. Many of these eccentric movements are not just unknown; they are fundamentally chaotic and unpredictable.[28] It is even possible that a large disk of dark matter in the Milky Way may have affected the earth's climate history and caused the extinction of the dinosaurs.[29]

Therefore, in another sense, the pedosphere also refers to the pedesis or errancy of the earth's motion. The word "pedesis" (again, from the Greek root *-ped*) refers to the motion of self-transport: the motion of the foot in walking, leaping, running, dancing unpredictably. Flows of matter have no straight lines because they move pedeticly.

The discovery of pedesis comes from two of the most critical kinetic discoveries of 20th-century physics, Einstein's kinetic theory of matter (1905) and Heisenberg's quantum uncertainty principle (1927). In the first, Einstein argued that all matter is a product of the stochastic or pedetic motion of atoms. For example, the atoms in gases move faster and farther than those in fluids. Those in fluids move faster than those in solids. All matter, Einstein showed, was not only in motion but in pedetic or turbulent motion. The form of matter is fundamentally kinetic but also fundamentally and irreducibly turbulent.

However, by showing that all matter was in turbulent or pedetic motion, Einstein introduced a fundamental kinetic uncertainty and unpredictability into the heart of matter. Although the description of kinetic turbulence goes back to the Roman poet Lucretius, science has been entirely unable to produce a successful predictive theory of turbulent motion beyond minimally probabilistic models. The precise kinetic structure of turbulence remains one of the last unsolved problems of classical physics, with a current million-dollar prize for its mathematical solution.[30]

The unsolved problem of classical turbulence, combined with Einstein's kinetic theory of matter, has enormous consequences. All matter is in motion, and that turbulent motion is not deterministically solvable. Heisenberg supposedly once said that he wanted to ask God two questions.[31] The first was, "Why is general relativity so weird?" and the second was, "How do you explain turbulence?" He then said that he was certain God would know the answer to the first question.

In the second kinetic theory, Heisenberg showed that there is a fundamental limit to the precision with which we can know the position and momentum of a particle at the same time. The more precisely the position of a quantum field is determined, the more it looks like a stable particle, but the more its momentum becomes uncertain. Conversely, the more closely we determine the momentum, the more its position becomes uncertain.

In other words, the precise extensive path of a particle from A to B is fundamentally uncertain. There are different interpretations of this experimental finding. However, one consistent interpretation is that the human observation of the position or momentum of an electron with light (photons) is only one influence among others in the world that can directly affect the momentum of quantum fields.[32] If this is the case, then humans may not be the only observers to produce uncertainty.

The importance of these discoveries is that matter and motion are indeterminate at both classical and quantum levels. If the earth is matter in motion, then it is also subject to a fundamental indeterminacy. We should, therefore, take this seriously as a methodological starting point for a materialist theory of the earth.

RELATIONAL MATERIALISM

Pedesis might be irregular and unpredictable, but it is not random. Through the pedetic movement of matter, metastable formations emerge. Terrestrialization is a metastable process. By contrast, the ontology of randomness is quite bleak. In such an ontology, all matter would be moving randomly at all times, unaffected by any other matter. Since fluctuations from disorder to order are physically rare, the likelihood that anything like the sun or even our galaxy would suddenly pop into existence would be unimaginably rare. And it would then likely immediately fall apart due to further random motion.

The very idea of purely random motion presupposes that it was not affected by anything else previously because, if it had been, its movements would not be completely random but instead caused in some specific way by something else. Randomness presupposes a first thing or motion before which there was nothing. This is a version of the internally contradictory hypothesis of ex nihilo creation: something from nothing.[33] Given the high level of order and complexity in the present cosmos, this kind of pure randomness is demonstrably not the case.

Pedetic motion, on the other hand, is not random at all but emerges from and is influenced by other motions, just not in a wholly determined way. Pedetic movement is challenging to predict not because it is random but because it is so entangled with all other movements. It is the interrelation and mutual influence of matter with itself that causes its unpredictable character. Over a long time, the pedetic movement of matter combines and stabilizes into specific patterns and relations, creating metastable structures of stability and solidity. Then, after a while, they become turbulent again.

Absolute randomness would require a motion to be wholly unaffected by and thus unrelated to any prior movement that might affect its subsequent action in any way. This would be a nonrelational object. Nonrelational randomness and relationality are incompatible with each other. The notion of randomness is a logical category, a pure mathematical idealism, as if we could abstract matter from its motion and kinetic relations. We have no evidence for anything truly nonrelational in the universe. If all matter is relational to some degree, then it cannot be truly random.

Pedesis, then, is simply a degree of more or less ordered relations of motion. Turbulence, for example, is matter in motion with a very high degree of unpredictability as a result of the number of continually changing variables involved in the process. Yet turbulence also has relatively ordered patterns, spirals, swirls, vortices, and so on, which begin to emerge from these continually changing relations.

Matter is pedetic not only at the quantum level of indeterminacy. It is pedetic also at the atomic, molecular, meteorological, and cosmic levels. Pedetic motions crystallize into relatively ordered patterns at every level of nature—even if it takes millions of years. Pedesis always generates some order.[34] Matter in motion is always related to other matters in motion. Therefore, there is never a case of absolute determination or absolute randomness. Pedesis gives rise to more or less stable configurations, like the electron patterns of the atomic elements, the speed of light,[35] or the rate of acceleration at which things fall on the surface of the earth.[36] Clearly, we do not live in a random universe. The mathematics of probability is our way of making educated guesses about the future based on past events.

There are thus two factors barring absolute determination and absolute randomness, and both have to do with motion. Absolute determination is not possible because we do not and cannot have a perfect knowledge of all the material variables in the universe, since the universe is continually expanding and changing. There is no meta-object called "the universe." And absolute randomness is not possible because since all matter is kinetic and relational, it cannot be truly random. If all of matter in motion were truly random, there would be no universe, and nothing would hold together for very long.

This relational indeterminacy is what produces the ordered metastable patterns we study in the earth sciences. Pedetic air currents whirl motes of dust around a room and pedetic orbits whirl our planets around the sun. Spiraled rings of smoke swirl up from a cigarette and thunderstorms swirl up around the globe.[37] The earth is a vast, metastable process of metastable processes. This is the earth I would like to study in this book: the earth of indeterminate flows of matter. This is an earth made of flows and folds, as we will see in our next chapter.

2   The Fold of Elements

THE FLOW AND FLUCTUATION of matter constitute the earth and its elemental body. The word "earth" refers to the planet Earth *and* its earthy soil, but also to the classical element "earth." The earth is elemental in the sense of being everything that is not air, fire, and water but also in the sense that it is composed of "elemental" or "primordial" components.

What is an element? If we are going to take the materiality of the earth seriously, we need to know what it is made of, its elemental constituents. This chapter aims to answer this fundamental question.

To begin with, the earth is elemental and elementary because it is of an elemental universe. The earth is burnt-out star-stuff from the early universe, from 13 billion years ago. Elements do not preexist the cosmos but emerge through it as it flows. An element, then, cannot be a fully discrete particle, atom, or fundamental building block of nature. Particles are merely the products of a more primary material flow that began to fold back on itself. An element is a fold or pleat in matter.

An element is not just a constitutive or simple component. An element is an emergent process that came to be through the iteration of fields of energy. So-called "fundamental particles" like fermions and bosons are not, technically, actually fundamental—their existence already assumes spacetime and energetic movements high enough to produce both spacetime and particles in the first place. This is why the first 10–13 seconds of the universe contained no fundamental particles. Particles emerged from processes.

During what cosmologists call the "Inflationary Epoch," spacetime fluctuated so wildly that it could not sustain any relatively discrete regions or elements. Due to these primordial quantum fluctuations, spacetime did not flow homogeneously but rippled with gravitational waves. As the waves moved, they cooled down into dendritic patterns, the way that a river sheds its water into tributaries and pools. These metastable pools of energy created the earliest and heaviest particles (quarks, hadrons, neutrinos, and leptons). Each particle unfolded out of the previous, like the petals of a flower from a bud.

It is a common misconception to think of particles as discrete objects, when they are nothing of the sort. They are vibrational patterns in the energetic topology of the universe. If spacetime in quantum gravity is like a continually fluctuating "foam," then elementary particles are the regions where this foam has coalesced and folded into giant metastable bubbles.

Elementary particles are excitations in fluctuating quantum fields. What makes them elemental is not that they were first but that they were emergent regions of an indeterminate material flow that interacted with itself. This self-interaction is what I call an elemental "fold."

The early universe was fluid. Quantum perturbations and gravitational waves rippled like the surface of a pond in a tropical rainstorm. What we now call "dark matter/energy" also emerged at this time and still composes 85% of the universe. We know this only indirectly, of course, because we cannot empirically measure the quantum fluctuations that make up this dark process.[1] Throughout this turbulent and chaotic inflation, dark matter was distributed and diffracted unevenly across the cosmos. As the universe cooled and became less fluid, these uneven distributions became the regions where visible matter accumulated into little vibrating and diffracted pools in the form of particles, atoms, molecules, stars, planets, and galaxies.

Inflation eventually made the universe almost entirely flat and homogeneous, like a folded-up blanket smoothing itself out, but with little ripples still left here and there. However, since inflation involved the unfolding of quantum processes, the precise spacetime in which chaotic inflation slowed down was indeterminate.

This resulted in small deviations from perfect uniformity in the inflated spacetime that now shaped the whole structure of the universe, including us and our planet. The earth was born from a tiny indeterminate quantum fluctuation long ago during inflation. Quantum cosmology is undoubtedly one of the most mind-blowing and revolutionary discoveries of the last few decades. It, too, should be taken seriously as part of the earth's deep history.

FOLDS

Matter tends to flows turbulently and to fold itself up into metastable cycles. These are what I am calling "elements," in a broad conceptual sense. Elements are thus both atmospheric processes and relatively discrete regions where larger composites emerge. Folding is nature's way of self-differentiation and self-affection.

These elements are flows that sustain themselves through iteration. The condition of earthly terrestrialization is, therefore, the periodic folding of material kinetic flows at different frequencies. The earth iterates itself in water cycles, rock cycles, and carbon cycles because the universe is already based on cycles.

The idea of folds helps to explain how metastable forms can emerge from the tendency of matter to flow and spread out. An elemental fold is like an eddy or whirlpool in the flows of matter. It is a relative stasis that is always secondary to the primacy of the flows and fluxes that compose it. As such, an elemental fold is nothing other than a flow of matter. A fold does not transcend or preexist the flow; it is merely the redirection of a flow of matter back onto itself in a loop or self-transformation.

It is a mistake to think of a fold as the mere product of a flow, as if the two were ontologically separate. The fold that moves already presupposes a more primary constitutive flow that composes and runs through it: the creative movement of the matter itself. Material flow and elemental fold are thus co-constituted in the same immanent kinetic process.

An elemental fold is the repetition of a differential process, or "cycle." It is a vortical process that continues to repeat in approximately (but not exactly) the same looping pattern, creating a kind of mobile stability or "homeorhesis."[2] In this way, elements constitute a point of creative self-reference, or haptic circularity, in matter that yokes a flow to itself. Particles have charges, and atoms have weights, all depending on how energy fluctuates in quantum flows. At the level of quantum field interaction, however, the term "cycle" does not have a classical definition. Instead it just refers to a quantum intra-action of energy with itself at a certain metastable frequency.

The elemental fold then acts like a filter, or sieve, that allows some flows of matter to pass through or around the recurrent attractor and other flows to get stuck in the repeating fold. The movement of the captured flow can then be connected to the movement of another captured flow and made into all manner of mobile composites, or conjunctions.

However, the joining of the flows into an elemental fold also augments them, not necessarily by moving them more quickly or slowly but by subordinating them to a cycle that begins and ends at the same haptic point or frequency of oscillation. All kinetic elements thus have two basic dimensions or aspects: their period and their cycle.

Period

I call the self-affective point at which a flow intersects with itself a "period." Although the flow of matter is continually changing and iterating itself, this haptic point appears to remain in the same place—like the eddy of a river.

In this sense, the point appears to absorb and regulate all the mobility of the yoked flow while itself remaining relatively immobile. The period is a "mobile immobility" that moves by the movements of others. The period is what persists only as a material oscillation or frequency. Particles of matter are frequencies of oscillation, not discrete substances. This, too, makes up the elemental earth.

Cycle

Matter flows and folds over itself. Once it returns and connects to itself again, it creates a cycle. These iterations do not just move in or through spacetime. They integrally interweave into the gravitational field. They are spacetime itself, rippling and folding over itself. Matter, we could say, "spacetimes" itself into elements. What I am calling a "cycle" is the oscillation of matter as it continues to affect itself. The point where matter intersects with itself is its "periodic attractor."

Cycles are not identities, substances, or essences. They are processes. The period of a cycle never creates a perfect regularity, equilibrium, or classical "identity." Rather, it creates a metastable state like a Lorenz attractor that tends to overlap with itself at irregular but frequent intervals (see Figure 2.1).

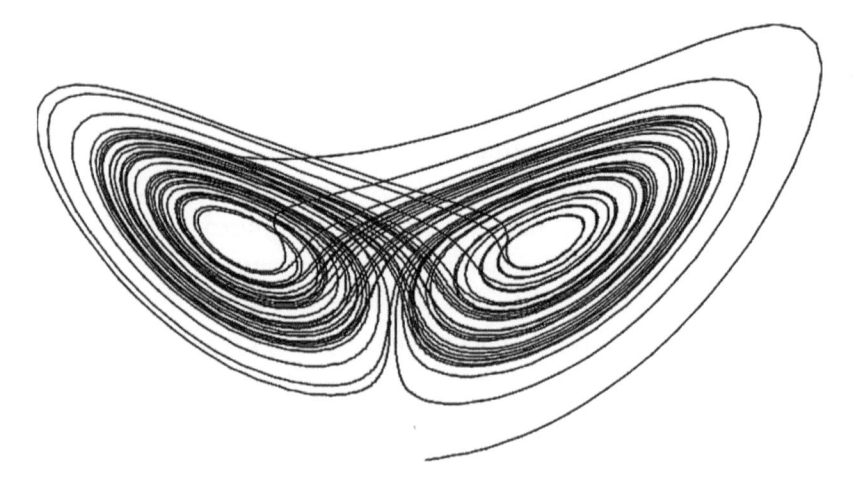

FIGURE 2.1 Lorenz attractor

Image from https://commons.wikimedia.org/wiki/File:Lorenz_attractor2.svg

Matter flows along but also creates eddies that begin to "cycle" around an area, just as in a river. The "period," in the diagram below, is the place where the cycle or eddy of flowing matter returns back periodically to itself. The reader will have to imagine the static diagram below in motion like a stream. Each time the flow cycles around and then moves out of the loop, everything is slightly different and not identical. What is stable is the pattern of the flow. The new influx of matter takes the place of the old and creates an "unstable stability." The quantum indeterminacies within the flow of matter never go away, but become increasingly metastable at the larger scales.[3]

The earth has elemental cycles. But this does not mean that the flow of matter has been arrested or rendered entirely discrete. The period is simply a slice or selection out of the whole recurrent process. When we mistake a periodic attractor for a simple static or fixed point, we lose the flow entirely. We see only an abstract product, without the motion that composed it. A process theory of the earth tries to overcome the prejudice that has us thinking of the earth as an object.

Even the circadian cells that are within all humans, animals, and plants do not cycle at precisely regular intervals but respond to the sun, the temperature, and the season. Without these influences, human circadian rhythms and body temperatures, for example, will produce all manner of different cycle lengths, ranging from 25 to 45 hours.[4]

Quantum cycles cannot be measured in the same way as circadian cycles because the level of indeterminacy and energy is so much higher. However, at least one feature is common to both processes: cycles, oscillations, and frequencies emerge only on the condition of a more primary flow and flux of matter.

The recurrent cycle produces a regional attractor or "identity" but only through motion, only through folding, habit, or synchrony. Folds are not only continually receiving a constant source of new motion from outside, but also losing some motion that passes through them. In other words, the universe tends to spread itself out into the cool. This is entropy.

A fold is only a regional capture of motion in a certain period. This is because when it intersects itself, it is crossing itself at a different point in the flow each time. There is no absolute "fixed period." There are only more or less dense periodic orbits or "limit cycles" that continue to shift around a metastable fold, like the 187-year-old Great Red Spot of Jupiter or the North Pacific Gyre. They cycle, but never in quite the same way. Folds are continually moving and self-differentiating processes that are never identical to themselves. As processes, they are indeterminate becomings.

Matter flows in a pattern of motion, and when the pattern returns, we say that it is "identical." But it never is, really. Matter persists over time, taking on relative stability, but it is never strictly identical. The periodic fold remains the "same," but only on the condition that others flow through it. As Heraclitus writes,

> On those stepping into rivers staying the same, other and other waters flow.[5]

For Heraclitus, each eddy in the river is like another river within the "same" river.

The flow from period to period is the cycle of a fold. A cycle is not a static unity but a fluid or kinetic one. Since the fold is only a fold in a flow that regularly enters and exits it, renewing it each time, its cycle can be said to be the unity not of an ideal identity but of a kinetic process. Just like a whirlpool in a river, a cycle is only a metastable unity of a differential process, refreshed each time with new water around a periodic attractor. The theory of geokinetics thus replaces the concepts of identity and unity with the concepts of a periodic attractor or differential cycle (see Figure 2.2).

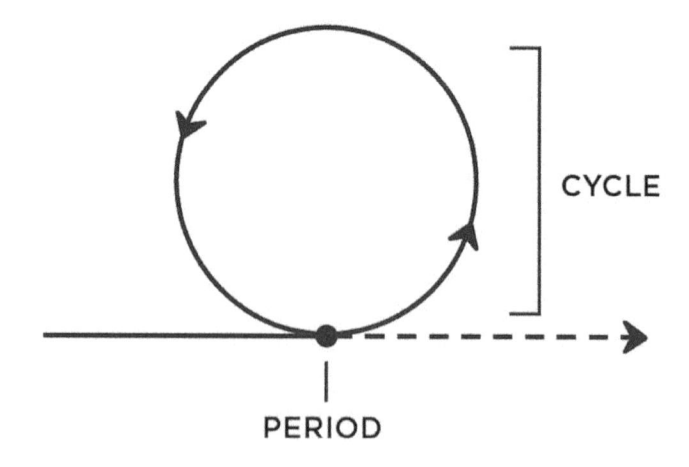

FIGURE 2.2 Cycle and period

The earth is a cycle of cycles, and they are all more entropic, turbulent, and chaotic than we ever imagined. When the universe and its elements are so iterative and indeterminate, why should we expect the earth to be immune to these instabilities, to entropy and turbulence? Uncertainty is not an accidental feature of our planet. It is a feature our earth shares with the rest of the universe. The matter of our cosmos tends to flow and iterate itself into metastable states at every level.[6]

CONJUNCTION

The earth is a process of terrestrialization. It continually destroys and remakes itself with every new experiment to reduce the energy gradient between the sun and space. It is one of many cosmic composites with integrated cycles of production and reproduction. From a materialist perspective, these conjoined cosmic cycles forbid any ontological division between mind and body, human and nature, earth and cosmos. The longer we study nature, the more cycles we find and the fewer oppositions. We find climate cycles, extinction cycles, lunar cycles, solar cycles, seasonal cycles, planetary cycles, galactic cycles, and even dark-matter cycles.[7] The same thing never happens twice but tends, instead, to cycle iteratively. A process theory of the earth is a theory of these singular cycles.

Conjunction is a cosmic process. A few minutes after the Big Bang, the temperature dropped enough for protons and neutrons to conjoin into nuclei, but it was still too hot for them to form into atoms. Perhaps one of the most critical events in the history of the universe occurred around 380,000 years after the Big Bang. After the emergence of early nucleus conjunctions, the universe cooled down enough for positively charged nuclei to combine with negatively charged electrons to form neutral atoms—mainly hydrogen and helium. These particular atomic conjunctions allowed for a metastable formation that released photons. These are the photons we can still see today as the cosmic microwave background—the oldest observation we have of the universe.

These are the major cosmic conjunctions from which all the known atomic elements were formed. The conjoined cycles of protons, neutrons, and electrons are what make up the vibrating earth's elemental body and all its planetary subcycles. This kinetic structure of conjoined cycles is what links the earth to the rest of the universe.

Everything cycles. Unfortunately, many humans have forgotten this. Capitalism, in particular, takes from the earth without giving back and ignores the metabolic balance among humans, nature, and society.[8] More on this in Part III.

In contrast, what we call "things" are merely conjunctions of cycles or folds. Every "thing" is supported solely by the flows of matter that move through the whole series of cycles. For example, living organisms are only relatively stable pools or folds in a continuous flow and transformation of energy running from the Big Bang to the sun, conjoined by the organism, reproduced in its offspring, disjoined in death, and radiated as heat out into space. The folds of life are only eddies in the kinetic stream of energy—the memorial traces of cosmic perturbations (see Figure 2.3).

Even the inorganic bodies of minerals are nothing more than relatively stable combinations of folds in the continuous transformation of kinetic energy. Igneous, sedimentary, and metamorphic rocks are three relative stages of a constant mutation and conjunction of the earth's liquid body—the rock cycle. Solid, liquid, and gas are the three corresponding stages of a continual conjunction in the earth's fluid body.

All organic and inorganic objects are conjunctions of smaller objects at the microscopic level. Quantum fields flow, conjoin, disjoin, and fold into particles on the luminous shores of existence. These particles then conjoin into molecules to create the visible objects of everyday life.[9] At the

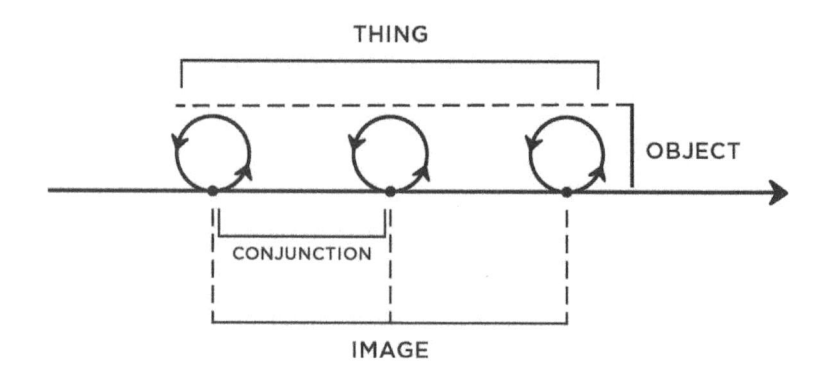

FIGURE 2.3 Conjunction, thing, object, and image

macroscopic level, however, all these objects do not produce a final state. They are all moving through an open and accelerating universe at incredible speeds. All things are products of kinetic conjunction without essence or identity.[10]

Extension, volume, and shape are nothing more than the products of a process of continuous and constant conjunction at the quantum and atomic levels.[11] Only after a series of qualities and quantities are added together in a conjoined structure of periodic cycles do "things" emerge. Only retroactively do things appear to have qualities and quantities by "necessity" or "essence." Necessity and essence, however, are only the *effects* produced by conjoined elemental cycles.[12]

Without conjunction, there are no things, only subatomic fluctuations: an opaque wall of quarks, hadrons, and leptons. Flows keep moving, folds keep cycling, but without conjunction, nothing holds together. Without the kinetic conjunction of folds, nothing stays attached to or part of anything else. Everything flows, but motion is not a thing. It is a process. Flows are vectors or trajectories in things, but not reducible to them.

For example, in a body of water, there is no "thing" called ice. The ice is water. However, as the kinetic waveform of the water changes, slows down, cools, folds, and contracts the cycles of its molecular orbits, ice comes into existence as a thing. Once the process of hydrogen and oxygen folding slows down below a specific velocity, there is ice. "Things" emerge through kinetic processes, but the processes are not separate or independent from the things. Flows are the processes by which things come into and go out of existence. They are the warps, woofs, and vectors by which things are woven, folded, and unfolded.

The conjunctions that compose the earth are, like the flows themselves, in constant motion and can always undergo a change and recomposition. Beneath the process, there is no stable foundation or substance. The determination of the qualities and quantities of the earth is never total, complete, or final because the flows that compose them always leak or connect to something else outside them. Therefore, as a process of flows, the earth is not reducible to any fixed set of qualities or quantities conjoined at a given moment. The earth is always becoming something else, hence the difficulty of theorizing it.

★ ★ ★

The kinetic structures of "flow" and "fold" are features of our cosmos. As conceptual terms, their purpose is to help us think across scales and earth systems. I hope that, throughout this book, they can help us identify some tendencies at work in all cosmic and terrestrial events. Precious little can be said of everything in nature. My argument here, however, is that, as far as we know, everything in nature *flows*, i.e., moves and tends to spread out. As the flows of matter spread out, they also tend to *fold*, i.e., to affect one another relationally in periodic cycles. Nature has no substances, essences, or forms. Nothing is pre-given. Everything emerges through process and motion. This is the conceptual basis for a theory of the earth.

There is no "reason" this universe of flows and folds had to emerge from the early state of cosmological indeterminacy. There is no principle of sufficient reason, hand of God, or external observer that can explain the emergence of our cosmos from quantum indeterminacy. The universe came from something that was "neither something nor nothing," like a generative void.[13] It was the most radical act of relational contingency: a cosmic clinamen.[14]

However, there is yet one last conceptual term needed to help frame another remarkable feature of nature, e.g., that it tends to *circulate* in at least four discernible patterns of motion. This is what we turn to in the next chapter.

3 The Planetary Field

THE THIRD CORE CONCEPT of geokinetics is the "planetary field." A "field," in my admittedly general definition, is a metastable distribution of periodic cycles. As matter tends to flow and fold, it also creates metastable feedback patterns that can sustain and reproduce themselves. I call this reproductive kinetic process the "circulation" of the field. This chapter provides a theory of how the conjoined flows I discussed in chapter 2 distribute themselves into four distinct patterns. These patterns of motion are at the heart of the kinetic theory of the earth, as we will see in the remainder of this book.

I use the concept of a planetary field to designate not only planets in the strict modern definition but all "wandering celestial bodies," to which the Greek word *planētai* initially referred. My purpose in expanding the meaning of this term is to make explicit the eccentric movements of all celestial bodies, of which the earth is only one. It is crucial to keep in mind that the earth is kinetically related to the cosmos and other celestial fields of circulation, including galaxies, solar winds, and even densities of dark matter.

My purpose in broadening the definition of planetary circulation is not to obliterate the distinctions among different degrees and orders of cosmic organization. Instead, I am trying to emphasize the connection of all earth systems to the broader astronomical context. I worry that if our theory of the earth remains too limited, we will miss the bigger process-kinetic picture within which the earth is embedded. I worry that we might think of the earth as a kind of object or as a self-enclosed autonomous subject.

I recognize that each field of circulation has its unique elements, relations, and capacities. I will explore these in Parts II and III. In this chapter, however, I want to establish a simple alternative conceptual framework based on a few shared tendencies that define material processes—what I call "flows, folds, and fields"—more broadly. I do this to avoid the problem of static, substance-based thinking that might emerge later on in our history and ethics if we did not start from a process perspective.

The history of Western thought has often sought ultimate grounds of being and knowledge in God, reason, space, and time. Rarely have philosophers considered the material conditions of their search itself in the geological or astronomical record. They assumed grounds to be stable because they lived during the Holocene.

However, if everything flows, then thought too is part of the earth. There is, therefore, a deep geological and cosmic dimension to our philosophical and scientific paradigms that we have not sufficiently investigated. Is human culture simply one geological and astronomical expression among others? The radio waves we use to broadcast culture on earth also radiate out into space and diffract with other cosmic waves in a silent symphony. It's time that we broaden our inquiry into the conditions of the present beyond human history and culture.

Human knowledge is a local cycle within much more extensive fields of planetary circulation that have no ultimate solid ground. The problem is that we have acted as if our particular geological situation were a universal feature of earth, wholly ignoring the material, kinetic, and deep historical conditions of our planetary being. Western philosophy has consistently interpreted the world upside down by falsely universalizing the geological particular.

The purpose of starting with an idea as broad as "motion" in this book is not to obliterate the specificity of the geological present. Instead, the aim is to show how the present emerged from much larger processes in which it is entangled. The earth is a specific region or field of celestial circulation implicated in others, like the diffraction of waves on the ocean.

People often call the earth the "lonely" planet because we have found no other humans or evidence of life outside earth. Even if it were true that there is no life outside earth, which is highly unlikely, this is a profoundly anthropocentric and biocentric way of thinking about the earth. It allows us to ignore and devalue our relationship of dependency on non-living processes here and in the cosmos at large. When we do this, we use a particular form of life to judge the world as a failed attempt to become that form.

By contrast, our whole solar system most likely emerged from the same nebular cloud of molecular gas and dust. We are, therefore, not externally related to the other planetary circulations in our solar system but internally entangled and intra-related with them. The earth is not simply among the stars, it is *of* the stars.

Celestial bodies are not separate bodies that come into relation but, rather, emergent regions of the same indeterminate movement of matter. So fields of planetary circulation are not the paths of mechanistic space rocks following predetermined laws but, rather, intimately interwoven processes.

What is required then, in my view, is a philosophical framework that begins with moving matter and builds up to more complex circulations like the earth and life. A theory of the earth should not start with the earth as a model, compared to which all other celestial bodies fall short. In other words, the universe is not a failed representation of the earth. Minerals are not life forms that failed, and we are not alone in the cosmos. A vast network of intra-relational dark matter, galactic gravity, and solar processes intimately supports us through patterns of vast kinetic circulation. Our movements are continuous with theirs and change all the time. We are dead stars blown on stellar winds.

Western culture has tried to deny its entanglement with and dependence upon matter, the earth, and the universe. This book is an attempt to flip the standard story of earth history in light of recent historical changes. Complete acceptance and understanding of our condition will not come from any "centrism," whether it is a geo-, bio-, or eco-centrism. There is no center of the universe anywhere.

Our "environment" is not limited to our planet. In reducing nature to the earth, much of environmentalism has suffered from a myopic metaphysics of substance. Nature is not, however, the recorded history of the earth. One of my hopes in *Theory of the Earth* is to provide a non-centric process-based perspective of our geohistorical conditions. It aims to be an analysis of circulations *without centers*.

We are living today, in the Milky Way galaxy, in our solar system, and on our earth, in a brief metastable bubble between regular 25-million-year cycles of mass extinction. Today we know much more than we used to about the incredibly eccentric and unstable nature of the celestial circulations that affect our planet.[1] It is, therefore, no longer possible to think of the universe as a mechanical process or even an organic unity. There is too much deep cosmic turbulence. It is important to remember that in archaic Greek myth, before the birth of Gaia, according to Hesiod, there was Khaos, a creative indeterminate void from which all else emerged. A theory of the earth should start with Khaos and end with Gaia.

THE FIELD OF CIRCULATION

What is a field of circulation? A planetary field is a flow of matter that has folded up into a series of ordered elements. A field provides a path of circulation that binds together and orders a regional distribution of elements. Conjunctions of elements are different because they add folds together into larger and smaller "things," or composites of qualities and quantities. A planetary field, though, binds together these conjoined groups of elements. It is a binding and ordering flow or pattern that moves through all the folds and subfolds and then repeats the process. Elements allow flows of matter to persist through cycling, but planetary fields distribute conjunctions through feedback patterns.

For example, around a hundred thousand years after the Big Bang, as the universe cooled and slowed down, atomic and molecular elements began to emerge. Before the domination of elemental matter, no structured or ordered circulation could emerge, because hot dark matter moving at near the speed of light was far too turbulent to sustain large-scale celestial structures.

However, at the transition point or gradient between hot and cold dark matter, metastable fields in the universe began to emerge. The violent perturbations that began with inflation seeded the initial turbulence that put a spin on this cooling matter—like stirring cream into coffee. The result was the first planetary fields: galaxies and exploding stars.

Hot dark matter pushed the universe apart, and cold dark matter pulled it together into dense clumpy regions. Stars, galaxies, and clusters of galaxies were all created from tiny quantum fluctuations and the uneven distribution of dark matter in the early cosmos. Even today, dark matter continues to support the galactic circulation of matter by pulling stars that might otherwise float away or be ejected by a supernova back into particular regions. We have dark matter to thank for keeping the heavy elements together that support our planet.

Early galaxies were not isolated islands. They circulated elements between and among themselves by converging and exploding massive stars that eventually broke down into stars that were more like our sun. Dark matter holds galaxies together and apart. As smaller galaxies evolved into larger ones and the matter between them drifted away, the process generated a network of high-density filaments that look like mycelia networks.

The shape of most galaxies is spiral due to the angular momentum they inherited from their formation from early perturbed gas clouds. Not only is the earth turbulent, so are galaxies, which are continually evolving, circulating, and transforming themselves.

The earth is a planetary field at the intersection of numerous other planetary and celestial fields, and the history of the earth is intimately enmeshed in extraplanetary processes. For example, the earth accumulated heavy metals like gold, lead, and iron because there was enough dark matter in our galaxy to pull them in from elsewhere. The sun runs on the same hydrogen made by the early post-inflationary universe. Even some of the organic carbon and amino acids on the earth came from meteoroids and asteroids. Nor is the moon just a passive hitchhiker: it helps stabilize the earth's orbit and climate. Even the stellar winds play a part, blowing away extraterrestrial debris that might otherwise smash into us. Other planets, like Jupiter, protect the earth from asteroids, and the heliosphere around the sun shields us from dangerous cosmic rays. All of this is also our environment.

The earth is not an autonomous, self-regulated, or organic whole.[2] It is a region of entangled and knotted celestial fields whose eccentric and pedetic contingencies participate in the major historical features of our planet, such as mass extinctions, climate change, the evolution of life, and technology. These are the deep material conditions of the earth, and they have important consequences for our theory of the earth.

The earth circulates, cycles, ebbs, and flows because gas clouds, galaxies, and dark matter also circulate, ebb, and flow. The idea that humans can geo-engineer the earth, as a new capitalist frontier, is based on an outrageous ignorance of the deeper history of the earth and its entangled planetary field of circulation. To reconstruct the earth, or to believe that it is reducible to a human-earth hybrid, is to ignore these larger processes.

The geocentric view also dramatically overstates the significance and power of human beings. The universe is not a human-nature hybrid, and the earth is very much part of the universe. Humans are simply a region of the universe with a highly asymmetrical relationship to it, as the universe conditions humans without humans having much control over the universe in return. Our prevailing idea of relationality as being reciprocal is just another particular mistaken for a universal. We think that we can affect things as much as they affect us. But that is not true.

Actually, since the past affects us, but we do not affect the past, all relationality is, by definition, asymmetrical. This is entropy, the tendency for the universe to spread out and move on.[3] The question is not whether all of being is related or not related to itself, but *how* it distributes itself in fields of circulation.[4]

I therefore define a planetary field as an asymmetrical order or arrangement of elements. A flow of matter moves through all the different elemental conjunctions, but at a specific limit, the flow recirculates and reproduces the arrangement of elements. The result is a metastable body that persists through iteration, such as a planet, a galaxy, a star, or an organism. Every field, however, is also subject to entropy.

Without a circulating field, elements are only partial objects, like a planet without an orbit. Kepler's laws of planetary motion (1609 and 1619), for example, describe a mathematical and geometrical circulation for the previously anomalous motions of the planets. Kepler began with a fragmented list of the observed magnitudes of planetary motions inherited from his teacher, Tycho Brahe, and after adding some of his own observations of Mars, discerned a consistent and repeatable elliptical circulation of motion in all the planets, which assumed the movement of the earth. Kepler thus discovered a planetary field of circulation for Brahe's anomalous observations.

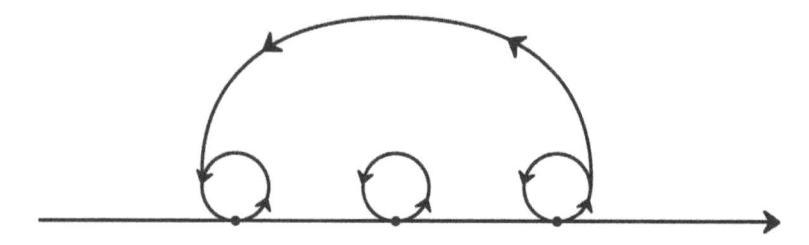

FIGURE 3.1 Field of circulation

A field of circulation is nothing other than the flows of matter and elements that compose it. The process of circulation is not a thing, any more than the irregular elliptical orbits of planets in our solar system are things. A field is a pattern that never precisely repeats. A field is an emergent order made by the collective iteration of material elements.

In other words, the kinetic condition, or "field," is immanent to what it conditions. This circulatory field is what gives folds a consistent and repeatable relation to one another. Without the differential repeatability of this relation, celestial bodies have no persistent arrangement or structure. Without fields, there would be elements with no discernible patterns of relation between them, as in the Quark Epoch, where there were not yet atoms (see Figure 3.1).

PLANETARY KNOTS

Fields can also combine themselves into larger and smaller networks, or knotworks, based on their shared motions. The earth is a planetary "knot" in that it is an intersection of two or more planetary fields sharing two or more of the same elements or folds (see Figure 3.2).

The knotting that occurs in the planetary field is what makes continuous intersecting fields of circulation possible. For example, in a geokinetic knot, each field remains in some sense distinct but also becomes connected at specific elements to other fields, making possible a series of shared or collective mutations.

One galaxy becomes another not by mimesis, metaphor, or representation but by literally sharing the same elements or folds with another field. Two fields are knotted together by their shared elements, but these regions of shared elements can also produce their own field in turn: a knot. The earth is a knot in several astronomical circulations. The circulation of the Milky Way, the sun, and the moon all braid together with the earth.

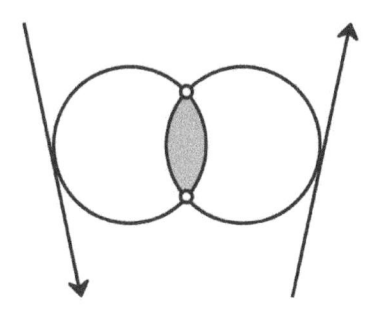

TWO-KNOT

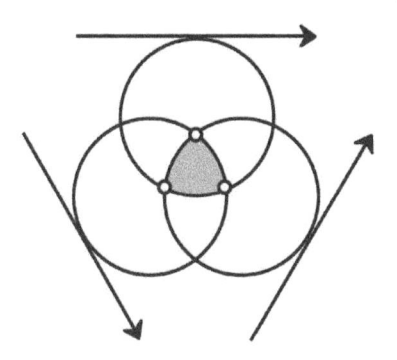

THREE-KNOT

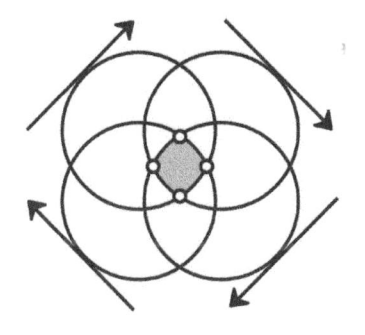

FOUR-KNOT

FIGURE 3.2 Knots

Nests

Nests are different from knots. Just as there are larger and smaller folds, each one containing the previous without any final one above or below, so there are larger and smaller nested fields of celestial circulation. Within any given field, only the elements appear; the field itself does not appear as an element because it is the pattern or trajectory of elements.

However, at the next larger level or nest, a subfield will appear only as a *thing* and not as another *field,* relative to the larger field that contains it. A knot is a nontotalizing overlap of one field with another that entails the sharing of at least two elements. A nest is a totalizing relation of one field completely inside another.

For example, our solar system nests in the Milky Way, but the earth is a knot between planetary orbits in the solar system. The point here is that nests and knots are entangled processes. Even minute differences in lunar and planetary orbits can affect the earth, its climate, and its creatures.

Therefore, the fundamental questions of the earth are material, practical, and kinetic: What can it *do*? How can it *move*? The earth has no essence, substance, or form. It is nothing but movements—flowing, folding, and the knotting together of different fields across shared elements. To describe the earth is to locate its kinetic elements and the field of circulation that orders it. The more folds it shares with other circulations, the greater the degree of similarity or "becoming" between them as an overlapping composite.

Fields of circulation braid together, but they also produce a knot, or new field, that occurs as the intra-action between the fields. A knot is like a dance where two fields directly coordinate their motions around a few shared objects. The knot is not an interaction between two separate individual fields, but rather what happens in between them: the entangled dance of their motion. Both fields thus undergo a mutual (not equal) transformation by coordinating their motions. Instead of remaining isolated, the knot allows multiple fields to become a single field with two or more dimensions or pathways.

Knots also make it possible for fields of circulation to morph or change their patterns of motion without changing the number of their shared elements or crossings. As long as the morphisms or movements in the circulations do not disjoin from the shared elements, the two fields remain knotted. However, as planetary fields change and move, their flows and folds may move closer to or farther away from one another, forming different "kinotopological neighborhoods" or proximities. Neighborhoods might change, but the number of shared elements will remain the same within the knot. In other words, knots are what allow composite celestial bodies and elements to persist in their composition without dissipating, even when they are moved around or morphed.[5]

The planets in the solar system, for example, are tightly knotted fields of circulation insofar as they share many of the same kinetic elements of the same nebular mass, but they also remain distinct in meaningful ways. They can also be moved around and changed to some degree without becoming unknotted from the solar system. Planetary orbits are always shifting slightly in response to one another. This shifting traces the knotted path of their circulation.

COSMIC FIELDS OF CIRCULATION

We circulate because the earth circulates, and the earth circulates because the universe circulates. The heliosphere, dark matter halos, and gas clouds all move, circulate, and fluctuate. Their continual circulation supports our terrestrial circulations (atmosphere, hydrosphere, geosphere). These circulations, in turn, support our biosphere, noosphere (the circulation of thought), and the new geological strata of the technosphere, circulating satellites and space debris.

The four major kinetic patterns of the earth described in this book were already invented by the cosmos. The earth is not the origin of these patterns but a particular iteration of them.

At the very beginning of the cosmos, however, there were no patterns, because there was no spacetime, no elements, and no fields of circulation. All the patterns we know today emerged relative to one another. In Part II, I show that the four main patterns of motion or fields of circulation invented by our universe were adopted and iterated by our earth in its own way.

Centripetal fields

The first type of field follows a centripetal motion. Centripetal fields circulate from the periphery toward a center, without necessarily creating one single center. By gathering elements from the periphery of a field, centripetal circulation captures the chaos of pedetic flows and orders them into a smaller but denser area. A centripetal pattern captures motion from the periphery and turns it inward, toward a central basin of attraction.

After the Big Bang, for example, the universe began to gather itself centripetally together into various turbulent eddies of gravity. As these gravitational eddies increasingly gathered more and more mass, they spun into particles and eventually into the first centripetally produced megastars, one hundred to three hundred times the size of our sun. As they burned out, they either exploded into supernovae, populating the early universe with newly fused particles, or they collapsed into supermassive black holes. Black holes then continued to contract spacetime and particles, centripetally, into early galactic clusters.

Gravity is the first centripetal pattern invented by the universe. It is the pattern that holds every relatively stable thing together. It is the kinetic pattern responsible for the origin of all our stars and their eventual implosion. Black holes are the most dramatic example of centripetal motion. In black holes, spacetime contracts in on itself and continually pulls matter toward a central region.

Centrifugal fields

The second type of field follows a centrifugal motion. Centrifugal fields circulate from the center of a field to the periphery. Centrifugal circulation occurs when a concentration emerges at the center of the field and begins to redirect all motion through this central fold. It becomes a radial point through which all flows move. It regulates and directs all the internal motions of the field. Movement is, in short, redirected from the center outward to the periphery.

In the universe, after the earliest stars accumulated enormous amounts of helium and hydrogen, they exploded outward in all directions under their weight. These centrifugal explosions then seeded other centripetal accumulations elsewhere. Even supermassive black holes accumulate energy centripetally, but then slowly radiate it back out centrifugally.

Tensional fields

The third type of field that emerged in the universe followed a tensional motion. Tensional fields conjoin their folds together through a system of rigid links. These rigid conjunctions keep the folds both together and apart.

In this way, rigid conjunctions decenter the movements of centrifugal circulations by connecting them. Kinetically, fields have their order and distribution of motion. However, when field movements are held together by rigid conjunctions, the motion of one fold becomes restricted by the motion of the other. Tensional motion occurs when there are two or more folds or circulations whose motion is constrained by others.

In the universe, after the centrifugal explosion and radiation of the first stars and black holes into the first galactic clusters, stars and planets began to settle down into metastable orbital patterns around one another within these galaxies. Orbital motion is a distinctly tensional motion because it distributes a gravitational tension between celestial bodies. If the earth were stationary, it would fall into the sun, but since it is moving at a great enough velocity, it runs parallel to the sun, though without escaping the sun's gravity completely. This orbital motion maintains a tensional balance, keeping the earth and sun together and apart at the same time. This is the same tensional movement that governs most orbiting celestial bodies.

Elastic fields

The fourth and final type of field follows an elastic motion. Elastic circulation conjoins the folds of a field together through a system of elastic links. These elastic links allow folds to move together and then return to their previous position after a contraction or expansion of motion. In contrast to the rigid conjunctions that constrain the movement of tensional fields, elastic conjunctions are flexible and allow the field to oscillate back and forth, expanding and contracting, without falling apart.

In the beginning of the universe, galaxies, stars, and planets were clustered together around enormous collections of dark matter. However, the remaining dark matter between galaxies began to pull the clusters further and further apart at an accelerating rate. Dark matter moves elastically, in the sense that it allows galaxies to expand and contract. Since dark matter makes up 85% of our universe, its movements and distributions cause some regions of spacetime to contract and then expand, depending on how these massive halos of dark matter converge and diverge. It stretches spacetime like the rubbery surface of a ballon.

* * *

This chapter concludes Part I of this book. My proposal so far is that recent events reveal the earth to be much more unpredictable and relational than we thought possible. It turns out that the material earth is much more like a process among other processes than it is a passive object or autonomous subject. The earth is not inert determinate stuff, but a kinetic process that prompts us to rethink our theory of matter and motion as well.

More specifically, I have suggested that cosmic and earth systems have three main tendencies that we can use as a historical-conceptual framework to make sense of the diverse processes of our planet. Cosmic matter flows and folds into elements that then circulate into celestial and planetary fields. The earth is part of this more massive material process. It flows, cycles, and circulates because the universe at large does.

The earth is, therefore, a continuation of cosmic patterns by other means. Human history, too, is a continuation of these patterns by other means. The patterns, however, are not predestined. They mix and may even produce new ones that we cannot anticipate.

My aim in emphasizing these patterns of motion across cosmic, terrestrial, and human systems is twofold:

> 1) To attenuate or overcome all notions of ontological division between the earth and the universe, as well as between humans and nature more generally.
> 2) To describe the material and kinetic conditions of the earth's motions as an interconnected knotwork of material circulations at every scale.

The aim of Part I of this book has been to create a theoretical framework for thinking about the earth as a mobile and metastable process rather than as a passive object of mechanical laws or an autonomous organic subject. The concepts of flow, fold, and field give us some tools with which to think about earth processes across multiple physical scales and to describe the coproduction, or sympoiesis, of the emergent patterns of motion that define the earth's history.

The concepts of this framework are important, but they are not enough. I have synthesized these concepts from history, and presented them here first to aid the reader, but they may feel a bit abstract. If we want to understand the immanent material conditions of the present earth, we also need a material history. Contained in our contemporary moment are the terrestrial traces of the universe's and the earth's deep history. If the past supports us and persists into the present, we cannot understand our present without understanding its history. This is the task of Part II.

In a basic sense, then, this book is not a theory *about* the earth but a process *of* the earth. When we think, we think through the earth and as the earth. Therefore, just as human history, society, and technology are crucial to understanding our contemporary condition, so too is terrestrial history crucial to understanding it as well. If we ignore the material conditions of the earth, we cannot possibly understand our present, which is nothing other than the kinetic structures of the past mixed and continued by other means.[6]

Only by profoundly shifting our frame and depth of inquiry will we be prepared to confront the most significant problems of our time.

PART II
HISTORY OF THE EARTH

A. MINERAL EARTH

4 Centripetal Minerality

THE EARTH IS HISTORICAL. If, as we argued in the last chapters, it is nothing other than matter in motion, then it is by definition historical. There is no matter outside history and no history that is not material. The deep historicity of the earth has at least two significant consequences for a theory of the earth, which will be the subject of Part II of this book.

First, because the earth is material, kinetic, and historical, it is possible for different, coexisting, and mixed planetary fields to emerge. In other words, matter can distribute itself differently over time into different patterns or orders of arrangement. There is no way to know what the earth is, I argue, without understanding its historical process of becoming. If this is the case, then it is possible to study this material history and to discern its planetary regimes or fields. What are the flows, folds, and fields that compose the earth? What are its minerals, atmospheres, plants, and animals when we look at them as processes?

The aim of Part II of this book is to study the four dominant regimes of planetary history during their periods of historical dominance. These include the Hadean mineral field, the Archean atmospheric field, the Proterozoic vegetal field, and the Phanerozoic animal field. Each of the following chapters defines the pattern of a primary geokinetic field as well as the concrete historical beings that constitute it.

The second consequence of the earth's historicity is that all these different geokinetic fields persist into the present. The contemporary earth is not a single geokinetic field or pattern of motion. It is composed of a heterogeneous mixture of everything that has ever been. Today all the dominant planetary fields that have ever existed persist and mix in various patterns. Deep history is not just the background for the present. It is the present.

The earth's history is not progressive, linear, or even zigzagged—it is compositional, hybrid, and additive, like the unfolding of a crumpled piece of paper that changes as it unfolds. Therefore, any theory of the earth that does not engage the coexistence and mixture of historical planetary regimes is not merely lacking background but is lacking a foreground as well. This will become clear in Part III, where all the historical labor of Part II will contribute to the ethical analysis of our contemporary moment.

Just as the earth is a concrete and singular historical expression of the four regimes of motion in the universe, so is human history a concrete historical expression of the earth's four main regimes of motion. If this is true, and I argue elsewhere that it is, then it means that there is no ontological division between the cosmos, the earth, and humans.[1]

What we call politics, ontology, art, and science are both historically singular and already present in the earth and in the cosmos in a more general sense. Humans and their unique histories do not emerge ex nihilo from a radical break with nature but are kinetically continuous with it. Humans are not on top of or separate from nature, as if it were a pyramid that supported them.[2] Thought, language, and culture are not present in humans and absent in nature. Rather, human thought, language, and culture are singular cases of general patterns in nature.[3]

In other words, the cosmos and the earth already invented and laid out the more general kinetic regimes of motion that iterate through human history. The earth's kinetic regimes are not just material conditions that support, but are separate from, human history. They are immanent processes contiguous with human history. Nature and culture are processes. However, not all patterns of motion are the same, nor it is the case that some patterns are copies of originals. Everything is singular, but tends to circulate in at least four main patterns of motion.

MINERAL GEOKINETICS

The first major planetary field to rise to dominance in the earth's history was the mineral field. This first type of field rose to dominance throughout the Hadean Eon, beginning with the formation of the earth, about 4.6 billion years ago, and ending around 4 billion years ago.

We often think of minerals as rocks or stone because these are its most reified and anthropocentric manifestations (the pebble, the monolith, the mountain).[4] However, this is a narrow range of minerality. Deep history offers a more robust view. Minerality is simply an ordered gathering of elements. It is atoms arranged in a crystal-lattice pattern.

The Hadean is perhaps one of the most beautiful and sublime eons of the earth's history. During the early formation of the earth, everything flowed. At extreme temperatures, all solids turned into liquids, liquids into gases. These, then, rained back down as liquids and turned back into solids.

Geochemists frequently define minerality by its highly periodic or ordered crystalline structure, but during the Hadean, minerals were highly fluid. This is a crucial insight: all that is liquid crystallizes into solids. Stone is not first something durable, like a fixed ground that humans and other beings live on, but something fluid and turbulent. Minerality is not something first ordered and fixed, that is then disrupted by volcanism and earthquakes. It is a kinetic pattern that emerges and submerges from flows of cooling magma.

The Hadean shows us that volcanism is not the exception to the rule of a stable earth but that the stable earth is instead the crystallized product of a continual volcanic process. The Hadean is not a relic of history but a continuous and contemporary process boiling underneath our feet. Therefore, when thinking about minerality, it is crucial to do so from a historical perspective. This history begins with the pedetic flow of rock, not the solidity of stone composites we find on the ground.

The Hadean is not the foundation or ground of the earth. It is the ungrounding of the earth. It is the demineralization of minerality. If our history of the earth is possible only because of the crystalline structure of minerality, then the Hadean earth is anti-history. The Hadean earth is the earth that is not and cannot be identical to itself. It is a fluid earth without memory.

In other words, at the historical and geological heart of the earth is a liquid history, real and fundamentally indomitable. Nothing on the planet could be as immanent and constitutive and yet at the same time as deeply ungrounding to humans as the earth's ongoing volcanism.

Minerality, in the Western conception, has long occupied the lowest rung on the great chain of being. Nothing is lower than the seemingly inert and passive stone we stand on and form with our tools. Minerality is the literal bedrock upon and above which all the other creatures of the world stand.

And yet, without the base materiality of stone and minerality, the whole chain of being would come crashing down. If minerals are the constitutive and mobile conditions of all earthly beings, then this has enormous consequences up the chain—throwing everyone back to the ground. In this chapter and the next, I would like to show that the movement of the basest of all matter, the mineral earth, is materially constitutive.

A genuine overthrow of the chain of being will not come by raising lower links of the chain to the fraternal level of life.[5] Instead, it begins by showing the dead and nonhuman movement of minerals as constitutive of and immanent to life and humans. Therefore, I do not want to extend a principle of vitalism or animacy to minerality. This would only tacitly reproduce the division between non-life and life that I wish to reject.

Minerals are not vital or animate. Rather, vitality is mineralogical. This inversion may seem trivial, but it is the subtlety that makes all the difference. Vitalist theories of the earth are backward because they assume that a given product (life) defines the essence of the process that produced it (minerality).

Hadean mineralogy, unlike life, does not survive on the energy of a solar economy. Unlike the processes in biocentric theories of solar economies,[6] minerals can move, order, and form themselves without the aid of the sun's energy. The sun's radiation is the source of life, but without the continuously molten Hadean earth that gave heavy metals like iron the time to sink to the center and create a magnetic shield around the earth, we would not be here, either. Minerality is, therefore, the condition for the sun to be a source of life on earth.

The earth, in this sense, is like a second sun, or rather, perhaps, a first sun. Liquid Hadean metals at the core of the earth are the immanent and base material conditions of solar vitality itself. More specifically, though, the earth's magnetic core and magnetosphere constitute an anti-sun. The mineral "soul" of the earth shields against the sun and insulates the earth. This is not the will of a living earth or Gaia to create and perpetuate life, but the movement of a dead minerality or Python to circulate a planetary field.[7]

The minerality for which I develop a theory in this chapter has three interrelated features: 1) it follows a centripetal motion; 2) it is a process of exteriority; and 3) it creates a material archive. After I lay out these core kinetic features of minerality, we will be able to recognize them more easily in the historical period of the Hadean Eon developed in the next chapter.

CENTRIPETAL MOTION

I define the "mineral field" first and foremost by the continuous flow of matter from the periphery toward a center. Any distribution or field of elements requires this centripetal motion such that two or more gathered heterogeneous elements can enter into some ordered relation that ties them together. Flows of matter without folds are insensible, while folds without fields are fragmented. Only when elements are gathered together into order do they become a planetary field.

Minerals are emergent patterns of matter's self-ordering. Flows of matter move centripetally into new planetary fields only at the expense of other fields. This is entropy. Every ordering spreads out and begins a reordering of matter at the same time. Elements from our early solar nebula were disjoined from the sun in heterogeneous flows of minerals around the periphery and conjoined into the centripetal accretion of the planets. In the process, an ordered mineralized planet emerged, mixed, and cooled the flows into dense crystalline patterns.

The accretion of mineralized spacetime occurred through centripetal motion. If the flows of matter and its folded elements did not move from a periphery to a center, the elements could never take on any mineral distribution. Everything would fly apart. Matter flows, but because the flow is pedetic it also curves and swerves back over itself toward a gravitational center.

The earth has no single center, only a liquid metal region with shifting boundaries.[8] The curvature of matter defines an area in which elemental folds can gather. As they gather, they produce a mineralogical order. Elements move into order only because they are gathered together into kinetic proximity with one another.

There is thus no fixed number of minerals, but like "species," minerals are emergent patterns of elements centripetally gathered together. As the molten earth moves and meteors hit it, new minerals are born and evolve—and are still evolving.[9]

EXTERIORIZATION

Minerality is the exteriority of the earth to itself. This is the second defining feature of minerality. Before there was an earth, there were nebular debris and meteors of various elements and minerals. The earth, like other planetesimals, was the vortical product of elemental and mineral composites swirling around the sun. Minerals thus emerged before the earth.

The earth is not something pre-made or fixed, but rather a process of becoming in which planetesimals and meteors accreted and destroyed one another in a spiral fashion around the sun. Each orbital cycle of debris around the sun produced new composites of rock and new mineral species and changed the orbital distribution of its mass. In other words, early planetesimal orbits were not circular or even elliptical, but spiral, since the whole moving mass was transformed through composition and collision each time.

There was no preexisting "little" earth that simply got bigger through accretion in a linear fashion. There was no progress. Matter accreted, collided, destroyed, re-accreted itself, and was absorbed into something much bigger.

The earth is an assemblage of singular material flows. It is the result of contingencies and swerves of solar winds and errant asteroids. All these contingencies make the earth a stranger to itself, without an absolute interior or exterior.

The earth is a cosmic exterior centripetally folded in on itself through the vortical motion of nebular accretion. It is the interior of an exterior. It is a fold. In other words, the earth is already profoundly alien to itself. It is a hybrid monster composed of all the motley stuff of nebular waste: a piece of shit stuck to a solar anus.

Even after the "early earth" formed through nebular accretion, it absorbed an enormous Mars-sized alien planetesimal (Theia) that crashed into the earth, liquefied it, and transformed it 4.5 billion years ago. How can we speak about "an earth" before this queer copulation? After the collision with Theia, there was a new earth, so mixed with another that the two could no longer be distinguished. The moon is not just a piece of the earth or of

Theia—it is a hybrid product of Theia and the early earth. The moon and the earth thus remain internalizations of cosmic externality.

The point here is this: there is no origin of the earth because the earth is not singular but a multiplicity of mutually extraterritorial or exo-territorial flows still coming together centripetally. The earth's center runs on an infernal and turbulent core, made partly from the alien metal heart of Theia.

Just as it shocked 19th-century readers to learn that humans descended from apes, so it should similarly shock us today to learn that the condition of our evolution (the earth) descended from alien minerals. The earth did not come from itself but is the product of a multi-planetary–mineral hybrid. There is no binary between inside and outside—only a continuous process of folding and unfolding.

Meteoroids, for example, composed our planet from the beginning and continued to bombard it throughout its entire history. All precious metals (gold, silver, platinum, and others) of the early molten earth sank to the core. They would have been enough to cover the earth with a twelve-foot-thick layer. Everything we find today on the surface has arrived from exoplanetary meteoroids. Our metal tools are made with alien minerals.

It was not just during the first 100 million years that the earth was strange and alien to itself. Alien matter continues to arrive and transform the earth. Foreign matter comes and begins to breed with terrestrial elements, thus producing new mineral species. These mineral species then, in turn, change the mineral components of living species and the atmosphere. Flows of meteoroids thus "precede and exceed" all planetary circulation. Our planetary body is the vortical production of wandering minerals.

The petrogenesis of spacetime

Minerality is not just the stones or rocks we see. Minerals, in the form of the accreted earth, are not *in* spacetime—they are spacetime itself. Mineral accretion from the early nebular cloud produced the spatiotemporal dimensionality of the earth. Minerality is why we have depth on earth.

Before there is a territory, there must first be a *terre*, which distributes and supports a spacetime. Minerality is, therefore, the condition of our spatiotemporal encounter with any particular rock or stone. A pebble is not just something we find lying around but a region of minerality that creates the spacetime within which such an encounter can occur.

Elements and minerals, for example, compose our bones and bodies. They give us a spatiotemporal externality. Minerals are processes of spacing and spanning. Cultural histories of stone are secondary to the more fundamental deep historical condition of kinetic minerality.[10] Rocks and mountains, for instance, are what support spatiotemporal relations. More fundamental, however, is the centripetal minerality that supports rocks and mountains in general.[11]

Spacetime does not preexist the centripetal motions of matter that gather to form it. The kinetic process is primary. The spatial difference between internal and external is, therefore, an emergent property of centripetal minerality. Minerals force us to imagine an "externality" more radical and primary than the difference between internal and external. They force us to imagine minerality as a process without an absolute inside or outside.

Furthermore, nothing stays buried in the earth. Extreme temperatures at the core and variable cooling rates for magma throughout mean that minerals solidify at different temperatures. Depending on their weight, they sink or float. Inside the earth, there are giant convection currents of magma that rise with heat, cool, and sink back down, like potfuls of boiling water.

Not only is the earth outside itself because of exo-planetary meteoroids, it is also outside itself because its outside crust is only a temporary externalization of a deeper internal process. The earth is a Möbius strip. It is externally and internally ungrounded. During the Hadean Eon, the difference between the earth, the moon, and the cosmos was a fluid process. Centripetal flows are not opposed to viscosity; viscosity, rather, is something that emerges through the folding of meteoric flows back over themselves. Matter slows and folds itself up through asymmetrical relations of accretion. These heterogeneous minerals merge and produce a viscous planet. However, just because there is viscous accretion does not mean that what accretes is homogeneous. The flows of matter are still pedetic and wandering. The emergent patterns of viscous relations then internally differentiate the earth and hollow it out through the swirling patterns of magma and volcanism.

The earth is porous from top to bottom. Inside, it is filled with lava tubes, venting systems, magma chambers, and magma dikes that burrow through the earth like worms.[12] Because the flows of meteoroids are pedetic and heterogeneous, so are their viscosities and burrowing porosity into the earth.[13] The earth becomes its own ungrounding by destroying itself and eating away at itself. It is precisely the earth's externality to itself that allows it to impede itself. It preserves itself as a metastable process of centripetal folding.[14] The earth is a kind of creation through self-destruction.

ARCHIVALIZATION

The third feature of centripetal minerality is its archival function. A common feature attributed to stone and rock is its ability to record ordered patterns of atoms into minerals, minerals into strata, and strata into recorded history. The 18th-century geologist Comte de Buffon called stone "the world's archives" and "ancient monuments from the earth's entrails."[15] However, minerality is not at all like an archive, in Buffon's sense, for two reasons.

First, because minerality is a kinetic process, it has no pre-given *arche*, or archive. Buffon knew nothing of the Hadean earth or of the material conditions of nebular and meteoric accretion that produced the earth. What is so radical about the Hadean earth and the pre-Hadean accretion process is that there was no archive. The earth does not begin with archival memories already in place.

Minerality is not an archive. It is a liquid history of turbulent flows, melting, and recomposing itself. Archivalization is, therefore, first and foremost, a kinetic process. Before there were minerals that could record, there were material flows of atoms and fluid minerality. Even after accretion, the Hadean earth was still liquid and left no fixed stone or crystal. The earth, therefore, as a kinetic process, is not an archive but a process of "archivalization" without a determinate beginning or end. It is a fluid history without recorded or fixed memory.

Second, the earth is not a mineral archive in Buffon's sense because even after the Hadean Eon, the earth remained fluid. Layers of rock (strata) are not chronological representations of the earth's history. Rocks erode, melt, collapse, and are reborn as different composites.

In other words, the earth is a recording process that transforms what it records. It is an "anti-archive," or *anarchive*.[16] For example, minerals need to be melted down and extruded from the earth in order to create ordered strata. However, the process of extrusion through magma also reconfigures the minerals that it records in the layers. The earth produces through recording and records through producing.[17] In short, minerality is what we might call an "immanent archive" that circulates without representing.[18]

Geolinguistics

Minerality is not an archive. It is a process of immanent archivalization that is the material condition for both the possibility and the impossibility of any given archive (including human archives). "Minerality," therefore, is not identical with minerals. Minerality, or mineralization, is the process that orders atomic elements into solid crystals. The ordered patterns of atoms are never pure but contain many impurities and pores. Each crystal pattern grows when more atoms coalesce through atomic bonding. Rocks, then, are composites of various mineraloids, and when rocks break or melt, they do so at varying temperatures depending on the melting point of the different mineral patterns and fracture points created by the impurities that compose them.

The English word "element" comes from the Latin word *elementum*, which has the same meaning as the Greek word *stoicheion*, meaning "an element," or "ordered letter in a row." In Greek, this process of ordering referred equally to a row of columns and to a row of letters in a word. In a general sense, then, the ordered arrangements of atomic elements (minerality) form the first written language. I do not mean this metaphorically. Minerality is not "like" human language. Rather, human language is a singular and historical expression of a much larger cosmic and geological process of minerality.

Human language is a metonymic region of a larger geolinguistics.[19] In no way do I mean to say that human language is identical to mineral patterns. That would be absurd. Human orderings are historically singular, as are plant and animal orderings, but all of them rely on a more general centripetal gathering of elements into ordered patterns. Without minerality and mineral order, there would be no humans or human language. Minerality is the general process of which human language is a distinct subset.

However, we often think of human language as separate from nature because it is supposedly "referential," i.e., it refers to objects outside the linguistic utterance. Nature, by contrast, simply *is*. But referential structures are not baked into language. They are the product of a long historical process of convention and anticipation (made possible by the earth itself). Language acquires meaning and reference through habits and coordinations of motion.[20] Language is just one of many ways that processes connect to create patterns of circulation. Meaning is fully material. This is the general case with mineral chains as well.

For example, oxygen and silicon accreted on the earth due to cosmic contingencies. Once they did, they formed new habits of kinetic coordination by floating to the surface of the earth and bonding to one another. In so doing, they made possible a whole new range of polysilicate crystal species. Once these formed, they created new "referential" structures such that putting oxygen and silicon together "means" or "refers to" quartz. Just as the sound "tree" triggers the chemical-neuronal habit of our (English-speaking) brains in a certain pattern (the "idea of a tree"), so the connection of oxygen and silicon triggers the chemical habit of electron bonding in quartz crystallization.

There is no a priori reason why the universe made these kinds of atoms (oxygen and silicon) or why they should end up in certain places, but once they do, emergent patterns of bonding can occur. Electron bonding is nothing but a habit (of accepting 2, 8, 18, or 32 electrons), just like the habit of linguistic references between certain sounds (phonemes) and objects. Both feel like laws because we cannot imagine the universe or a given language without them, but ultimately both are emergent historical patterns of motion—not ontological destinies. Mineralization, however, is the more general and deeply historical pattern, of which human language is but one regional expression.

Mineralization, like language, is not a mechanistic process, but responds creatively to its environment. The cosmos creates chemical reactions, directs meteoroid paths and volcanic turbulence, and mixes impurities of all kinds. Mineralization is even capable of inventing new minerals and reorganizing itself into new compact patterns deep under the earth. All the pedetic variables of the planet (temperature, pressure, erosion, etc.) push minerals into new arrangements with unusual results and novel mixtures. We are discovering new ones all the time. Geolinguistics is an open process defined by centripetal addition, "and, and, and . . . ," in ordinal series.

Geomnemonics

Mineralization, then, is not a fixed archive but the *process* of archivalization. This is a critical difference. An archive is the result of a process of archivalization that centripetally gathers matter into an ordinal series or territory. Mineralization, therefore, does not occur on a surface or territory but constitutes the territory and recording surface itself. Order is not made all at the same time, like an eternal Platonic geometry, but rather bit by bit, in an open geokinetic series. However, as an entropic flow of matter, this process will also eventually unravel all archives. Ultimately, every trace of the earth will break down into vibrating fields at the end of the universe.

The "anti-archive" of mineralization is the basis of all memory in plants and animals. Mineral memory is not like an animal or human memory, but the other way around. As ordered elements gather and iterate in a series, they create inorganic and organic patterns. Living matter remembers not merely by looking to external stone markers but because its mineral and elemental body is already a recording surface. Biological bodies are made from elements and minerals, as are all their recording tools (stylus and tablet).

Memory is a material process. There are no permanent files or filing cabinets in the brain or the earth. There are only cycles and circulations of vibrating atoms ordering themselves into series: minerals in a row, DNA in a row, synapses in a row. To remember is to re-perform this ordinal series. Each memory is, therefore, a "re-membering," or reassembly of the ordinal series. There is no static or fixed memory waiting to be recalled, reproduced, or represented.[21] There is only a continual activity of re-ordering and re-mineralization.

The deep historical and geological conditions of memory on our planet, therefore, have enormous consequences for a theory of memory and history at every level. The earth's process of centripetal mineralization is a moving and creative ur-memory. If human memory follows the same material-kinetic pattern, then it cannot be the recollection of an objective past situation either. It must be materially creative as well.

The past itself is in motion. Since this motion is also transformative, matter cannot store the past flawlessly. The past is something produced, destroyed, and recomposed, again and again.[22] Eventually, the whole geological and biological record will be melted back down into molten, disordered flows of magma. Human history is a subset or region continuous with natural history and is thus mineralogically ungrounded. History and memory in their "general sense" are transformative and re-performative. Each recording or archiving (human and nonhuman) is a non-neutral action that adds something to what it is recording. In short, human memory is much more like mineralization than mineralization is like human memory.[23]

* * *

Centripetal motion, exteriorization, and archivalization are the three defining kinetic features of mineralization. These are not, of course, a priori features of nature. They define only one significant historically emergent pattern that shaped and continues to shape the earth: centripetal motion. The theory of minerality above, however, is still too general on its own. I now turn to the historical and geological conditions of its emergence.

5 Hadean Earth

ALL THE MAJOR EVENTS of the Hadean Eon follow a centripetal pattern of motion. In the beginning, there was no division between the earth and not-earth. There was only a continuous flux and flow of matter moving in increasingly centripetal patterns of mineralization toward a central region. The early earth, for instance, had no atmosphere and directly touched its exterior: outer space. Because the surface, core, and exterior were all completely fluid, the difference between inside and outside was in continual flux for the Hadean earth.

The Hadean Eon is named after the Greek god Hades and is associated with hell. However, the real historical Hadean earth is much more radical than the metaphysical imagination of hell. The historical distinction between heaven and hell, above and below, form and matter, is itself conditioned on a much more radical geological and liquid hellscape. The Hadean process occurred before the difference between inside and outside, immanence and transcendence. Terrestrial difference was therefore an emergent feature of a more primary process of grounding and ungrounding.

Humans like to take credit for the theoretical invention of heaven/hell as a defining feature of their powers of mental abstraction. However, the material and conceptual divisions put forward by humans are only products of a much more primary historical process of mineralization. The distinction between inside and outside is already present in nature as an emergent property. "We never invent anything that nature hasn't tried out millions of years earlier," as the great science fiction writer, Arthur C. Clarke, wrote.[1]

The difference between the earth and not-earth, for example, is an effect of the emergent historical formation of the earth's crust. But since the crust itself is only a temporary crystallization of a difference between the depths of the earth and the heights of the cosmos, we can see that this division is, first, a material process in nature, and only then one in human minds.

This chapter argues that centripetal patterns of motion emerged and prevailed throughout the Hadean Eon. I argue that this deep history of mineralization is the condition of terrestrial motion for all subsequent eons up to the present. I will look closely at the kinetic patterns produced by four major geokinetic phenomena that define the Hadean earth: meteors, the moon, water, and lightning. I argue that each of these major phenomena contributes to a distinctly centripetal pattern of motion.

THE DIALECTICS OF METEORS

For the first billion years of the earth's history, the earth received a continuous bombardment of meteors and asteroids, some of them up to 100 kilometers in diameter.[2] But the distinction between meteors and the earth is a retroactive one. At some point, the earth was also just a meteoroid among other meteoroids. The mineralized spacetime of the earth emerged from a centripetal accretion of meteors. Instead of the bombardment *of* the earth, we could just as easily call this process the earth's bombardment of itself by itself: meteor-earth.

Furthermore, all the early planetesimals were similarly meteoritical. The process of accretion was not a linear development from small to more massive but rather a nonlinear process of accumulation, fragmentation, and re-accumulation in a material dialectic. In the early solar nebula, there was only a nebular and meteoric materialism without a goal. There was not even a strict ontological distinction between the planets and "alien" asteroids.

Thus, before there was an earth, there were meteors that accreted into thousands of planetesimals. But what actually accreted what? It is arbitrary to say that the more massive meteor or planetesimal accreted smaller ones. And yet the planetesimals bombarded meteors just as much as the planetesimals were bombarded by meteors. The bombardment was mutual. However, as mineral accretion increased, some planetesimals eventually got larger and more diverse, with increasingly dense, metal-rich cores, lighter silicate mantles, and thin, brittle crusts. This process gave birth to the first three hundred mineral species we know. Inside the planetesimals, the pressure increased and helped melt and mix minerals. This alien sex produced the diverse orders and internal layers of minerals that now populate our home.

Every planet today retains traces and memories deep in its inner layers and mineral species of what we could call "the great exteriorization event." As thousands of planetesimals collided, their mineral archives were also scattered and fell to the surface of the early earth and other planets (as they have been falling ever since). The blasted iron cores of early planetesimals fell to the earth and merged with it.

Deep space and time are thus always reemerging, here and now. Our bodies and tools are made of them. The ancient surfaces of early planetesimals now fall from skies all over the solar system. Meteors are heaven on earth. What a mistake that we ever thought of fixed grounds and heavens. The two continually weave together by the threads of the mother of mineralization: meteors.

The word "meteor," from the Greek word *metéōros*, meaning "raised from the ground, hanging, lofty," originally described any object in the sublunary world (below the moon and above the earth). Meteors and weather, for the Greeks, existed in between gods and men, heaven and earth. This is why the term "meteorology" today, following in this tradition, is the study of weather and climate. We are not made from gods, nor from the earth, but from meteors. They are the process-objects of the cosmos.

The kinetics of meteors and mineralization, however, go much deeper than this. They utterly overthrow our conventional sense of what climate is and where climate happens. Meteors are other grounds from other bodies with depths and surfaces whose collision with the earth has brought us all the minerals that now compose our atmosphere and climate. In other words, meteors are the centripetal conditions of the earth's depths, surface, and exterior. Meteors are not just in between in a spatial sense. They are kinetically constitutive of the whole field of planetary circulation more generally.

We distinguish meteoroids (in space), meteors (in the atmosphere), and meteorites (on the earth), but these are only three relative aspects of the same material process. If the earth is already composed of meteorites then how can it truly distinguish itself from those outside it?

Our planetary space is the product of kinetic flows of meteors. Things flow on the earth because flows became things on the earth during the Hadean. It is a geocentric bias to divide meteors by their location on a surface, in the atmosphere, or in space. All three are the same swerving material.

The movement of meteors weaves the ground and sky together. "Above" and "below," "matter" and "form" are regional effects of the same process of meteorization. There is a similar kinetic process of mineral crystallization between gas, liquid, and solid. The difference between the ordinal patterns found in gases, liquids, and solids is simply a matter of degree of purity in the order and of centripetal contraction. There is no ontological difference in kind.

This is strikingly clear in the earth's Hadean history. During the Hadean Eon, as noted above, the earth was hit by a Mars-sized planetesimal called Theia. After the collision, the two bodies sent mineral debris of various sizes, in gas, liquid, and solid form, exploding in all directions. This mineral cloud surrounded the two bodies as they circled one another in close orbit. During this time, they shared the same radiant silicate vapor atmosphere at temperatures of 10,000 degrees Fahrenheit. The collision blasted the whole distinction between inside, surface, and outside into a crystalline cloud. The distinction between inner depth, surface, and exterior involved only the degree of centripetal density. There was no ontological distinction between the earth, the moon, and a meteor.

Eventually, this rocky gas cooled and condensed into droplets of magma that rained down on the twin celestial bodies. Dense metal from the cores of both planets mixed and cooled back into liquids that sank to form the new larger core of the earth. The cloud of vaporized rock then reconstituted a new mantle. On the Hadean earth, all that was solid melted into air. The ground became the sky, and the sky became the ground. Everything remembered its meteoritical birth: its fundamental condition of kinetic ungrounding and process.

The first weather on the earth was thus mineral, meteorological, and meteoritical: rock-rain. It was crystalline weather that became the new surface and eventually, the depths of the earth and the moon. The old earth died so that a new earth could rain down and cool rapidly into the first metastable crust, floating on an ocean of magma. These are the first and oldest recorded crystals we have recovered from the earth's earliest centripetal archive.

The earth is, therefore, not a self-identical object or subject. It is a kinetic process. It is a centripetal collection of exteriorizing flows continually folded together. The spatial division between inside and outside, ground and sky, is an emergent and metastable kinetic pattern, not an a priori structure of reality. Meteors are the material ungrounding and historical deconstruction of all geophilosophy. To modify a line from the French poet Arthur Rimbaud, "the earth is an other."

LUNAR ALTERITY

We cannot fully understand the earth without its co-emergence with the moon. Had the moon's angle of collision been slightly different, it might have become another life-hosting planet, similar to the earth. The earth would surely never even have become been the living earth as we know it without Theia. The inclination or angle of Theia's collision altered the entire history of the earth. The earth today is, in part, the result of a few degrees of declination: a swerve in the flow of matter.

Most importantly, the collision allowed for a double centripetal accumulation of minerals. According to the prevailing and well-evidenced lunar impact theory developed over the last forty years, the earth accumulated most of Theia's iron core. Most of Theia's volatiles, on the other hand, were blown away in the impact. This allowed the earth to become larger, increase its core pressure and temperature, and thus develop a large, convection-driven iron core strong enough to produce a powerful magnetic field. Without this strong magnetic field (the magnetosphere), the sun would have destroyed life before it had a chance to evolve. Furthermore, without Theia's collision, the earth would not be tilted on its axis. There would be no seasons.

Just as importantly, the moon orbited closely around the earth (just 15 thousand miles away), pulling gravitationally on the earth's surface. This produced enormous equatorial waves of rippling crust. Every five hours, the earth's surface bulged more than a mile outward toward the moon, creating incredible internal friction and adding more heat to the earth. The Hadean earth was not a sphere but a spiraling kinetic process of centripetal accretion and elliptical inclination.

The moon's orbit also prolonged the molten state of the earth's crust, thus helping it to cool faster than it might have otherwise. Just as you might rapidly stir a cup of hot coffee to cool it, the faster the moon spun around the earth, the faster the earth's crust cooled. Theia's collision, iron contribution, and the moon's ultrafast orbit all contributed to an increasingly rapid crystallization of the earth's crust.

If the earth is still defined by mineralization and centripetal accretion today, it is in part because the moon continues to support us as it moves farther and farther away. Even as it drifts away, however, the moon will not allow the earth to remain still. Recent studies show that the moon's contributions go far beyond tidal effects.[3] The moon also pulls on the earth's mantle and liquid core, continually deforming it into a turbulent spiral and generating billions of watts of power that continually fuel the earth's geodynamo. It is precisely the moon's eccentric elliptical orbit that pulls and twists irregularly on the earth's core. Without this gravitational pull and pedesis, the earth's core would have cooled a long time ago. The earth would be a cold, dead place.

Western cultures have often described the moon as an inferior replica of the sun. The moon was the *becoming* to the sun's *being*. From a deep-historical perspective, however, the situation is quite the opposite. The moon is the cosmic motor of the earth's inner sun. The moon is the sun of the earth. However, the moon is also a black sun, deeply internal and external at the same time. The moon is an internality so inner that it is exterior, so exterior that it is interior. It is the movement that twists our core into knots and drives our inner heat.

Hell is moonbeams and star-stuff. The flux of our mantle keeps the magnetosphere in flux and affects volcanism and even earthquakes in ways we do not fully understand. The so-called "D double prime" region of the earth, between the outer core and lower mantle, remains one of the most mysterious and unmapped inner regions of the earth, in part because of its pedetic and eccentric fluidity, related to the moon. Fluctuations at this level are also related to the movements of magma plumes that travel up to the surface.

We often describe the moon in terms of its phases of light reflected from the sun. However, even more important is the kinetic structure of the moon's eccentric flux. The moon is the inner "black sun" of the earth.

The magnetosphere fights back invisibly against the sun's damaging rays, cosmic radiation, and meteoroids.

The moon orbits and spins in such a way that it centripetally gathers and helps crystallize the earth gravitationally. It also helps define and protect the earth that it gathers. This centripetal crystallization, in turn, helps differentiate the inside from the outside on earth.

Once the earth developed a surface, the largest body in the sky was not the sun but the moon. The moon was at one time so close to the earth that it blocked out the sun. The solar economy was thus secondary to the lunar economy, when the sun was much weaker than it is now and the moon much stronger. The distinctions between ground and sky, matter and form, being and becoming are all related to the formation of the moon.

The primary error of all geophilosophy has been its geocentrism and lack of engagement with deep history. Geophilosophy often begins with the division between ground and sky, mortals and gods, the being of the sun, and the becoming of the moon. However, these divisions are all products of more primary processes of centripetal accretion.

The moon is the coordinated deformation of the earth and vice versa. Both continue to stretch each other's bodies into elliptical wobbling eggs. The moon stirs up the liquid metal geodynamo inside the earth that drives its convection cycles and numerous geomagnetic reversals. Their co-motion ensures the impossibility of spherical perfection. The earth is not round because it is locked in a process of continual deformation with the moon.

The effects of the moon were not and still are not trivial for earth's integrated systems, including human systems. This is why a *cultural* history of stone or the moon is still inadequate and anthropocentric. The moon is not like us or our ideas; our ideas are more like the moon. Humans are a regional and historical expression of a more general lunar pattern of centripetal movement. The human idea of "gods" began with real celestial bodies like the moon and only later became more abstract. Many humans like to think that their abstract ideas make them different from nature, but all our ideas are only iterations of patterns from nature. We have "gods" only because we have a moon and planets, not the other way around. We do not project our humanity anthropomorphically onto the moon; the moon projects its "lunomorphism" on to us. Perhaps anthropomorphism, then, is really just our way of trying to reciprocate with the moon for our moonlike natures.

What new systems of thought and being might emerge if the celestial conditions were different? What if there were two suns or several moons? Without a single sun and single moon, would we still think in terms of metaphysical dualisms? These questions are speculative, of course, but my point here is that the moon plays an active role in all earth systems, including human culture. There is no Gaia without Theia.

LIQUID EARTH

The Hadean earth was a liquid earth: liquid rock and liquid water. The two flowed and co-formed one another in a cycle of mutual heating, cooling, deformation, and centripetal concentration. Water was most likely present on the earth before the collision with Theia, at perhaps a hundred times today's levels, but Theia's impact blasted it into space. Some of this blasted water even remains trapped on the icy poles of the moon. After the collision with Theia, asteroids continued to blast the earth's volatiles, such as water, into space for half a billion years.

The earth's water is fundamentally mineralogical. The elemental components of water (hydrogen and oxygen) came to the earth centripetally, from meteors. Recent experimental research has shown that "the most common of minerals—olivine, pyroxene, garnet, and their denser deep-Earth variants—may be able to incorporate a small amount of water at mantle conditions."[4] These minerals do not typically incorporate hydrogen atoms, but under the extreme pressures of centripetal accumulation, they can. Since these deep minerals also contain oxygen, they effectively contain the mineralogical equivalent of water.

What this means is that centripetal mineralization is the condition for the protected storage and eventual release of water into the earth's atmosphere. The earth produced its water mineralogically, and only half of it ended up on the surface of the earth as ocean.

Recent research also shows that there is at least as much water under the earth, trapped in rocks and minerals (in the mantle transition zone), as there is in all our oceans combined.[5] Ringwoodite, in particular, can hold one percent water. It sounds like a tiny amount, but it adds up. Rocks and minerals hold an incredible 99.9999 percent of the earth's oxygen. The movement and slippage of these deep interior "oceans" plays an essential role in the earth's volcanism and internal motions. Centripetal mineralization thus acts as a crucial archive of basic ingredients for all the earth's systems.

The earth is not just its surface. We cannot understand its cycles and spheres as self-enclosed loops. For example, in addition to its surface water cycle, the earth also has a deep water cycle. Water not only arrived from elsewhere, but it is also an emergent metastable product of the earth's own body. It is a product of volcanism, as well as of the meteorological and lunar energies that formed, pulled, and heated the earth's body and continue to drive its turbulent volcanism.

We tend to think of water, oceans, rain, and hydrology as happening *on* the earth or *to* the earth, but shockingly, water is nothing but the earth itself. Hydrology is entirely immanent with geology. Liquid water is archival material that becomes unarchived and disordered. There is no ontological division between surface and depth or mineral and water. There is only a change from crystal to liquid as hydrogen and oxygen depressurize through volcanism.

My theoretical conclusion here is that the material conditions of the earth are immanent to moving the bodies conditioned by it. In other words, the earth is nothing other than the material patterns that shape it. It is a process. For example, volcanism both produces water and is produced by a watery slippage and turbulence in the mantle transition zone.

This sounds like a minor gesture, but understood theoretically it is a radical inversion of transcendental philosophy, which is always looking for stable "grounds." The consequences of my argument are that earth has neither stable ground nor a total absence of stable ground. It is the continual formation of multiple metastable grounds or regions of kinetic circulation. There is no ahistorical ground or unground of the earth but rather many metastable historical grounds of the earth.[6] Centripetal circulation is the first and principal regime from which the others emerge.[7]

Geology and hydrology are thus poorly understood if looked at in terms of a division between surface and depth. Instead, they have knotted circulations that enfold the surface into the depths and unfold the depth into the surface. They are both centripetal circulations.[8]

Meteoric, lunar, volcanic, and hydrological processes are thus continuous folds or sub-cycles within a much larger regime of motion. As mineral flows move through this regime, they continuously transform from one to the other. There is no ontological or transcendental difference between inside and outside, depth and surface, matter and form. There is only crystallization, decrystallization, recrystallization. What I am calling a "kinetic transcendental" is not limited to surface and depth but is the more fundamental process of circulation. Surface and depth are only regional differences in the more extensive process of centripetal mineralization and crystallization.

After the collision with Theia, global volcanism, fueled in part by hot water mixing with rock magma, blasted through to the earth's surface and created a superheated mineral soup. Volcanoes spewed billions of tons of hot water vapor, nitrogen, carbon dioxide, and noxious sulfur compounds into the heavy, sun-blocking atmosphere. Deep-earth gases were also released in enormous explosions at the surface. Hot water dissolved and concentrated rare elements locked in the earth's mineral body—beryllium, zirconium, silver, chlorine, boron, uranium, lithium, selenium, gold, and others—and redistributed them through the crust. Huge waves, giant geysers, and roaring rivers became the primary sculptors of the earth's crystalline surface.

The emergence of water and its liquid history completely overthrow the division between inside, surface, and outside. Water is inside the earth as a solid, on the surface as a liquid, and in the atmosphere as a gas. The earth's early liquid atmosphere was dense with water vapor but also mineralogically continuous with the surface. The primary movement of water was, therefore, still mostly centripetal insofar as it almost immediately turned to mineral raindrops and fell back down to the earth.

Torrential rains and superstorms across the entire earth gathered exploded minerals from their air, accumulated them into drops, and redistributed them back to the earth in the form of a thick mineral soup. In short, the earth's surface underwent a vast archival redistribution of its minerals using superheated liquid-mineral rain. This soupy mix of mineral liquids then leaked back down into the cracks and fissures in the earth, causing it to cool faster and faster, in a feedback loop.

The earth's early water was not only extremely saturated with minerals but was also highly saline and acidic, causing it to dissolve the surface further and release even more minerals. The oldest known solid bits left over from the Hadean earth are tiny zircon crystals discovered in the Jack

Hills of Australia. They are are more than 4 billion years old and attest to a relatively cold and wet late-Hadean earth during this time.[9]

The faint sun paradox

The sun, however, was not responsible for this superheated state on the Hadean earth. Most current estimates suggest that the young sun, 4.4 billion years ago, was 25 to 30 percent less bright than it is today. If the sun had been the only heat on the earth, the whole planet would have been covered in ice and produced no ocean at all. The leading hypothesis is that the earth stayed warm by creating extremely dense concentrations of carbon in the atmosphere—producing a radical greenhouse effect with ten times as much pressure as our current atmosphere.

There is an important philosophical point here. Perhaps we have credited the sun with such a large role in the earth's history because of a biocentric, Holocene-centric, and solar-centric perspective. But the earth is not the slavish terrestrial dependent of the sun. Indeed, the sun has played an important role, but the centripetal gathering of minerals played the first and most foundational role in sustaining the earth's heat and dynamism. Like a clam or beetle, the earth birthed a protective shell. Instead of calcium, though, it was made of electromagnetic waves and atmospheric gas.

As we will see in the next chapter, the earth's emerging atmosphere was continuous and immanent with its geology and hydrology. There is no need to invoke Gaia or vitalism to understand the kinetic pattern of circulation and feedback at work here. The hydro-volcanic activity produced a thick protective atmosphere, which in turn made possible higher temperatures and more hydro-volcanic activity. without deference to the "life-giving" sun.

Life on the earth was made possible by a densely centripetal contraction of minerals on the earth's surface. The material conditions of early life are thus more directly related to these centripetal conditions than to the creative power of a feeble sun. This weak sun would not have been enough on its own to sustain the process of centripetal mineralization required for life. Instead, the earth's first ocean surface was formed by a mineralogical rain.

This extended process of heated fluidity also made possible an archival layering of heavier and lighter minerals in the earth. Most notably, the earth's continuous heat during a time of weak solar contribution made possible the emergence of relatively lightweight granite from heavier volcanic basalt that had sunk to the hot ocean floor. As matter moved from the periphery toward the center, it differentiated itself by weight. As the lightest among the rocks, granite accumulated on the surface, while denser rocks sank back down.

This granite is the foundation of the continents we currently stand on. Granite was not a gift from the sun. It was not even a centrifugal expulsion from the earth. Instead, it was a differentiated layer of a more fundamental process of centripetal accumulation. Paradoxically, we perceive granite as hard, fixed stone but it is really because of its light weight and lower melting point relative to other minerals that it has the position it does, allowing us to think of our floating continents as so firm.

The flow of water, with its spiraled storm systems, river eddies, and turbulent whirlpools, was the first kinomorphic process that shaped, cooled, and concentrated the earth's folded surface. Hot water expanded and burst open rock to cool it down along fluid lines.

We tend to think of geological processes as collisions between rock strata, but the earth was and continues to be shaped by a confluence of fluid dynamic processes. The earth was formed by *flows* of liquid rock, water, and air. In this sense, the earth is not a sphere. It is a fluid and vortical spiral. It is whirling storm systems, cycling convection patterns of magma plumes, vortical currents, and ocean whirlpools.

THE LANGUAGE OF LIGHTNING

The earth centripetally gathered and layered its minerals from meteors that dissolved in the sky, from heavy clouds that rained them down, from volcanoes that spread them out on the surface, and also from *lightning*. The late Hadean began to produce an early cloud-covered atmosphere of gases, vapor, and mineral dust from asteroids and volcanoes. Carbon dioxide and dinitrogen dominated this hazy atmosphere, which was filled with volcanic and atmospheric lightning. These electrical blasts of lightning ripped apart molecules and produced nitric oxides and ozone. Raindrops then carried these molecules back to the earth, where they eventually provided the essential biomolecules of life and the atmosphere.[10]

Lightning is, therefore, a crucial condition of centripetal accumulation on the earth. Even today, most of the earth's nitrates that we rely on to support life come from lightning. The earth re-mineralizes itself through lightning. Valuable nitrogen and oxygen that might have escaped its surface are brought back down in a "fixed" crystallized form.

But just like meteors, the moon, and the ocean, lightning is fundamentally pedetic. Lightning is not merely an aerial descent from the sky to the ground. The movements of centripetal mineralization are not single and unidirectional but collective and indeterminate. They are like the fluid dynamics and liquid topology of the magma flows of the "D double prime" zone.

Clouds are the product of extraterrestrial meteors and volcanism. When these mobile oceans of negatively charged electrons race over the surface of the earth, they create a positive charge on the surface. The earth thus confronts its externalized self as an "other." Hadean clouds were as dark as the skin of the earth, moist like its oceans, and aerial like the fall of meteors. Lightning thus formed a material and dynamic synthesis of all the major Hadean events: leftovers from Theia, vapor from the ocean, and meteors from space. Lightning allowed the earth to become what it is by accumulating itself as an "other." The earth extruded itself as clouds, then transformed this material into nitrogen oxides, and finally re-accumulated this material as a new skin on its surface. Lightning thus allowed the depths to metamorphose into the sky and then birth themselves again on the surface. As the sky reached down to return centripetally to the ground, the ground reached up to centripetally pull the sky back down.

Before the actual lightning bolt, the lightning sends out "stepped leaders" that stretch out pedetically in various directions. The leaders flow for a few hundred feet, then pool up at a particular region in the sky, pause, and move in another direction for a few hundred feet, pool again, and so on in a zig-zagging fractal pattern. Within each zig are numerous zags and within those zags, many zigs, and so on. The pedetic movement of the stepped ladder still remains a mystery today.[11]

When one of these leaders gets close to the ground, stepped leaders from the earth move upward toward the sky in the same fashion. Once two leaders get close to one another, a massive bolt of lightning is discharged—but from the ground up. The part of the channel nearest the surface will discharge first, then the parts that are higher up. This process repeats several times, in a strobe-like effect. The children of this queer copulation then fall from the sky in the form of life-giving nitrogen and oxygen.

Centripetal accumulation through mineralization is thus not unidirectional or predetermined. Rather, it is a tendency of flows to find and transform one another from the periphery to a center. These indeterminate and mutually exterior transformations then layer themselves in an archival and ordinal pattern on the surface of the earth. The earth is, therefore, not a mere recording surface for events, but it is also transformed by the recording process itself.

In short, lightning demonstrates quite dramatically the performative and transformative nature of the earth's mineral archive. The electrical difference does not resolve the difference between the earth and itself (as depth, surface, sky, and space) but experimentally explores and transforms it. The difference between ground and sky is not absolute or pre-given but is an emergent process of exploratory differentiation made possible by lightning's queer and centripetal collection.[12]

★ ★ ★

Centripetal mineralization was the first dominant kinetic regime invented by the earth. This first movement inward toward the center from the periphery along differentiated layers continues today. It remains the immanent condition of planetary life and all mineral-based structures. Meteors continue to fall and weave into the earth; the moon continues to pull the charge of the earth's core; the earth continues to make and rain down water onto its surface and regenerate itself with lightning. The whole process is neither grounded nor ungrounded, but circulated.

Centripetal motion does not disappear after the Hadean. Instead, it becomes the condition for a whole new regime of centrifugal circulation. This new circulatory pattern is what we call "atmosphere," and it created life.

B. ATMOSPHERIC EARTH

6   Centrifugal Atmospherics

THE SECOND MAJOR geokinetic field to rise to dominance in the earth's history was the atmospheric field. This second type of field became increasingly prevalent throughout the Archean Eon, from about 4 billion years ago to about 2.5 billion years ago. Three significant events define this transition: the end of heavy meteor bombardment, the emergence of living organisms, and the rise of a highly oxygenated atmosphere.

These events constituted a dramatic historical shift in the earth's pattern of motion from centripetal accretion to centrifugal expansion, respiration, and reproduction. This shift to a more centrifugal motion was not an exclusive break from the previous pattern by any means. All the older centripetal movements that first defined the earth continued into the Archean period, but to a lesser degree. For example, there were still meteors, lunar cycles, volcanism, lightning, and oceans, of course, but by the Archean, these processes had settled down into a much more stable pattern of motion. By the end of the Hadean, the earth had cooled and crusted over. This provided a surface that the earth could vaporize and grow.

ATMOSPHERIC GEOKINETICS
The atmosphere is not part of the earth; it *is* the earth. We tend to fetishize what we call the "climate," as if it were an object or region that could be

studied and manipulated on its own. However, the climate is not a separate part of the earth. Climate thoroughly entangles all earth systems. Focusing on the "climate" might allow us to ignore the rest of the earth's entwined movements.

Thinking of the climate as a manageable and visible stand-in for the earth might also allow us to forget the deep historical and non-constructible regions of the earth. We do not like to think of the deep earth, planetary volcanism, meteor strikes, solar rhythms, and cosmic events as ecological issues because of an implicit human bias that privileges visible, "human-level" objects and events, especially those that we can directly manipulate.

We have fetishized the surface of the earth and the near-surface climate. However, these make up a mere one percent of the planet. What environmentalists call "deep ecology" is thus, in this sense, in fact profoundly shallow. Environmental ethics is essential, but it also risks subordinating a theory of the earth to the area that humans can see.

Just because we cannot help or hinder solar flares, meteors, or volcanism, does that mean that they are ethically and philosophically irrelevant events? Because of their emphasis on human-centered ethics, ecology and environmental philosophy have tended to ignore geology and cosmology. Ecology has implicitly defined the earth by an extremely tiny region of time (the Holocene), space (the surface), and activity (human ethics).

We desperately need a broader theory of atmosphere that takes deep time and deep space seriously. This is important not only in order to squelch the techno-constructivist impulse of "climate management," but also in order to properly situate human ethics within this deeper spacetime. The atmosphere is not a place above the earth that holds a bunch of gas. The atmosphere is the whole planetary process of vaporization (evaporation and condensation). Just as minerals are not *on* the earth, the atmosphere is not *above* the earth.

After a long process of mineral accumulation and relative stabilization, vaporization is something the earth began doing quite rapidly. The atmosphere is the earth's own vaporized body. It has layers of sedimented cloud formations, airborne rivers and oceans, and convection cycles that are metabolically continuous with the deep earth's heating and cooling rhythms. It is the most mobile of the earth's geological strata.

The atmosphere is, therefore, neither a substance nor a place. It is a process that comes from the depths of the earth, expands out into space, and then cycles. Gas, vapor, or volatile elements are present in every region of the earth. We do not merely breathe the atmosphere. We *are* the atmosphere, breathing itself. Our bodies are made almost entirely of the volatile elements of our atmosphere (oxygen, carbon, hydrogen, and nitrogen). Every time we breathe in, we add something to our body, and each time we breathe out, we lose something from our body. This goes on until we completely rebuild our entire body. Every living being, and many non-living ones, on this planet came from airborne matter that once gathered in the sky billions of years ago and rained down to the surface. Let us not forget that lighting made the nitrogen in our bones.

The atmosphere is the *medium* of the earth. It is the immanent material process by which mineralization is broken up, transmitted, and reordered. The earth communicates with itself, as and through itself, in the atmosphere. The atmosphere both is the medium of light and sound and is itself *made* of light and sound (photons and pressure waves in gas).

This deep historical point has radical consequences for all theories of communication. We should not divide communication into sender, message, and receiver. Atmosphere is the wholly immanent medium of mutual penetration within which these three emerge as metastable patterns. Communication is the radical transformation of the whole atmospheric medium itself.

From a materialist perspective, "the message is the medium," not the other way around. The sender, receiver, and message are all the medium. This is because language is not a symbolic representation. It is an atmospheric change in the whole vibrating sonic medium. Language does not happen in our brains or our symbolic systems but in a coordinated kinetic medium: *the* atmosphere. Humans have language in the same way that the atmosphere has humans. When we perform well-coordinated sonic and gestural acts, this is nothing other than a performance of the atmosphere itself. Language is something that the atmosphere does *through us*. Sound vibrates us in the same way that it vibrates the atmosphere that we are.

This chapter develops a new and expanded theory of the atmosphere as defined by three interrelated features: centrifugal motion, respiration, and reproduction. These are the core kinetic features that emerge from the Archean historical events discussed in the next chapter. The purpose of putting the conceptual product before the historical process in this chapter is to help the reader identify the general pattern of motion that developed in the historical events. Once I lay these features out, we will be in a better position to identify the role played by the centrifugal patterns of motion in the Archean Eon.

CENTRIFUGAL MOTION

The "atmospheric field" is, first and foremost, a flow of matter from the center of the earth to its outer periphery. Once planetary accretion had created a surface, vapors began to circulate across this membrane. Minerals and atmosphere are immanent and continuous with one another. However, vaporization assumes mineralization as its necessary material historical condition. Before a movement can take place from a central region to a periphery, there must first of all *be* a central region. The centripetal mineralization of the Hadean earth provided precisely this.

Centrifugal motion is not a necessary or a priori consequence of centripetal motion; it is an emergent feature of it. On the earth, the more flows of matter that gathered, the more the distribution of matter became uneven. As heterogeneous things gather into a pile, some end up at the bottom of the pile. Some also end up on the top, others at the periphery, and others still closer to the center. All the flows are equally heterogeneous, but they become differentiated and ordered topologically by the pedetic nature of the gathering process. The earth is a heap. And a heap has an emergent and relational topology: a "structured criticality."[1] Minerals sink, float, and settle in different places as they are heated and cooled.

An incredible amount of energy was required to gather the earth together centripetally. Once it was gathered and heated, the heat began to radiate from the center toward the periphery. The earth's geodynamo is continually energized by the orbiting moon pulling on its molten core, but eventually, the moon will leave its orbit, and our core will cool.

But the centrifugal movement of the earth is far from linear. The centralization of heat inside the earth's core produces a convection flow of matter that moves to the surface and into the atmosphere. The atmosphere, in turn, has convection cycles, which also recycle some of that heat and release the rest into space. Energy flows and matter cycles to kinetically sustain a metastable state.

Therefore, we can see that the centrifugal motion of the earth is a historical and emergent feature of a prior centripetal collection, and not any absolute logical necessity independent of history. Entropy itself is not absolute but rather a particular feature of our universe at this time. If, for example, there had been a "big bounce," where the universe continually expanded and contracted, then an opposite law would be true.

Furthermore, centrifugal radiation is not opposed to centripetal accumulation; both occur at the same time. For example, the earth's convection cycles comprise a double circulation. Matter at the deep core of the earth heats up and rises to the surface. As it rises, it cools and sinks back down to be re-accumulated, reheated, and circulated again. The same thing occurs above the earth as hot air rises into the sky, cools, and returns to the surface as rain. Thus, centripetal mineralization and centrifugal vaporization form continuous and asymmetrically linked convection cycles above and below the earth's surface.

Once a central accumulation of matter/energy occurs, it begins to radiate and evaporate this energy outward and away from its core. In other words, it becomes possible for an atmosphere to emerge. In short, every great center requires an even greater periphery to support it.

RESPIRATION

There are at least two major interrelated kinetic features of the atmosphere: respiration and reproduction. The word "atmosphere" comes from the Greek word *aetmós*, meaning "breath, vapor, steam, or smoke" and associated with "life and soul" or *psukhe*. This is a revealing historical etymology, but not in the way we normally think. Typically, the conclusion of this etymology is that the immaterial idea of the soul is historically derived from a material idea of human respiration. However, this only begs the question: where did the human idea of respiration itself come from in the first place? Before there were breathing humans who could imagine their material breath as an immaterial soul, there was already a breathing earth. Humans breathe because the earth breathes. Human breath is composed of two movements: a centripetal movement in and a centrifugal movement out. The product is a release of heat and vapor.

This is what the earth does, as atmosphere. The atmosphere is the breath of the earth. This is not a metaphor—or perhaps it is an inverted one. The earth does not breathe as humans breathe, but the other way around. Anthropomorphism is already a *terromorphism* that has forgotten itself.

Human breath cycles and respires because the earth cycles, circulates, vaporizes, and respires. The earth already invented the centripetal/centrifugal vapor cycle so that life could participate in it. In short, humans live, and believe that they have souls, only because the earth itself materially inspires them to do so.

Furthermore, the whole metastable existence of planetary geo-, hydro-, bio-, and atmo- *spheres* is only possible because of the more primary kinetic process of circulation. The earth's spheres are fluxes, circulations of centripetal and centrifugal rhythms. The earth has no parts and no whole, only processes of inspiration, vaporization, and respiration.

One consequence of this is that there are no perfect or static spheres in space. Our spatialization of the earth's spheres is a terrible mistake. Sphericity is centripetal and centrifugal flows circulating in and out in dynamic kinetic stability.[2]

Starting in the late Hadean, the whole planet began to respire. Let us look at four significant kinds of respiration.

Volcanic respiration

After the earth's crust cooled, volcanoes continued to off-gas volatiles, which accumulated in the sky. Hydrogen, oxygen, carbon, nitrogen, and other elements followed thermal convection movements that began deep in the earth and radiated centrifugally upward, higher and higher into the periphery of the earth, contributing to the atmosphere. Volcanoes breathed in cooled mantle and crust and breathed out magma and airborne vaporized volatiles. When the volatiles settled back on the surface, the earth breathed them in and out again.

Oceanic respiration

This volcanic process produced the earth's first oceans in the late Hadean and gave birth to a new terrestrial breath: the hydrosphere. The earth's hot crust and volcanoes heated the oceans, causing them to evaporate volatiles like hydrogen and oxygen higher and higher into the sky. Once airborne, they accumulated, rained back down, evaporated again, and so on, creating a second floating ocean of mobile water and clouds. Through the alternating process of centripetal accumulation and centrifugal evaporation, the oceans became more mineralized and the sky more vaporized.

Atmospheric respiration

This enlarged atmosphere, in turn, also started to breathe. As lightweight hydrogen and oxygen rose higher and higher, these gases were increasingly affected by their expanding convection cycles, periodic high-speed solar winds, and extreme ultraviolet radiation. In this way, the whole outer atmosphere (thermosphere) began to rise and fall with the cooling and heating fluctuations of the sun. As it contracted, it put pressure on the lower atmosphere and thus changed temperature and convection cycles on the surface as well. The entire atmosphere began to breathe in centripetally and breathe out centrifugally, ultimately releasing hydrogen and electrons into space.

As the earth's skies began to clear up from all this respiration, the rhythmic alternation between day and night also became more pronounced than in the previous hothouse atmosphere of the Hadean. As the earth cooled, it was increasingly affected by the rhythms of solar radiation, which in turn affected ocean water evaporation rhythms and weather patterns. As the earth revolved around the sun, its atmosphere heated and cooled in response to the fluctuating rhythms of photons and electrons. This also produced "breath" in the atmosphere. The whole planet began to breathe to the increasingly dramatic rhythm of its solar revolution.

Biological respiration

Eventually, all this breathing made possible a new kind of breath: microbial respiration. Microorganisms, possibly all over the earth (in the skies, on the surface, underwater, and deep in the earth),[3] began to release various volatiles and gases from the earth's accumulated minerals—most significantly oxygen. During the late Archean, huge amounts of free oxygen were released into the atmosphere, in what is called "the great oxygenation event." This free oxygen even created a new aerial sedimentary layer called ozone, which shielded the earth and early life from ultraviolet solar radiation. Thus, oxygen captured deep in the earth eventually made its way to the surface and skies through the centrifugal action of microbes.

Kinospherology

The geosphere, hydrosphere, atmosphere, biosphere, and sun all contributed to the centrifugal respiration of the planet. Volatiles of all kinds radiated outward away from the earth, supported in large part by the entropic flow of heat from the core. The atmosphere should therefore not be understood as a region or sphere but, better, as an integrated process of respiration. Respiration is defined by a centrifugal movement outward and centripetal movement inward. All the so-called "spheres" of the earth are static spatializations of an internally dynamic kinetic process. What appears to be a spherical region is only a respiring metastable state between an inward and outward motion. Perhaps they should be called "kinospheres" instead.

There is no single isolatable region that we can call "the" atmosphere. There are gases below, on, and above the surface. They continually radiate out into space and connect the inside of the planet with its outside. The atmosphere turns the whole planet inside out. Unfortunately, we tend to think that the atmosphere is just what we breathe, and so separate it off from the deep earth, ocean, and outer space. But deep-earth gases are atmospheres for other forms of life, just like water is atmosphere for fish. The ocean respires through fish.

It is precisely the anthropocentric separation of human and biological respiration from the earth's respiration and metabolism that has allowed us to justify our pollution of it—as if the sky were a giant trashcan where things "disappear." But in the atmosphere, everything moves around, in, and out. Nothing gets destroyed, only transformed.

This chapter and the next argue that we need a new, non-anthropocentric kinetic theory of atmosphere based on respiration and centrifugal circulation at the planetary level. The specialized fields of geology, climatology, and biochemistry require a more general kinetic physiology of the earth to bring them together. The earth sciences are not separate parts of a whole system but wholly continuous and entangled aspects of the same process.

Scientists have recently discovered a whole microbial world living in the deep earth, dubbed the "dark biosphere." Today, we need a study of *all* the "dark spheres": the "dark atmosphere" and the "dark hydrosphere." What I am calling the atmosphere is continuous with all the dark and light spheres of the earth. Air weaves its way through the entire earth, creating all manner of bubbles and oceans at every scale. Ultimately, it is gas and heat that are the radically centrifugal vehicles that will carry out the earth's entropic destiny. They are the radical exterior of our interior, whose escape both maintains and slowly destroys our planet.

REPRODUCTION

The second major kinetic feature of the earth's atmosphere is its reproduction. The earth reproduces itself through respiration. Breathing, in the general sense, is an alternating centripetal and centrifugal motion that creates and reproduces a cycle. Energy continually flows into a region from the periphery and then releases a portion of that energy outward. It then re-accumulates enough energy to repeat the process. This is what I am calling *kinetic reproduction*. In physics, it is called a "self-organized" or "dissipative" system.

Not only do kinetic cycles respire, but the movement of respiration also helps reproduce the respiration cycle itself. Every metastable kinetic fold or cycle follows this basic pattern of motion. It forms the basic material foundation for increasing the complexity of mineralogical, atmospheric, and biological order on the earth. The earth thus reproduces or re-plicates (refolds) itself through all its folding cycles and spheres, in what we could call a general "inorganic metabolism," "geometabolism," or "atmospheric metabolism."

For example, all the earth's respiration/convection cycles discussed in the previous section not only release energy centrifugally but also help return energy to the earth, which facilitates the further centrifugal release of energy. This is how the earth reproduces itself across a range of metastable states. If energy only flowed out without cycling, nothing would exist. The earth would be a cold, dead place. Thus, matter at every level either reproduces itself or returns to its indeterminate quantum state.

Unfortunately, we often conceptualize reproduction in a much narrower sense, as strictly biological, sexual, and mimetic. However, what I am interested in here are the more general and fundamental material kinetic conditions that make possible this more narrow biocentric definition. It is not enough to merely posit the existence of narrowly defined biological reproduction. Kinetic reproduction has the benefit of describing a much broader spectrum of iterative ordering motions, of which biological and highly mimetic reproductions are only a small and emergent portion.

Every reproduction or replication is an iteration that requires a thermodynamic difference between the first iteration and the second iteration. This difference means that there can be no strict identity between model and copy. There is only a continuum of processes that cycle in more or less similar ways. Reproduction itself is a process that matter does, and thus something that changes the "original" in the process of replication.

Mineral reproduction

For example, minerals reproduce themselves without life, sex, or mimesis. Minerals are composed of atoms that vibrate back and forth within certain physical limits that define their shape. They preserve their crystalline order not because they are static but because their atomic lattice moves in a dynamic kinetic state that maintains order, usually with very little energy input. We often treat minerals as classical equilibrium states; however, at the atomic and quantum level, they are not. They, too, lose energy and become disordered. All minerals radiate heat and are continually losing photons. Eventually, entropy will win, and they will break down. In the meantime, they slowly reproduce themselves by kinetically repeating a given pattern of motion. Each iteration is slightly different because of the pedetic movement of their atomic vibrations. Thus, mineral reproduction does not need life, sex, or mimesis.

There are also, for example, dozens of chemical cycles like the citric acid cycle, in which one of the waste products of the chemical breakdown process is a necessary input for starting another cycle.[4] There are also many different chemicals whose products catalyze other chemicals whose products, in turn, catalyze others, in a large autocatalytic network that ultimately further catalyzes the original chemical reaction again.[5]

Fluid reproduction

Fluid cycles reproduce themselves through dissipative structures like vortices, whirlpools, and eddies. These patterns are efficient ways to reduce an "energy gradient"—a difference between low and high entropy states. There are energy gradients between the deep earth and the surface. There are also gradients between the surface and the sky and between the warmer and cold regions of the ocean. Spiraling storm systems emerge in the skies, whirlpools and eddies emerge in the oceans, and convection cycles emerge under the earth, all to reduce energy gradients. Before chemical and biological metabolism, there was already a fluid metabolism (inflow and outflow of energy in a system) of the earth.

The ocean, in particular, produced several major emergent reproductive orders that made other metabolic forms possible. Regular seismic activity shook and mixed the ocean of the Archean earth, producing a large energy gradient that dissipated through ordered waves, storms, and vortical movements. The strong pull of the moon also created enormous wave and vortical currents that mixed hot and cold waters.

These and other movements of water brought elements (atomic, molecular, and mineral) together at every level. It is not only water that is the historical material condition of chemical and biological metabolism but, more specifically, the ordered and reproductive *movement* of water. Molecules and minerals were not brought together randomly in the ocean but in particular kinds of kinetic cycles. Water shaped chemical and biological relations through vortical motions.

Chemical and biological metabolism is the shape of water. Chemical reactions tend to follow the form of the self-affective spiral that brought them together. Flows of water slow and curve as they encounter rocks, divots, and topological differences along the ocean floor. As water currents slow, sediment drops out and falls in the same place, again and again, accumulating, reacting, expanding, and possibly catalyzing one another. The ocean floor and eddies of water do not mix elements randomly but rather based on very specific relational criteria such as the weight of the elements, the exact topology of the floor, and the vortical fluid dynamics of water. All of these produce regular periodic structures that are the basis of all other metabolic cycles. Metabolism is a subcategory of kinetic cycles more generally, which form a "general material metabolism."[6]

If we assume that the Archean ocean was merely a *random* prebiotic soup, then the statistical odds of emergent chemical and biological metabolism are outrageously improbable. In my view, if we want to understand where emergent orders come from, we should first look at the more general material kinetic orders already at work on the earth. For example, because water is a polar molecule whose oxygen atoms have a negative charge and whose hydrogen atoms have a positive charge, water molecules bond together. Water, therefore, already polarizes and orders itself into a simple alternating kinetic pattern that shapes and orders how elements move and interact through it. Furthermore, molecules that are hydrophobic or hydrophilic, or both (like lipids with hydrophobic tails and hydrophilic heads), will arrange themselves in emergent orders because they are in water.[7]

Experiments like the famous Miller-Urey attempt to simulate life in a flask of mineral-rich liquid with an electrical spark fail to answer the deeper kinetic question of how the minerals were gathered together in the first place. The laboratory flask is an abstraction from the Archean ocean and thus already assumes the nonrandom relational gathering of specific mineral concentrations. In other words, order has already been assumed and added in the shape of the flask. The chemical reactions that occurred in the Miller-Urey experiment were not random at all. But what is the kinetic and topological equivalent of this experimental process? This is the kinetic question we are trying to answer.

Multiplication
Multiplication is a distinct subcategory of this much larger domain of kinetic cycling and respiration. Multiplication occurs when a kinetic cycle not only continually maintains or reproduces itself but also duplicates or produces new kinetic cycles.

During the Archean period, for example, hundreds of new minerals evolved and multiplied all over and through the earth. This multiplication was the result not merely of centripetal meteoric addition but of geochemical and biochemical respiration and reproduction. Minerals became airborne and liquefied in the prebiotic soup of the earth's ocean. Then, as elements mixed along the vortical patterns produced through convection, they crystallized into new orders. Older crystals could grow by the addition of new atoms. Crystals could even divide by breaking into two pieces and regrowing. Each divided crystal retained an archival memory of its past and could continue to grow in the same pattern. Novel crystal formations could emerge with variations and impurities in response to the changing environment. They could even mix with other atoms and minerals to create minerals never seen before.

Even convection currents can multiply. Within the larger convection cycle of hot fluid rising and cooler fluid sinking and reheating, smaller self-organizing nonlinear convection plumes emerge and create turbulent eddies and vortices.[8] Additionally, convection cycles heated from below spontaneously form into convection cells along the bottom surface. Heated fluid thus self-organized into multiple cellular membranes well before biological systems began to follow this same pattern. Which direction the convection cycle moves (clockwise or counterclockwise) is indeterminate and cannot be known deterministically in advance of the cycle. Moving matter thus creates cellularity, reproduction, and novel variation.

However, crystals and convection cycles are not biological—even if their patterns of motion are continuous with life. Living organisms are particular types of self-organized dissipative structures. They do not spontaneously self-organize into vortical patterns but emerge from preexisting kinetic cycles. Through a complex process, which we will cover in the next chapter, organisms retain a material archive of the kinetic cycles and patterns that formed and that can sustain, reproduce, and multiply them.

In this way, the organism does not have to reinvent the metabolic patterns of kinetic and chemical transformation each time that are required for reproduction. Living organisms, just like a self-organized convection cell, take in energy, release energy, and cycle some of that energy to reproduce themselves. However, they can also store energy internally and, like minerals, retain an archival pattern. Instead of the atomic series in minerals, organisms use nitrogen-based chemical series and other biomolecules to multiply themselves.

Center of orientation

The centrifugal movement of respiration and the kinetic cycle of reproduction also create a center of orientation—a spatial region from which movements go out and return. Every cycle has its limits or cellular membrane—a point beyond which flows of energy do not cycle back. However, there is also a metastable region where flows enter and exit, which sustains a general metabolism. This central region is a center of orientation in the sense that once it emerges, the system will continue to maintain itself through positive feedback. It may move to areas where there is a higher energy gradient or may grow larger or multiply from this initial center of orientation.

Kinetic cycles, therefore, are the metastable basis from which growth, dissipation, multiplication, and so on all move and develop. Development and variation all assume the emergence of a prior metabolic center of orientation from which anything can develop. This center is established first and foremost as a centrifugal result of respiration and reproduction.

★ ★ ★

"Atmosphere" is the general process of planetary respiration and reproduction. It is not just a region of the earth, but the earth itself as a process of breathing and centrifugal cycling. This includes mineral, aerial, aquatic, and biological processes during the Archean Eon of the earth's history up to the present. These are the immanent historical conditions of the present.

However, this conceptual outline is merely an extract from the real historical conditions of the Archean earth, which we must turn to now in order to ground this theory of centrifugal motion and show all the historical events that produced it.

7 Archean Earth I

Pneumatology

DURING THE ARCHEAN EON (4 to 2.5 billion years ago), the entire planet began to move in an increasingly centrifugal pattern of motion from the center out to the periphery (and back). This movement and the birth of the atmosphere produced an enormous material transformation of the earth, affecting every region.

As the earth began to settle, cool, and extrude new sedimentary layers, these new layers of mineral, water, air, and biota could accumulate and develop increasingly complex reproductive structures and cycles without being so bombarded by asteroids.

The Archean earth emerged from the Hadean, but also marked a move toward much more numerous metastable formations. This was due in large part to the emergence and multiplication of iterative kinetic cycles alternating between centripetal and centrifugal movements. In particular, the respiration of volatile gases and liquids played a crucial role.

It is impossible to overemphasize the importance of atmospherization on the earth, as the deep material and historical condition not just for humans but for all semi-stable emergent orders on the planet. As the Hadean earth was slowing down the flow of rock, the Archean earth began increasing the flow and circulation of gas. This produced an enormous energy gradient between a hot energetic earth and a cold exterior. Just as *hot* smoke flows, plumes, and spirals off into *cold* air in beautiful vortical patterns, so the atmosphere was born as the epic exhalation of the earth. Complex orders of mineral, gaseous, liquid, and living matter emerged through the pedetic disorder between this enormous difference in energy.

This radical move is the historical and material basis of all planetary orientation in general. Centripetal accretion made possible a distinction between a central region and a periphery, but centrifugal respiration introduced a relatively stable center of orientation compared to a mobile periphery. The kinetic structure of the atmosphere is the second key material condition for all planetary order.

The birth of the atmosphere is what gives central orientation to the planet in general and to everything on the earth in particular. Existential orientation is not unique to biological life. In its most general structure, atmospheric convection cycles are what orient all planetary existence up and away from the center of the earth.

This chapter argues that the emergence of a prevailing centrifugal pattern of motion occurs increasingly throughout the Archean Eon and that the deep history of the atmosphere is the material condition of terrestrial motion for all subsequent eons up to the present. In this chapter, I look closely at the kinetic patterns produced by two major geokinetic phenomena that define the Archean earth: sky and clouds. I argue that each of these historical phenomena follows a distinctly centrifugal pattern of motion.

ON THE HEAVENS

The first historically significant event of the Archean period is the emergence of the sky from the centrifugal movement of the atmosphere. As the earth's crust hardened, volcanoes subsided, and meteors slowed, something revolutionary happened: the sky began to clear. The carbon dioxide and sulfurous gases that darkened the earth's greenhouse atmosphere began to dissipate. Flows of light and radiation from the sun, the moon, and the cosmos increasingly reached the earth's surface.

What had for millions of years been a relatively sealed and turbulent surface became a *medium* from which the tripartite distinction between interior, surface, and cosmic exterior emerged. The earth no longer mainly recycled its dense low atmosphere but increasingly released its gases higher, outward, and into the cosmos, raining heavier gases and minerals back down. This process was not a sudden event but a slow transition of increasing volatile evaporation.

The atmosphere is a material medium that orients the earth as a center, which radiates heat into the periphery. The centrifugal motion of radiation transformed the earth into an oriented center with a peripheral cosmos. Centered orientation is, therefore, a kinetic effect—a by-product of a pattern of motion.

Is it such a surprise that so many ancient cultures believed that they lived at the center of the universe? The atmosphere is the material-historical condition of this idea. The centrifugal cosmologies of the ancient world are not ex nihilo ideas but real descriptions of the earth's atmospheric orientation. Humans see themselves at the center because the earth sees itself at the center—as an effect of its clear atmosphere.

History of the eye

The atmosphere is also the medium through which the earth absorbs, reflects, and circulates cosmic, solar, and lunar heat/light. The earth is the first eye—a central zone or surface of luminous accumulation and orientation. All other planetary eyes emerged from this basic kinetic pattern invented by the earth. The earth's black basalt surface is its light-absorbing pupil, its clear sky its cornea, its blue oceans and swirling clouds its iris. Again, we have an inverted metaphor. Biological eyes are material metaphors of the earth itself, not the other way around. History transports kinetic patterns asymmetrically from the past to the present, not the other way around.

Geoaesthetics

The earth's atmospheric medium invented the primary kinetic structure of sensation, communication, and aesthetics on this planet. The surface of the earth became a stage for the performance of light, form, and motion. Relative to the changing starry periphery, the earth was a calmer surface for the play of emergent forms (waves, clouds, whirlpools, and convection cells). The aesthetic distinction between figure and ground has its deep historical origins in the Archean earth, with its transparent atmosphere.

This division between figure and ground is not pre-given. It is an emergent distinction in which figure and ground kinetically distinguish themselves not by a break or division but through a semi-transparent medium.

The history of aesthetics has forgotten air. Atmosphere is treated only metaphorically, as a human feeling about our surroundings, when the historical conditions are precisely the opposite.

Human sensation is part of the atmosphere. The atmosphere is the deep transcendental condition for everything we usually call environment and background.

Figure and ground are regions of the atmosphere that have crystallized *as ground* and evaporated *as a figure*.

Cloud figures float on a stage made of rock and sea. The distinction between figure and ground that we usually make in art and aesthetics is made possible by the material condition of the atmosphere.

The earth is the first artist, the principal Narcissus, whose oceans painted the first image of the clouds and sky in their reflections. All painting is atmospheric. It is nothing other than the use of pigment to reflect images of light through a semi-transparent medium. In this way, the earth paints and sees the cosmos in the reflected surface of its oceans. The oceans and pools give the earth multiple perspectives. All other terrestrial mirrors have this same tripartite distinction between a darkened background, or tain; a semi-transparent surface; and a luminous medium.

The ocean was the first camera and cinema screen. As sunlight moved across the oceans over a day, it catalyzed chemical reactions in minerals, leaving a trace of its season, temperature, and luminous motion. Rock records from the great oxidation event still record a geophotographic image of a red earth covered in iron oxide pigment. The moving ocean surface was, and still is, filled with moving images of celestial and meteorological bodies. The earth was the first performance artist, whose act of recording was itself a self-transformative act without representation.

Light and sound are, therefore, not *in* or *through* the atmosphere but immanent transformations *of* the atmosphere itself. The earth sees and hears itself through itself as an atmospheric medium.

Being and becoming

The deep historical material foundation of the distinction between being and becoming is already performed by the sun moving over the sea. The light of the sun leaves traces of its radiation spiraling around the earth. This ring of fire draws and writes a spiral around the earth that distinguishes a central arch from its periphery. The sun remains relatively unchanged, compared to the changing sea.

This process is the real material historical foundation of the human idea of eternity. As the French poet Arthur Rimbaud writes in his poem "L'Éternité":

> Elle est retrouvée.
> Quoi? L'Éternité,
> C'est la mer allée
> Avec le soleil.
>
> It is recovered.
> What? Eternity,
> It is the moving sea
> With the sun. [1]

If Rimbaud is right that the idea of eternity is a material structure of the earth itself, this presents the possibility of a whole new geological epistemology and geopoetics. Humans know, not because they are different from nature and animals, but precisely because they have iterated the terrestrial drama of the sea and sun in their bodies. The earth knows itself *through* the human body (and other bodies) not because of any logical or ontological necessity but because of the practical, historical, and material way that the earth has come to iterate itself, billions of years before their existence.

How could we have forgotten the real material conditions of our geo-ontogenesis: the sun over the sea through the medium of atmosphere? This is not a metaphor. The sun does not rise over the sea because humans have delusions of eternity, but rather the other way around. Human thought does not represent the terrestrial drama. The earth is not an external passive condition for our cultural projection. Rather, the sun over the sea is an immanent and performative condition continuous with the human body itself. Human thought is the terrestrial drama played out and performed through us and other living creatures, each in its way. As another French poet, Paul Valéry, writes, "The core of man does not have a human figure—"[2] it has a cosmic and geological figure.

Solar rhythms

The clearing of the Archean sky also increased the rhythmic or periodic flow of energy from the sun relative to the irregularities of volcanism and lightning.

The kinetic periodicity of energy flow from the sun, made possible by the earth's new atmospheric medium, was one of the largest high-energy patterns of motion on the earth. Duality, symmetry, alternation, balance, and so on all have their terrestrial origin in this simple kinetic structure. The entire atmosphere began to pulse between hot and cold as day turned to night. Convection patterns on the earth began to sync up with solar radiation cycles. Temperature and light-sensitive chemical reactions began to sync and reproduce to the new rhythm, creating a visual archive of alternating matter and form. In other words, the general metabolism of the earth, with its short solar days and longer solar seasons, became increasingly periodically ordered around temperature and pressure changes related to solar cycles and fluctuations.

Geomusicology
This first and simple periodic structure of solar exposure shaped every significant emergent order on the Archean earth. The flow of matter/energy from the sun began to sync up the existing kinetic cycles on the earth and produce new integrated ones. All matter vibrates differently, depending on its composition and temperature. Sound is the vibration, resonance, and diffraction of pressure waves through an atmospheric medium. The basic structure of all sound on earth, therefore, comes from the regular pulse of the atmosphere's sonic qualities synchronized by the sun's orbit.

This is a simple but profound geohistorical point. Rhythmicity begins not with human walking or with animal rhythms, but with the earth. The earth vibrates and diffracts sound waves as it synchronizes with the movement of the sun. The earth itself not only plays music but is the instrument as well. The historical emergence of the atmosphere transformed the earth into a giant resonant membrane.

Differentiated layers of the atmosphere, for example, produced a metastable sonic cavity between the earth and the ionosphere where low-frequency Schumann resonances were generated and excited by lightning. As solar energy expands and contracts with night and day, the walls of the cavity and their frequencies modulate.

Essentially, the Archean earth became a guitar whose resonance chamber was this atmospheric cavity and whose strings were bolts of lightning. Thunder from the lightning accentuated the solar beat with cymbal crashes.

The earth thus became *musical,* in the general sense that it produced periodic sonic patterns through its newly formed atmosphere. The earth became a sound system playing its music out into space.

This is the geology upon which the whole history of "heavenly music" is based. The idea that the cosmos has an order and thus a musical or harmonic character to it has its real foundation in the actual musicality of the earth. Humans look for order in the heavens precisely because there *is* order there, oriented by the centrifugal atmospheric productions of the earth.

Humans could see this starry order because the atmosphere was transparent. The atmosphere connected light and sound and ordered them with the sun. The entanglement of sight and sound is, therefore, fundamentally atmospheric and not exclusively human.

CLOUD FORMATION

The second main event of the Archean was the emergence of clouds: the geological sediment of the sky. Clouds are the immanent geo-atmospheric formations from which figure and ground emerge. The atmosphere clarifies itself to become the background medium of its cloudy figuration. The atmosphere both grounds and figures.

It is a simple but profoundly humbling idea to think that the earth has been producing extremely dynamic, emergent, and novel orders and forms for around 4 billion years *in clouds.* Over 4 billion years, how many distinct forms have clouds made? Did the earth already sketch out all the forms produced by plant and animal activity before their emergence? Was the alphabet written in the sky before humans traced it by hand? Are clouds linguistic, textual, graphic, in a language of convection and respiration? We have general typologies of clouds (cirrus, cumulus, stratus, and so on), but we do not have a database of the specific shapes produced by the ingenuity of clouds.

Anti-form

Cloud forms, however, are a strange kind of form. They are forms made of matter (ice, minerals, microbes, water, gas). They are little ecosystems that tend to resist fixed forms. Clouds are forms-in-motion, continually shifting and deforming between forms. Atmospheric form reveals the emergent structure of form itself as a momentary and fleeting product of matter in motion. Instead of movement causing deformation or imperfection in form, the earth shows us that the basis of terrestrial form is continual topological deformation.

Furthermore, clouds are anti-forms because their boundaries are fractal and not well defined.[3] The edges of cloud formations continually dissipate such that it is impossible or at least arbitrary to decide what vapor concentration defines the "end" of the cloud. Clouds also have "self-similar" forms at the macro and microscopic levels. This is in part because they are metastable structures formed by iterative processes and not geometrical calculations. Clouds are thus emergent forms defined by ongoing but patterned irregularity.

Clouds lack edges. They have no points, lines, or planes as in geometry. As such, they resist all discrete geometries and topologies, because they are processes. They are folded and iterated bodies without surface or depth. They are neither wholly formal nor merely material because they are continuous gradients of more and less vapor without a hard boundary.[4] Clouds are not forms *of* something else, because they are radically singular, without model or copy. They are woven together through layers of iterated sedimentation and crystallization. Technically, when water turns to solid ice in clouds, it becomes a mineral.

Clouds are emergent patterns in a centrifugal flow of heat from the core of the earth into the atmosphere. They are not reflective copies *of* the earth, or of anything else, but rather products of a centrifugal expansion of airborne matters formed into distinct bodies. They are novel experiments in anti-form. Clouds produce all manner of spontaneous vortices, funnels, and other curved structures that gather elements together into new orders.

Ice crystals

Early clouds were home to some of the first ice crystals on the earth. Ice crystals produce emergent forms like hexagonal plates, hexagonal columns, and dendrite formations, as the result of pedetic flows of matter and the molecular structure of water. Flows of dust, gas, and vapor ride atmospheric convection currents into the sky and begin to cycle inside clouds. This constant iteration of cycling heterogeneous elements slowly produces metastable patterns of ordered crystals. Clouds thus become ephemeral archival memories of the atmosphere through their crystallization.

Once water becomes mineralized/crystallized, however, it falls centripetally back to the earth. There, its mineral archives are not entirely lost but are integrated in precise ways back into the air, ocean, and crust. Each drop of rain contains a melted-down mineral memory that it shares with the surface. Clouds also move through the sky and transport these mineral ideas all over the earth. This historical transmission and deposition resulted in new breeds of evolving mineral species.

Ice crystals were some of the first quasi-geometrical forms, alongside the cooling mineral concentrations on the crust. What they show is precisely the historical and material emergence of kinogeometrical form and order on the earth. Geometrical forms do not transcend matter and history but are emergent products of it.

Before there were bees to make hexagonal hives, clouds had already figured out how to make hexagonal plates and hollow columns. If there are hexagonal beehives, it is because Archean clouds were already their immanent material condition. Atmospheric ice crystals also invented the dendritic patterns we find in trees, in watersheds, and in the veins and neurological fibers in our brains. Could we have had any of these without the creativity and material conditions of clouds? A precise genealogy of this question would be impossible to write, and yet it is hard to ignore that the same dendritic patterns emerged across numerous earthly scales.[5]

Clouds and brains are different, and yet they move in similar ways that reveal the geo-material conditions common to both. Crystals in clouds grow dendritically, intersecting one another in some areas more than others just as neurons fire together in the brain in some regions and patterns more than others. The brain is capable of producing new firing patterns and so are clouds. Without a deep genealogy of their transmission, it is still necessary to say that hexagonal and dendritic forms are common because the same matters move through them. Human thought, of course, is neither mechanically determined by these processes of dendritic mineralization nor entirely free from them. Dendrites remain the asymmetrical and immanent geo-material conditions of thought.

Clouds are not like brains, but brains *are* clouds in the general sense of being dendritically organized minerals, water, and gas. Humans are made of the same atoms that were produced by clouds billions of years ago. Our bodies also retain a molecular memory of hexagonal geometries, also found in basalt columns, quartz crystals, water bonding, ice crystals, and the hexagonal carbon atoms rings that structure our biology. Matter has a memory that traverses and exceeds the human brain.

We were born from clouds. They produced the water that we are. When we look up to the clouds, we are looking at our immanent origins. We can see in them the six-sided symmetries and dendritic patterns shared by our bodies and brains. Atmosphere and clouds gave birth to oceans and life—as so many oral mythologies have described.[6] When Aristophanes satirically wrote that "from [clouds] come our intelligence, our dialectic and our reason," that was geohistorically closer to the truth than he realized. Clouds did invent the material kinetics of our material intelligence.

Weather formation

Wind, clouds, and weather are deeply nonlinear processes, such that there is no complete set or "phase space" that can include all the variables and trajectories at work. This is not just a problem of data collection. It is a fundamental problem for all moving systems. At each moment, the whole phase space changes such that it was not what it was before. It is inaccurate to treat the atmosphere as a totality of "particles with trajectories," because there is a real emergent novelty in weather and cloud formations.

This is the kind of phenomenon scientists usually do not like. Weather is neither purely random nor is it wholly determined or determinable. It is relational, pedetic, and so diffuse that the actual measurement of weather with various instruments may affect the weather one is trying to model.[7]

The imagination of the earth

Clouds are the imagination of the earth not because they are fleeting like the imagination of humans but vice versa. Because the atmosphere is centrifugal and creative, it can externalize heat, elementals, minerals, and biota up into the sky and reform them into new shapes. Clouds are new earths, new continents, metastable worlds. The earth takes its mineral, liquid, aerial, and biotic beings and invents unique ecosystems and microworlds that do not mimic the earth, but use the heat and elements of the deep earth to make and multiply peripheral earths around the earth.

Clouds are sketches and traces of worlds that could be. Rain sediments and seeds the cloud images back down to earth. One result of this creative rain is that the earth becomes more liquid, flexible, and vortical as weather transforms it.

If the oceans produced the kinetic cycles and patterns that created life, they got the idea from the clouds whose vortical patterns and fluid images preceded their own. The creation of cloud plumes and cloud systems are also like dissipative versions of the mineral earth's volcanic islands. Both create a "new earth." The two even share minerals through the water cycle—seeding and reseeding one another.

The noosphere, or sphere of ideas, therefore, does not originate with human minds or their technologies. It is already there in the atmosphere, where images multiply and forms circulate all over the planet, affecting one another and producing emergent novel patterns never made before on the earth. The noosphere is social, creative, and global because the atmosphere already was and remains the immanent medium of all ideas, imagination, sensation, and communication.

In the next chapter, I look at one more important kinetic event in the earth's Archean history: the emergence of life.

8 Archean Earth II

Biogenesis

THE THIRD MAJOR HISTORICAL EVENT of the Archean Eon was the emergence of living organisms with metabolism, genetic multiplication, and natural selection (prokaryotic bacteria and archaea). Organisms, in my kinetic view, are dissipative or vortical systems that have the distinct ability to remember and reproduce the material patterns that produced them. Life, as I will define it in this chapter, is a fundamentally atmospheric phenomenon in the sense that it came into being with and as the atmosphere itself. Life does not live *in* the atmosphere but is immanently something that unfolded *from* the atmosphere.

Prior to life, the atmosphere produced all kinds of membranes, metabolisms, multiplications, and emergent orders (hexagons, dendrites, and so on) that were not alive. Organisms adopted these patterns and continued them by other means. I argue in this chapter that the emergence of organisms followed a centrifugal motion similar to the emergence of the atmosphere itself.

VORTICAL LIFE

The accumulation of prebiotic materials necessary for the emergence of life assumed, first of all, a centripetal kinetic pattern of motion capable of bringing them together in the first place. Once enough of the necessary elements and molecules came together enough times, a biological habit of this iteration emerged. Since no one has yet experimentally produced a living organism from prebiotic chemicals, the precise physical process and materials remain unknown. However, there are many ingenious theories and new experimental data developing all the time.[1]

All of them, though, presuppose the accumulation of certain prebiotic materials (lipids, citric acids, carbon atoms, amino acids, proteins, and so on). But what was the material kinetic structure necessary for the collection of these matters in the first place? Entropic dissipation is too general of an answer to be of much help, although certainly, life, along with everything else, involves the physical process of entropy.

I propose that life is not merely entropic, but vortical, and continually reformed through a material process of folding. This is an idea with many notable precursors.[2] For example, the 19th-century French naturalist George Cuvier (1769–1832) wrote that "Life, then, is a vortex, more or less rapid, more or less complicated, the direction of which is invariable, and which always carries along molecules of similar kinds, but into which individual molecules are continually entering, and from which they are continually departing."[3]

The English biologist Thomas Henry Huxley (1825–1895) described life's "manner of association" as a "whirlpool," whose stoppage or broken form we call "death."[4] The English neurophysiologist Charles Sherrington (1857–1952) described the cell as "an eddy in a stream of energy" and as "a stream of movement which has to fulfill a particular pattern in order to maintain itself."[5] The great American biologist Carl Woese (1928–2012) wrote that "organisms are resilient patterns in a turbulent flow—patterns in an energy flow."[6]

Life is a metastable formation that is also able to store enough energy to keep the vortical pattern going even when environmental conditions may be insufficient. The continual reproduction of a semi-permeable membrane allows for this regulation and stabilization of the metabolic intake and outflow. Life is also distinct from its vortical beginnings, however, insofar as it developed highly differentiated internal functions and orders not found in waterfalls or whirlpools. Vortical systems do not necessarily lead to life, and they are much simpler, but there is a clear kinetic continuity between the two.

Life and its prebiotic components, I argue, must have emerged from these metastable patterns of folding and iteration. Ocean topology, iterative and diffractive waves, and vortical spirals in fluids probably helped sustain certain patterns of accumulation. Similarly, the bipolar and hexagonal structure of water molecules, planetary convection cycles, atmospheric turbulence, and cloud formation all likely played a role in sustaining patterns that may have helped life take hold.

Almost all the big theories of abiogenesis suggest that life originated at the site of a large energy gradient such as deep-sea volcanic vents, lighting strikes, or solar radiation. However, from my perspective, equally important to the energy of the gradients were the *kinetic* patterns that emerged from the gradient *reduction* process: vortical formations, eddies, and foldings. The most probable origins of life also all involve water, whose vortical formations are well known. Unfortunately, biology has spent much more time studying the so-called "mechanical biobits" of life than it has the kinetic patterns of the atmospheric conditions for them.

Accordingly, the narrow and fetishizing perspective of mechanistic biology risks postulating either the improbable event of random assembly, or the spontaneous appearance of autonomous membranes, metabolisms, RNA, or DNA out of nothing.[7] To think this way is not only a failure to understand the kinetic material and historical conditions of life but a misunderstanding of what life is and how it works. By starting with ex nihilo substances, i.e., biobits in a flask, we end up thinking that life is an autonomous machine composed of unidirectional parts without any relational or kinetic precursor. Mechanical and substance-based biology is, therefore, an implicitly anti-relational approach.

Unfortunately, this substance-based mechanical metaphysics defined most of twentieth-century biology, and the consequences have been disastrous. How we frame the problem of the origins of life has significant implications for how we understand what life is. Before moving on to the kinetic theory of life, let us take a short detour to clarify its contribution vis-à-vis a critique of substance-based biology.

AGAINST THE MACHINE

The reigning ideology of molecular biology since the mid-20th century, appropriately called the "central dogma," is defined by a reductionistic appeal to both randomness and mechanism in biological phenomena.

According to this dogma, first, there is DNA that contains discrete genes or "information" that is directly transmitted and recombined through generational heredity. DNA is said to contain a genetic "program" or "blueprint" for the precise reproduction of the organism.[8] This is the coded information for the creation of RNA, and RNA creates the proteins that do the functional work of the organism. In this story, the transmission of information is one-directional and linear.

The central dogma or "neo-Darwinian synthesis" is also explicitly reductionistic in that it defines life strictly by the operation of supposedly simple, discrete, machinelike, basic building blocks (amino acids, proteins, RNA, and so on). Richard Dawkins's concept of the so-called "selfish gene" has done much to popularize the idea that genes are reducible to individual units whose organisms are mere "vehicles" for the transmission of individual information. Population genetics then analyzes the statistical distribution of this information in large groups over time.

At the heart of this paradigm is also the idea that DNA *randomly* mutates its information to create phenotypical differences that are more or less successful in the survival of the organism. For example, in the 1970s, the French molecular biologist Jacques Monod put forward a hugely influential account of the substance-based and reductionistic model of molecular biology. Monod described DNA as a universal machine that governs the production of the whole organism. Life, according to Monod, is an entirely mechanistic process that follows ironclad laws, indifferent to human values and interests. On the other hand, Monod said, DNA also mutates randomly. However, once mutated, life follows the necessary laws of linear, mechanistic biology. Hence the title of his 1970 book, *Chance and Necessity*.

Another influential French biologist of the time, François Jacob, described DNA as a "dictionary" of sixty-four genetic letters, where the twenty amino acids in the cell are like "synonyms."[9] Crucially, for Jacob, the genetic program does not learn from the organisms and does not change with the world. Occasionally, however, DNA makes "copying errors" that lead to phenotypic differences that are "put to the test" in the population and either selected or not for survival and reproduction.[10] In this way, DNA is reworked in a "roundabout way," as random errors are either reproduced or not.

The central dogma described above has constrained the problem of the origins of life to the search for machinelike bits of preassembled prebiotic particles. It has also led to a complete misunderstanding of how organisms work. Life is not a deterministic machine composed of discrete parts or substances. DNA is not a blueprint or set of instructions. It is not information that unilaterally copies itself or regulates or "selfishly" controls the organisms. Changes in DNA, I will argue, are not random mutations or "errors," nor are they deterministic adaptations either.

Instead, the organism, as the Austrian biologist Ludwig von Bertalanffy (1901–1972), wrote, "is in reality a momentary cross-section through a spatio-temporal pattern" of "a continuous stream."[11] Or as the Australian biologist Paul Alfred Weiss (1898–1989) wrote, "Life is a dynamic process. Logically, the elements of a process can be only elementary processes, and not elementary particles or any other static units."[12] The origin and continual operation of life are much more like a dynamic system, with unique emergent orders and micro orders forming like "nodal points in a vibrating string."[13] As the American biologist Richard Lewontin put it, "First, DNA is not self-reproducing, second, it makes nothing, and third, organisms are not determined by it."[14]

ORIGINS: FOLDING-FIRST

Life is profoundly historical and kinetic—not random and discrete. Life emerged from a practically incalculable number of processes, each with their own incalculable conditions of emergence. Calculations done by biologists from Fred Hoyle to F. B. Salisbury all have a different way of determining the "likelihood of the emergence of life." Still, they all agree on one thing: 12 billion years isn't enough to produce a single enzyme by chance alone. Formal equations of the probability of life look something like this:

> the number of building blocks on planet × 1/(average [mean] number of building blocks needed per "organism") × (availability of building blocks during time t) × (probability of assembly in a given time) × time.[15]

There are at least three problems with formulating the origin of life in this way. First, there is no complete set or phase space of "building blocks" on the earth. The building blocks are themselves metastable and emergent processes on the earth and are continually moving and changing. The idea of a complete set of all blocks is an abstraction of processes that are continually shifting and changing. Second, there are no building blocks.

Every discretely identified block is itself only a fleeting dynamic pattern in matter. It exists only by continually iterating and folding itself like a river eddy. Third, matter does not move randomly. Matter, down to the smallest quantum levels, is self-affective and relational. Its relations are never isometric, but there is nothing that is absolutely unrelated and thus unaffected by something else.

The idea of randomness is just that, an epistemological abstraction that imagines movement without any other relations because scientists do not know the precise kinetic relations involved. Randomness is an epistemological marker of our ignorance, not of absolute laws. In short, the known laws of physics do not predict the likely emergence of life. There is no necessity of life from physics, nor is there any total set of variables from which to deduce the so-called randomness of life. Life emerges neither from chance nor from necessity.

Matter creates the emergent conditions for its reorganization through what I call *sympedesis*, or collective pedetic emergence. Emergence is not random, but like walking or wandering: each step comes after the other. The previous step does not determine the next step, but it does shape its range of action in a shifting context. Once a step occurs, it irreversibly directs but does not determine the next, and so on iteratively. There is no discrete origin or end of the universe. Rather, there is pedetic emergence, sympoiesis, or sympedesis.

Theories of the origin of life all tend to include the word "first" in their name, as if their origin required no further or relevant precursors: "DNA first," "RNA first," "Protein first," "metabolism first," "lipids first," and so on. What most biologists agree on at this point is that chemical or mineral metabolism came before proteins and RNA, which all came before DNA. The problem is that no experiment has produced DNA from RNA, or RNA from proteins, or RNA from chemical metabolism.

It seems to me that the fetishization of the idea of building blocks has rendered the kinetic transition and emergence between blocks obscure and mysterious. The main biomolecules responsible for creating proteins, amino acids, have been produced by the Miller-Urey experiment, but scientists have also discovered amino acids in the Murchison meteorite. This means that amino acid synthesis has already happened outside the earth, and may have come to the earth from space. An under-studied question is then "what must the material conditions have been for the production of amino acids?" Whatever they were, they must be common to space and the Archean earth.

The critical conclusion of the last forty years of "DNA *not* first" theories and research is that DNA did not emerge out of nowhere as a divine "master molecule" containing self-authored information with which to dominate the slavish organism. DNA is not a selfish god but a metastable form emerging from other, more primary sympoietic and kinetic processes. In particular, I suggest, DNA, RNA, protein, and metabolic chemicals all involve a process of folding that is essential to their reproduction. Folding is not a thing, object, or building block—it is a kinetic process. Folding is immanent to and constitutive of all life's processes. Cell membranes emerged from folding and are themselves folds. Genetic reproduction occurs through the folding of DNA, RNA, and proteins.

Thus, instead of asserting a linear developmental order between discrete building blocks, or even a simple coevolution between blocks, life should be understood, I propose, as without a first block or final goal. That is, we should think of life as an emergent process of *folding* "first." However, since folding is not a thing nor a bio building block, it cannot be a chronologically first thing or substance. Instead, since the process of folding, as we have shown in previous chapters, is continuous with a much longer material and historical process, it is "first" in a historical-ontological sense. Folding is how matter transforms itself through motion.

From this kinetic perspective, life is not a unique, random, ex nihilo mystery of the universe that makes us and our planet so special and cosmically significant. Instead, it is a regionally emergent feature of a much larger flow, fold, and circulation of cosmic matter. Life may have singular features, but it is not separate from the flux and flow of the universe.

DNA, for example, is a unique stabilization and refinement of a folding process that RNA had already been performing long before. We have known since the 1980s that RNA must already have been capable of transmitting genetic information by folding itself over into unique three-dimensional shapes and knots called ribosomes that produce base pairing and thus transmit genetic material. However, RNA only has one backbone and is thus much less stable and regular in its transmission. DNA is a more stable version, using two RNA strands folded together into a double helix.

The point is this: DNA is a highly ordered and extremely accurate duplication, with only one difference in a billion base pairs (after self-correction). However, this accuracy was an emergent feature of an increasingly refined process of RNA duplication through folding. RNA is not just a messenger for DNA but the immanent historical condition of DNA itself. DNA does not do anything kinetically or structurally different from what RNA had already been doing.

In other words, DNA is not a historically autonomous ruler over RNA; instead, DNA got all its complexity from RNA in the first place. RNA is not an inferior version of DNA; instead, DNA is a more stable topology of RNA. RNA is not a "copy" of DNA; instead, DNA is a topological version of RNA transmitting itself to itself.

We can say the same of proteins. Proteins are hundreds of folded strings of amino acids less complex than RNA's nucleotides and thus were probably more stable on the Archean earth. Proteins are capable of folding themselves into novel structures and functions without the aid of RNA or DNA. However, RNA and DNA can also produce new proteins. Recent theoretical models have shown that it is possible to create self-reproducing RNA strands from self-organized proteins. If current experimental efforts can show the emergence of RNA from protein folding, then again, RNA's duplication would be an emergent and possibly coevolutionary property of protein folding.

None of these three are autonomous processes. They are historically emergent orders that built themselves up pedetically, one step at a time, through countless iterations of folding. Each biomolecule learned to fold from the previous ones, but all of them learned to fold *from water*. Water's polarized structure brings some amino acids together and pushes some apart, depending on whether they are hydrophobic or hydrophobic. In a general sense, then, water was already folding itself over itself through electrical attracting and bonding before other molecules began to attract and repel it.

Water flows and folds. It is the fluid atmosphere and oceans that taught biomatter how to fold over itself—and thus how to move. The movement and function of DNA, RNA, and proteins are directly related to the fold of their topological shape. The precise method by which folding occurs remains perhaps one of the greatest mysteries in biology, rivaling and related to that of abiogenesis.

Hydrophobic kinetics is not something a single molecule possesses like an autonomous essence. It is an emergent dance of polar and apolar molecules entangled together. The whole division between inside and outside, particle and environment, is profoundly challenged by the process-driven kinetics of folding. This is not a special case. All form, I would submit, is the product of a kinomorphic process.[16]

Folding is the immanent and primary process for the higher-order emergent structures of proteins, RNA, and DNA. This is why there are no blueprints, only continually woven patterns. Life is, therefore, kinetic and performative. Whatever "there is" in a cell must be performed and reproduced through the drama of folding. This is what I would like to emphasize in the Archean history of life: the central neo-Darwinian dogma is ahistorical, retroactive, and wrong.

MULTIPLICATION

Biological reproduction, multiplication, and mutation are not random. However, neither are they entirely determined by a series of bio-machines defined by essence, function, or mere replication of DNA code. Macromolecules like DNA, RNA, and proteins are not substances but kinetic processes that emerged historically through folding and continue to rely on the folding process for their reproduction and multiplication. The performative and creative structure of life has kinetic folding baked in.

This is in contrast to the central dogma, which holds that DNA contains the complete set of information for reproducing every organ and function in the cell. Depending on the kinds of organs and proteins needed, the story goes, only part of the DNA code is active in each place. In short, DNA is the master molecule, and all the others merely follow orders directly from DNA.

However, there are three significant challenges to this idea: externally caused mutations, epigenetic mutations, and nonrandom internal mutations. While the first two can be and often are admitted to the growing list of "exceptions" to the "rule" of DNA dominance, the third challenge fully overthrows the substance-based unidirectionality of DNA.

External transformations
DNA mutates due to a variety of different influences, including chemicals, ultraviolet light, radioactive decay particles, viruses, and even cosmic rays.

Biologists often treat these as merely external or random effects on DNA, but they are highly relational processes. For example, they mean that chemicals, sunlight, and cosmic rays that were present on the Archean earth likely played a lasting role in the transformation of DNA and, thus, in the evolution of life. If solar flares, radioactive elements coming up from the deep earth, and cosmic rays can influence DNA mutation rates, then life is not quite as autonomous and autochthonous as is often thought. Ultraviolet light may have provided the early energy to produce life's macromolecules in the beginning, but may also have continued to play an ongoing mutagenic role as well.[17]

Viruses also played and continue to play an active role in the transformation and evolution of DNA and life.[18] Viruses are ancient Archean biomolecules (made of proteins and amino acids) that attach to host cells and use host DNA and RNA to reproduce more viruses, with high rates of mutation. The relations between cells and virions are incredibly diverse, and viral genes are often recruited for cellular functions and contribute directly to the evolutionary development of organisms.[19]

From the beginning, viruses thus played a central role in reproducing and mutating life through coevolution. Viruses are the ultimate symbionts whose nonlinear and highly mutational expression strategies directly introduce new changes in the folding patterns of DNA, RNA, and proteins. Viruses change the way life folds and unfolds itself—and thus how it multiplies itself. Viruses also use an enzyme called reverse transcriptase to create DNA from an RNA template, a process called reverse transcription. The virus thus repeats the historical performance of the creation of DNA from RNA, reversing the coding of DNA.

Viruses, therefore, blur the distinction between life and non-life because they do not have cell membranes or reproduce on their own and yet they can direct the folded structures of the organism, which define what life is. Viruses participate intimately in the supposedly autonomous reproduction of the cell. Furthermore, cell structure and the mutational folding process that define life may also have been the result of virions *before* the division between cells and virions in the first place.[20]

Epigenetic mutations

DNA does not transcribe itself. The multiplication of cells is the result of a complex host of folding processes through a liquid environment (cytoplasm and water) and can be altered by various environmental or "epigenetic" processes. This is not a Lamarckian point, however, because there is no direct influence of the environment on DNA.[21] Instead, there is an indirect process called methylation, by which gene sequences inside the DNA string can be turned on or off to produce different folding patterns in the RNA and proteins. Numerous environmental factors, such as starvation, disease, and pollution, can cause gene methylation—and then be passed on to offspring. Genes can also be turned on or off, depending on how they fold around histone proteins. This can also be affected by the epigenetic factors mentioned just above.[22]

In both cases, DNA does not make itself but emerges through a process of modified transcription and folding that is not a mere tracing of a model or preexisting form. DNA is continually modulated and tweaked creatively through the environment, niche construction, and folding. The process of transcription is thus genuinely creative. "The problem is not only that the music inscribed in the score does not exist until it is played, but that the players rewrite the score (the mRNA transcript) in their very execution of it."[23] It is the process of transcription itself that is primary and within which DNA is only one of many agents that must perform the highly plastic event of transcription each time. The process of transcription is not at all random or predetermined, but rather highly relational, since it includes the environment, methylation, histone folding, and the folding of RNA and proteins through the dynamic cytoplasm. In short, transcription is not representation but performance.

However, external and epigenetic mutations do not necessarily overthrow the central dogma. Only internal transformations can truly challenge the primacy of DNA.

Internal transformations
We saw above that biological multiplication works not like a machine but more like a musical performance or choreographed dance. Multiplication includes all kinds of so-called "external" factors that can become part of the internal life of the cell. However, the central dogma can and has still widely tolerated external and epigenetic factors as "modifications" to the unidirectional DNA process.

As long as the above changes are merely changes to DNA transcription, and not relational changes internal to DNA itself, then the central dogma can still hold on, even if by a thread. Viruses and methylation are not enough to overthrow the idea that DNA contains the totality of all expressible genetic information—whether that information is turned on or off.

However, DNA is not just Platonic information to be transcribed. The information in DNA is not a priori but is inseparable from its kinetic and topological performance of folding. DNA folding patterns and structure are some of the most critical but enigmatic aspects of reproductive life. DNA does not merely contain information but performs this information through its movement and folding around itself. Its double-helix shape renders it highly collapsible and foldable around itself and around the histone proteins that are themselves made of folds. We currently have no map of this intricate folding pattern, but recent research shows that it plays a role in shaping what DNA is and does.

For decades, experiments suggested a hierarchical folding model in which DNA segments passively spooled around protein particles. However, biologists based that model on structures of DNA as they were after the harsh chemical extraction of cellular components. Now, new 3D images of DNA inside the cell nucleus of *intact* cells show that DNA is extraordinarily flexible and plastic, with fluctuating diameters, contrary to the textbook models.[24] These folded structures expand and contract in a wide range of configurations that affect what DNA is and does. DNA is not like a database or selfish gene, then, but like a dancer whose pattern of motion is in a continuous process of becoming.

The radical idea here is that the process of DNA folding affects gene expression. Scientists have counted the genome, but the three-dimensional shape of DNA remains mostly mysterious. Contrary to the central dogma and textbook biology, DNA folding is not just an efficient storage method for information (like a database). When DNA folds over itself, it touches itself in certain places that change and regulate gene expression.[25] This is neither random nor "merely" an external modification of a preformed substance. This process is a kinetic and performative act of internal self-affection. Critically, this means that there is no isolated information "contained" inside DNA. DNA is what it *does*: its being is a doing.

Furthermore, DNA mutates in nonrandom ways. DNA mutation occurs when base pairs do not align properly. The precise mechanism of this misalignment was discovered recently, in 2011, along with the fact that misaligned protons play a role in the self-mutation of replicating DNA.[26] But why would protons move to the wrong position in the first place? For a proton to move, there needs to be a sufficiently strong thermal energy to cause it. However, there is not such sufficient energy available in the nucleus nor in surrounding water molecules.[27]

It is possible, though, for protons to tunnel quantum-mechanically across the energy barrier and cause a mismatch in the connection of DNA base pairs.[28] In quantum mechanics, subatomic particles are neither determinately here nor there but have an indeterminate or fuzzy position that allows them to leak across energetic thresholds. If quantum tunneling were a nonrandom cause of mutation, we would expect to see an increase in mutations when the DNA was read and copied more frequently. Since each reading and writing of DNA constitutes a performative act of quantum "measurement," then these would cause an increase in perturbations, leading to higher rates of mutation.

This is indeed what several recent studies have all confirmed without a doubt.[29] The more genes are read and transcribed, the more mutations occur. It turns out that the more cells struggle to survive and the more they try to replicate their DNA, the more this replication leads to mutations, which then lead to an increased chance of survival in hostile conditions.

This is an incredible result that directly contradicts two fundamental tenets of the central dogma: that mutations are random; and that information flows only one way, from DNA outward to proteins and then to the environment of a cell or organism. If mutations lead to an increased chance of survival stimulated by the environment, they are not random but rather relationally "adaptive mutations." This is, again, not technically Lamarkian, however, since the environment does not determine a specific mutation but merely provokes the organism to increase its rate of self-mutation and natural selection.

Some studies have shown this outcome to be consistent with the act of quantum tunneling and measurement—since DNA measures itself in the act of replication.[30] However, since other biochemical reactions can also cause mutation, and since indeterminate quantum positions are challenging to isolate, scientists have not yet been able to isolate quantum effects as the *sole cause* of mutation. In either case, experimental results strongly suggest that mutation rates are directly related to environmental changes. This point alone is enough to overturn the central dogma of linear causality.

To conclude, DNA folding and quantum tunneling both offer us a new vision of DNA as a performative process of relational and nonrandom self-transformation. Strictly speaking, there is no isolated, autonomous DNA substance, code, or genome in real living cells—only in biochemistry labs. There is no such thing as abstract genetic information in living cells. DNA is always vibrating, flowing, folding, and measuring/transforming itself in a more extensive fluid medium or atmosphere. The movement of life is neither random nor mechanically determined but pedetically choreographed and profoundly entangled with its environment or atmosphere.

PNEUMATOLOGICAL MATTER

Life, therefore, is not *in* the atmosphere but *of* the atmosphere, in the deepest sense. It is a fundamentally pneumatological matter. Biochemical metabolism and respiration consume volatiles from the air, transform them, and release them in a continual centrifugal circulation up and out. Life does not live on the earth, as a spatial dimension, but is thoroughly penetrated by it. Living bodies comprise minerals, liquids, and gases from the sky that they circulate, transform, and then use to reproduce themselves. DNA is not the blueprint for the organism.

Life is a continual act of com-penetration, multi-folding, or plication—not a mere juxtaposition in space. The difference between acting and being acted on is atmospherically indeterminate. RNA, proteins, and viruses, just like DNA, must perform their information through folding; atmospheric life *is* what it *does*. Life remakes the atmosphere, which in turn becomes the material of its being. Being is making.

The atmosphere completely overthrows the division between the material conditions of life and the conditioned existence of that life. What emerges also transforms its condition of emergence. Quantum and environmental factors work together alongside DNA. The three coevolve with one another and make possible the transmission of modifications from one generation to the next. While minerality is the archival sequence that life internalizes in carbon, the atmosphere is the fluid immersion that connects these archives through the sky and ocean. Together, minerals and the atmosphere produce a kind of ecological inheritance, or bioculture, that builds pedetically on prior modifications.

Microbes, in the Archean, fed on dissolved iron from the ocean. They oxidized it into hematite and, in the process, released enough energy to support an entire ecosystem. We can still see massive banded iron formations in Australia and South America today. They are the geo-bio-cultural memory of an epic feast that lasted tens of millions of years. Mineral topologies and curved centripetal enclosures provided the surfaces, shelters, and vortical templates where early microbes could concentrate, fold up, accumulate, build, and respire.[31] In turn, populations of mutating microbes produced new metabolic products that catalyzed energy-producing reactions in the mineral-rich ocean—producing a gradual increase in carbon, sulfur, nitrogen, and phosphorus near the surface of the earth, along with hundreds of new minerals.[32]

To live is to make an atmosphere. Life breathes the breath of other beings. The atmosphere is the circulation of life and death—ultimately radiating heat centrifugally out into space. The clouds are, therefore, alive not only because they brim with microbes but because they are the dying breath of microbes. The atmosphere is not just the medium of exchange and communication between microbes. It is itself the structure of all material language. We say that the wind whispers, howls, and speaks because speech, in a general sense, is the act of pushing air over a modulating surface. The atmosphere is thus the immanent condition of all speech on the earth. It is the real breath, spirit, and vision of the earth.

What we call life and mind are merely reifications of the centrifugal circulation of breath radiating from a site of central orientation. But the kinetic structure of this site is something already drawn up and formed by the earth, as a central body visibly surrounded by a periphery of heavenly bodies through the medium of a clear sky. Mind = spirit = life = breath = atmosphere. Atmosphere and respiration are the material conditions that precede any distinction between mind and body, subject and object, self and other.

Living organisms are part of the same dominant planetary regime of centrifugal motion because they are continuous with the airborne release of volatiles and heat from the deep earth and atmosphere. Even genetic transfer follows this same prevailing historical motion. Genetic material and energy move from a central region of storage to a peripheral region of protein multiplication. This is not a matter of model and copy but rather, like the atmosphere itself, a matter of transmission, mutation, and experimentation. Genetic matter moves from center to periphery, which in turn maintains and reproduces the cell and its central nucleus. Volatiles move from the earth's core to the periphery, from where in turn they rain back down to maintain and expand the earth with new mineral strata, organisms, clouds, and ozone. The cell is shaped like a tiny earth because the same regime of motion circulates through both—heat moves from center to periphery.

* * *

During the Archean, the entire earth erupted into centrifugal motion. Volcanoes blasted themselves into the air, the ocean evaporated into the clouds, and organisms released an incredible amount of volatiles and stored energy. However, by the end of the Archean earth, around 2.5 billion years ago, a new form of life emerged that would change the motion of the planet yet again: plants. This is the subject of our next historical section.

C. VEGETAL EARTH

9 Tensional Vegetality

THE THIRD MAJOR geokinetic planetary field to rise to dominance in the earth's history was the vegetal field. The Proterozoic Eon, from about 2.5 billion years ago to 541 million years ago, was the longest eon in the earth's history. Over the course of this eon, three critical events occurred: the emergence of eukaryotes (cells with a nucleus and organelles), the emergence of multicellular organisms (such as protozoa, fungi, and plants), and the arrival of life on land.

All these events followed a tensional motion inside, between, and through living organisms. However, this new pattern of movement was defined by a system of "held contrasts" that was not limited to life alone. Life, like mineral and atmospheric flows, was not just one isolated region among others. Vegetal life completely saturated and transformed *all* planetary processes.

The earth is what it is today because eukaryotic and multicellular organisms have been affecting the earth's weather, ozone, mineral evolution, and even the chemical balance of its oceans for a long time. Life is what ultimately resurrected the earth from its Proterozoic cycle of cryogenic snowball phases. What I would like to show in this chapter and the next is that the principal historical events of the Proterozoic Eon all share a unique pattern of motion. During this time, the earth invented a new tensional pattern of movement through organisms, which in turn reorganized the entire earth according to this same pattern.

Before we get into the historical specifics of this transformation in the next chapter, I would like to show in this chapter the basic outline of what tensional motion is and how it works. Although a bit abstract, I hope it will help to more easily identify the existence of this pattern in the historical events described in more detail in the next chapter.

GEO-PHYTO-KINETICS

Vegetal life does not emerge *on* the earth; instead, the earth *becomes* vegetal. The most significant error in thinking about vegetality, or plant life in general, is to think of plants as distinct or separate from their deep historical emergence as cosmic and earthly bodies. Plants are fallen stars, minerals, air, and water. They are made wholly of the stuff of the cosmos and the earth.

Vegetality is something that the earth does—not something that merely occurred on the earth. Vegetality is an expression of the earth's own inner tensions and pressures relative to its atmospheric and gravitational orientation. Without the memory of minerals and the fluid dynamics of the atmosphere, vegetality would have been impossible. Therefore, there is no ontological difference between plants and the rest of the earth—only a difference in kinetic pattern. The question of this chapter is then, what is the distinct kinetic pattern that defines plant and vegetal life?

Just as the atmosphere is not *on* the earth, so plants are not *in* the atmosphere. Spatial prepositions inadequately describe what are ultimately dynamic and kinetic intra-penetrations. Just as atmosphere is the continuation of minerality by other means, so vegetality is the continuation of atmosphere by other means.

By treating vegetality in isolation and outside of its deep historical evolutionary context,[1] we may be tempted to look to "vital powers" to explain its inner movements.[2] In my view, this is a mistake. A theory of vegetality requires deep material history.

There is no Newton

Kant once said that there could be no Newton of life because life is so radically creative and unpredictable that it resists all the tools and formulas of classical mechanical physics.[3] Complex life, it is still often said, is entirely unlike either the randomness or the lawlike structure of physical particles.[4] Therefore, the story goes, there must be some hidden creative force inside life that dead matter does not have (vitalism) or some divine creator who created it (God).

These explanations, I believe, are just another way for humans to tell a story about nature that makes a tiny fraction of the universe (themselves as life) more special or consequential than the rest. We think our life is so extraordinary that only a physical or spiritual miracle could have produced it.

The assumption here is that non-living matter is not creative—but this is wrong. Non-living matter is just as unpredictable and creative as living matter, if not more so: because it is non-living matter that produced living matter in the first place. Matter, as I have argued in this book and others, is neither random *nor* fully determined. Matter is not the passive stuff that Newton and others thought it was. Therefore, not only is there no Newton of life; there is no Newton of matter, either. Chaos and quantum theory have sufficiently attenuated Newtonian mechanics. Today, we need a different kind of theory of life that is consistent with the indeterminate mobility of matter. This is what this book tries to provide.

Phytohistory

The first step in this direction is to take the deep history of matter seriously. This means, first of all, that nature has no taxonomic "kingdoms" or substances but rather emergent patterns of motion. So instead of defining vegetality by what it *is*, this chapter and the next try to define it by what it *does* and how it *moves*.

There are two problems with the idea of biological kingdoms. The first is that it fetishizes life by conceptually distinguishing it from the non-kingdoms of inorganic matter, viruses, and prebiotic proteins. The second and related problem is that thinking of life in terms of "kingdoms" is an attempt to purify and hierarchize various forms of life, where there are in fact no natural hierarchies.

Philosophically speaking, the idea of kingdoms comes from the substance-based metaphysics of types, kinds, or categories that goes back to Aristotle. The assumption was that matter is made up of ontologically divisible stuff with essential properties and forms. Darwin showed that

Aristotle was wrong about this in biology, and yet we continue to use the idea of kingdoms even while admitting that they are just a conventional notation. It is thus important to remember that organisms are patterns in motion and that kingdoms are abstractions of real historical processes.

In contrast to the ideas of kingdoms, the history of geokinetics that I have been trying to trace in this book argues that minerals, gases, liquids, and life are not kinds of substances but metastable processes and patterns. Each pattern emerges from flows that have folded up into fields.

For example, symbiosis shows us that all life forms historically emerged from the coordinated activity of microbes. Biological complexity emerged from diverse symbiosis—not from kinds or types. Every life thus contains the history of all life and matter. *Ontogeny recapitulates cosmogony*. Without this ongoing process, we would not be alive today.

But even the idea of symbiosis does not go far enough. It stops at life and goes no further. We need another, more encompassing, idea that gets at the immanent contribution of non-living matter as well. This is what I call *symkinesis* or *kinopoiesis*: the co-emergence of matter through and as motion.

Life, in other words, is thus mineral and atmospheric. Plants arrange minerals that came from deep space and deep earth. They contain gases and liquids made possible by volcanism, rain clouds, and lightning. Plants are not a type of substance but a new *pattern* of material motion. The concept of "kingdom" privileges static forms of life over processes and thus tends to lose sight of the bigger geokinetic picture. Just as it is absurd to think about plants without their fungal symbionts, it is ridiculous to think about either without their symkinetic relationship with the cosmic, mineral, and atmospheric patterns of motion that sustain them. This is the idea I will try to summon in this section on vegetal motion.

In particular, a deep historical perspective allows us to see a kinetic relationship between beings that we often treat as separate substances or kinds, such as protists, plants, fungi, and animals.

At one extreme, philosophically, all of being is just one continuous substance; at the other extreme, every being is singular. However, I think that monism, pluralism, and even the combination of monism and pluralism are metaphysical and idealist positions that turn away from matter, history, and motion. Being is neither one substance nor many singularities. It is neither identity nor radical difference, but rather a multiplicity of moving flows and historically emergent patterns. Vegetality, I argue, is the historical name of the tensional pattern of the earth.

Approaching the history of life from a kinetic perspective allows us to see more clearly how specific differences emerged in the first place. The theory of biokinetic patterns is more primary than the theory of kingdoms. In particular, biokinetics shows that bacteria, protists, plants, and fungus are all defined predominantly by what I am calling a tensional pattern of motion. The differences between them are not differences of kind but rather differences in degree of tensional motion. The historical rise of this specific biokinetic pattern does not emerge out of nowhere, of course. It is a way of extending and transforming the centripetal and centrifugal patterns of planetary respiration discussed previously.

This deep historical perspective also overturns the retroactive *fallacy of extension* that humans frequently impose on their study of nature—as if animals had "human-like" qualities, plants had "animal-like" qualities, and so on.[5] Deep history teaches us that the process is precisely the inverse. Later historical formations emerge from earlier ones. This is why a cultural history of plants that is limited to humans and human history is insufficient.[6]

Natural histories of the earth have so far been plagued by the "great chain of being," with form on top and matter on the bottom. If one believes that history is progressive, then the present is by default at the top of a hierarchical pyramid. The great chain of being is thus the effect of an arbitrary philosophical assumption. The chain of being also assumes that the links are discrete beings, rather than processes. The deep historical approach of this book differs from the analogical structure of the "chain of being" in that it treats each "type" of being not as a substantial rung or link but as a regional pattern of kinetic flux.

It is therefore not enough to show that the plant "link" of the chain is much more animal-like than we thought and thus deserving of our moral consideration. Nor is it enough to invert the chain of being such that plants are superior to animals because they were historically prior to them.

The whole structure of the ladder of being and all its rungs has got to go. Nature is not analogical, teleological, or merely composed of singularities. It is radically kinetic and patterned. There are no types, rungs, or substances, only emergent and dissipative patterns here and there, now and then, without origin or end.[7] Each of these patterns, including plants, is not like something else but is the continuation of the cosmos and its patterns by other means. There are no substances—only kinetic processes. This is what the theory of vegetality or phytokinetics in this book aims to show.

Therefore, in using the term "vegetal," I am talking not only about plants or the kingdom *Plantae*, but about a historical structure of geokinetic tension produced by all forms of life with cell walls, including bacteria, protists, plants, and fungi. These walled and pressurized cells in tension with themselves and with one another saturated the Proterozoic earth—effectively cellularizing the earth itself.

TENSIONAL MOTION

The vegetal field has three kinds of kinetic tension or pressure: an intracellular tension within cells, an intercellular tension among cells, and a trans-cellular tension through the environment.

The historical emergence of tensional motion did not replace centripetal and centrifugal motions. It emerged immanently from them. Centripetal mineralization continued as the earth gathered and distributed rocks and minerals, producing terrestriality. Before plants or animals came on land, the land came onto the land, centripetally. Centrifugal atmospherics also continued to circulate and respire volatiles into the air and water during this time. This produced the vortical patterns through which life emerged and gathered together again and again.

It is precisely the continuous operation of these two prior patterns of motion that gave birth to a third new motion. Living matter continued to fold over itself through centripetal and centrifugal convective respiration patterns. However, this process slowly introduced a new relation of kinetic tension between distinct bacteria that were vertically thrown together but also held apart by their cellular walls. Single-celled bacterial life, for example, was increasingly thrust together by specific mineral topologies (holes, cracks, shorelines, and energy-rich mineral sources) as well as atmospheric vortices (current patterns and cycles in air and water, convection cycles in volcanic vents, and weather). This is not just symbiosis but a whole kinopoiesis that allowed bacteria to gather together.

Terrestrial flows threw organisms together. The cellular tension within and between organisms was like the birth of a new ocean composed of a multitude of tiny oceans. The emergence of life was like the birth of a new sky consisting of a variety of atmospheres, or the birth of a new earth composed of a multitude of new earths. Each cell contained a liquid ocean, mineral earth, and metabolic respiration. Accordingly, this made possible a new kinetic tension between cells and the larger ocean, sky, and earth. It also introduced a new kinetic tension between cells and other cells.

Ultimately, as bacteria saturated the ground, ocean, and sky, the entire earth developed a kind of kinetic tension with itself. The membranes of cell walls, filled with hydrostatic pressure, were like little earths that made the earth manifold with and against itself. Vegetality is the earth in tension with itself, through the use of walled membranes.

During the Proterozoic, the earth became membraned and enfolded like a single fluid skin or surface folded up into little compartments, rooms, or labyrinthine vacuoles—to which the Latin word *memebrum* refers. It is, therefore, not just plants, in the restricted sense, that are tensional, but the vast proliferation of walled cellular life that continuously "membranized" the earth into a "vegetal earth."

There is no opposition or development between centripetal, centrifugal, and tensional patterns of motion. Life is the complex cellularization and "membranification" of the earth itself, holding itself together and apart kinopoietically. Vegetality is the earth turned inside out and folded up like a cellular foam or a vortical labyrinth woven from the flux of matter. Each cell was centripetally gathered from the cosmos, centrifugally breathed out from the earth and ocean, and manifolded with and against itself. Vegetality holds the earth together and apart from itself through the use of pressurized metastable membranes: a pomegranate earth.

The historical rise of this new pattern of motion followed three main kinds of tensional motions: within, between, and through cellular life.

INNER-CELLULAR TENSION

The first tensional motion of the Proterozoic occurred *within* living cells, in the form of cell wall pressure, cytoplasmic strands, and organelles.

Living cells are nothing other than flows of earthly matter folded back over themselves. They are the product of a continuous vortical movement of biomolecules swirling around a shared region. Most scientists today believe that the first cell membranes began as chains of self-organized lipids and proteins that folded over themselves to form a sphere. They speculate that at some point, RNA must have been trapped inside the sphere and was able to flourish under the protection of the walled-in area. Eventually, this RNA produced some DNA that could reproduce the protective membrane—and thus ensure continued security.

This story raises an important question: why did lipids and proteins self-organize this way, and how did RNA get inside them? The hydrophobic theory, as we discussed in the previous chapter, is a step in the right direction but still not enough. The hydrophobic theory still assumes a *random* encounter between lipids. It thus assumes that lipids were the sole agents of order—and therefore leaves unanswered the real kinetic conditions of this order in the first place. Instead of looking only at the molecular structure of water, we need to look also at its kinetic structure. For example, the movement of water is easily capable of organizing into vortices and eddies that could have brought the hydrophobic lipids together in the first place. The molecular polarity of water can also explain why the lipids stayed together.

Just as bubbles emerge from the turbulent waters of eddies, so did lipids likely gather in much the same way. Just as a bubble of water can trap air, so was RNA probably swirled up into a vortical lipid "bubble" structure. The so-called mysterious origins of life are, therefore, only mysterious if we ignore the agency and self-organization of non-living matter. If we focus only on the biomolecular building blocks (lipids, amino acids, nucleotides, etc.), we miss the kinetic patterns that underlie them.

Over enough iterations of self-organizing structures in nature, new formations can emerge without leaving a genetic or linear biological trace behind. The most significant "leaps" in evolution are precisely where genetic and biocentric explanations break down and where, perhaps, a kinetic account may fill in the gaps. We may also be able to explain the leap from lipids to membrane-enclosed RNA or from simple cells to eukaryotic organelles in this way.

From the perspective of a linear "bio-substantialism," however, everything looks random and miraculous.

Cell wall pressure
The first primary source of internal tension within the cell emerged alongside the rigid but porous cell wall. Early cells used osmosis to transfer energy into and out of the cell. By pushing the cell membrane against the cell

wall, the cell could regulate the flow of water and nutrients with its environment.

The cell wall introduced the first major kinetic tension between the inside and the outside of the cell. The wall held the cell together and apart from the ocean. It also introduced a new internal hydrostatic, or turgor, pressure inside the cell that affected how the cell itself was organized and was able to move. For example, cell expansion through pressure can contribute to cell growth and mobility. As the cell expands and contracts (respires), it becomes round or flaccid and thus able to break out of confining areas or squeeze into smaller ones. The roundness of a cell also affects how other cells will interact with it and how much its internal components (RNA, DNA, etc.) will move.

The cell wall thus allowed cells to produce, for the first time, their own variable atmospheric pressures in tension with their surroundings. However, the life of the pressurized cell is one of constant tension and pressure with its surrounding environment or atmosphere. The cell wall turned these new cells into their own little atmospheres. Accordingly, the earth's atmosphere was multiplied and turned inside out into a billion little bubbles.

Cell walls thus produced an internal tension through pressure that structured the inner growth, mobility, and organization of the entire cell.

Cytoplasmic tension

The second leading source of internal tension in the cell is the use of cytoplasm. Cytoplasm holds the inner parts of the cell together and apart from one another. The evolutionary origins of cytoplasm are not known and are under-studied, but most scientists believe that cytoplasm likely originated from the ambient seawater of the ocean. Cytoplasm began as seawater trapped, just like RNA, in the vortical eddy of the lipid membrane. It is thus interesting that recent studies of cytoplasmic streaming in vegetal cells show the same vortical motion of ocean water that first gathered the cell together, now continued by the vortically circulating cytoplasm.[8]

Cellular cytoplasm is not simply a passive goo or jelly in which all the "real" parts of the cell float around. The cytoplasm is exceptionally active, circulating in the cell in a vortical pattern within which we can see subvortices spinning with cellular elements inside them.[9] These vortical motions are produced by the tensional osmotic flow of fluid through the cell.

One of the fastest ways to reduce an energy gradient is with a spiral.[10] Thus, the vortical movement inside cells is the most efficient way to take in nutrients and expel waste at the same time. It also happens to be an excellent method for efficiently mixing inner-cellular material, including symbionts, into new composites and relations.[11] This vortical motion also maximizes photosynthesis in plants and is the source of mobility in slime molds.[12] However, without the tensional pressure maintained by the cell wall, the energy and the osmotic gradient would collapse. Nothing would whirl or mix.

In addition to this vortical flow gradient of cytoplasmic streaming, there is also an inner-cellular tension produced by the cytoskeleton—interlinking protein filaments that connect the cell nucleus to the cell membrane. These cytoplasmic strands that compose the cytoskeleton are not passive supports, like the frame of a building or a body. Instead, they are active vibrating microfilaments that crisscross and weave together the inner structure of the cell. Cytoplasmic strands carve up the entire interior of the cell into a manifold of sub-cellular vacuoles. These strands can also transport transcripts, proteins, and signals from the nucleus to the periphery of the cell. The cell is internally *cellularized* by these strands.

Furthermore, these strands and the moving microfilaments of the cell also sustain a critical tensional structure. They hold all the vacuoles and organelles together and apart. Cytoplasm and the cytoskeleton are the self-organizing tapestries that weave all cellular components into a coherent matrix. Sub-cellularity is, therefore, a product of vortical kinetic eddies.[13]

Cytostrands are not a static tapestry or fixed order but rather a highly dynamic and mobile fabric, a "living labyrinth" in which all the threads and strands can shift around in tension with one another and in response to the global pressure made possible by the cell wall. It is precisely the coordinated pushing and pulling of these microfilaments that helps sustain the vortical circulation of cytoplasm. As the strands expand and contract along with their microfilaments, they spin the cell into vortical patterns. In this way, tensional motions and circulation produce the inner order of the cell.

Cytoplasm, contrary to what many of us learned in high school biology, has never been the passive medium or stage for cellular order and organelles. Rather, it is the kinetic and tensional condition for the possibility of cellular and sub-cellular structure itself.[14] For a long time, the standard story in biology has gotten the order backward, thinking that the organelles of the first eukaryotic cells were merely centrifugally "pinched off" from the nucleus.[15] This is wrong.

It was tensional motions, not centrifugal ones, that created organelles. First, there was a vortical pattern of cytoplasmic movement; then, subvortical patterns emerged within that; and only then could various materials and symbionts gather together into organelles.[16] Once biomolecules and bacteria entered the cell, they were vortically mixed over and over again. Eventually, some of these could become symbionts or even, through DNA exchange, permanent organelles of the cell and bound by the microfilaments into a tensional whole, moving together under the conditions of hydrostatic pressure.

Symbiosis and organelles

This leads us to the third leading source of internal tension within the cell: the introduction of other cells that eventually became organelles, such as mitochondria and chloroplasts, which are what define eukaryotic cells. The jump from prokaryotic cells with no internally distinct organelles to eukaryotic cells with distinct organelles is one of the most profound leaps in evolutionary biology, after the emergence of life itself. The origins of this leap have also remained shrouded in mystery because there is no linear genetic trail showing the precise evolution of organelles from something prior. Physicists love to point out that the odds of organelles emerging randomly from particles are astronomically low.

However, the theory of endosymbiosis, developed by Lynn Margulis, shows how bacteria could have been captured inside the cell membranes of other bacteria and continued to live together. Those relationships that had an advantage tended to survive longer, transfer their genes horizontally without sexual reproduction, and eventually reproduce as a single whole organism.

For example, about 700 million years ago, a blue-green alga (*glaucophyte*) swallowed a single cyanobacterium (*Cyanophora paradoxa*) and began extracting energy from its photosynthetic process. The two organisms shared pieces of their DNA, and the cyanobacteria eventually became a part of the algal cell called a chloroplast. Other ancient bacteria that breathed oxygen and produced energy from it allied inside other bacteria, providing energy, disposing of cell waste, and eventually becoming cellular mitochondria.[17]

The theory of evolution is incomplete without the story of symbiotic encounters, trans-species cooperation, and the biokinetic tensions inside the cell. For example, the fastest and most superior method for a cell to change itself and adapt to its environment is not through DNA mutation and centrifugal cellular reproduction. Instead, the quickest way for cells to change is by increasing their inner tension with other cells. These other cells are vortically gathered in, swirled around, and recombined through horizontal gene transfer within the original cell. Outside cells can be sucked in and given a chance to cooperate or not. If they do, then they are reproduced as part of the cell, much faster than random mutation and evolution could ever take place. However, such a "tensional symbiosis" also leaves little or no vertical genetic trace of its process or origins.[18] This is why it looks like a "leap" has occurred in the genetic record—where kinetic symbiosis has done all the work.

Symbionts and organelles thus introduced a biokinetic tension inside the cell. They did this by holding themselves together and apart from one another in the same cell. Each directly fed off others in a linked tension. Evolution is vertical and centrifugal, but symbiosis is horizontal and tensional. Both harness the pedetic flows of the environment that continuously support them.

We can see now that the cell has at least three kinetic patterns of motion. First, there is the centripetal accumulation of energy from the periphery. Then, there is the centrifugal emanation of DNA from the center. And then there is the vast tensional and manifold cellularity of walls, strands, filaments, organelles, and symbionts that structure the whole balance between centripetal and centrifugal motion.

INTERCELLULAR TENSION

Cells are also in external tension with other cells. Thus, we can see a second tensional motion *between* living cells in the form of mucilage, filaments, and the thallus.

Mucilage

As we have seen, some of the earliest many-celled structures were gathered together by the vortical patterns of mineral and atmospheric flows. But why did they stick together? Cells held together and apart in tension with one another because of mucilage—a sticky, sugary fluid secreted by certain protists and bacteria.

When these cells whirled together, they formed what biologists call "colonial" structures. Some had shapes like plates and others like spheres. Some developed in dendritic, others in rhizoid patterns. Some colonial structures contained moving cells whose flagella began to move together in collective mobility. This movement also produced internal vortices that would eventually become the sub-cellular interior.

Each of these colonial cells is appropriately called *coenobium*, the Latin word meaning cell, cloister, convent, or monastery. Just as the individual cells of the monastery surround the empty center garden (*garth*), held together and apart in linked tension with one another, so the coenobia surrounds the hollow center of the biological cell.

In the case of the spherical colonies (*volvox*), coenobitic cells eventually began to exude not only mucilage but cytoplasmic strands that connected them more firmly. These strands bound the inside of each cell with the exterior of other cells, without dissolving the cell walls that kept them apart and without reproducing the whole assemblage as a single multicellular organism.

Filaments

When this process of mucilage and strand connection was extended linearly between cells, it formed a long tensional filament that could extend vertically, supported by the water, toward the sunlight to photosynthesize. This is the basic structure of many algae that form threadlike chains of cells held together and apart in linked tension with one another. Importantly, each cell still functioned and reproduced relatively independently from the others in these pluricellular filaments.

Thallus

The emergence of the thallus, or single body, combined the inner tensions of multicellular bodies with the external tensions of filamented and branched bodies. This produced a single multicellular organism without any central organization. The structure of the thallus is fascinating because it is still not divided into specialized tissues or organs. Even without specialized organs, though, different regions of the thallus can always take on different morphologies. Kelp, for example, is a thallus and still has at least three distinct areas: the holdfast (the anchor), stipe (the blade support) and the blades (for photosynthesis).

The differentiated structure of the thallus allowed for a new type of tension. For example, thalloid plants, fungus, and lichen could grip onto a surface with one part of their body while using another portion to increase photosynthesis. This allowed for a new macroscopic tension between bottom and top, rock and water, and eventually rock, water, and air.

The differentiation of the thallus into select regions thus put the plant in tension with itself. The locations of energy accumulation, conversion, storage, and reproduction were different yet coordinated. Even without organs, regions of the plant acted in linked tension with one another and had to share and exchange energy between other regions in tension.

Furthermore, this was not just a kinetic tension in plant and fungal bodies. It was a way that the earth and the cosmos were put in tension with themselves, as light, air, water, and minerals were gathered, held together, and kept apart inside the manifold body of the vegetal earth. Sunlight and solar rays, for example, touched and mixed with water, air, and dirt without entirely dissolving into one another inside the thallus. The thallus kept them together and apart in the same body.

Kelp, for instance, fill parts of their bodies with air and float to the surface. Fungi take solar energy from the sky and bury it under the ground. The kinetic cellularity of vegetal life, thus, is designed precisely to gather the earth together into pressurized cells, store it away, and release it in a linked tension with the world.

TRANS-CELLULAR TENSION

The third and final tensional motion occurred *through* vegetal cells in the form of capillary, communicational, and cosmic tensions. Vegetal motion is in tension with itself, other plants, and the material cosmos more broadly.

Capillary tension
Trans-cellular vegetal motion is a tensional movement of matter through cells via osmotic pressure. After vegetal cells began adhering to one another and forming composite multicellular bodies (thalli), they produced a new kind of kinetic tension that ran through their cells, connecting them. Capillary

tension or surface tension is the flow of liquid through narrow spaces without the assistance of, or in opposition to, gravity.

In plants, this occurs in two complementary ways: 1) the movement of water evaporating out of leaf pores creates a transpiration in the water tubes that pulls water up through the cell walls, using surface tension, and 2) the osmosis of water and nutrients moving from the roots upward. This is how many plants and vegetal bodies move fluid from outside their bodies throughout their cells. Water adheres to the inside of the narrow cell tubes, coheres with other water molecules, and is pulled up by the surface tension of water pressure.

First, however, the cell wells become connected as new cells grow out of old ones. Cells are no longer held together and apart by mucilage but by a sticky inner membrane of their cell wall made of pectin. As new cells grow, they adhere to one another tightly except for a series of pores or passages (plasmodesmata) between the cells. These plasmodesmata, or intercellular junctions between plant and fungal cells, enable the transportation of hormones, enzymes, sugars, amino acids, ions, and proteins between cells.

This tensional flow between pressurized cells is regulated by a cytoplasmic sleeve that can open or close to adjust the size of the sieve and thus control what can flow through. Fluids can travel between cells inside the cell wall or through these cellular junctions. Furthermore, these junctions internally connect cells through cytoplasmic strands adhering to the cytoskeletal filaments within each cell.

These junctions and their tensional linages make possible a trans-cellular motion through the cells of the vegetal body. As water pressure increases at the roots, it is moved upward in a continuous cellular tube toward the sky. As fluids move upward and through the chain of cellular junctions, they can also produce vortical eddies around the edges of the junction (the septum), creating kinetic sub-compartments. In fungal cells, these eddies capture cell nuclei and accumulate them to stabilize flow-stressed pores.[19]

In short, capillary tension introduced an enormous trans-cellular movement of fluids through huge numbers of cells.

Communication
Trans-cellular vegetal motion also occurs through or across other vegetal bodies via signaling and communication. Tensional vegetality is a dramatic

way of being "together-apart." Communication between plants is kinetically tensional because it allows them to act together at a distance via chemical, electrical, and mechanical tensions above and underground.

However, vegetal communication is not a linear, representational, code-based language. Vegetal communication is kinetic, haptic, performative, and dynamically netted in a vast mineral and atmospheric network of intersecting material flows. Vegetality is not reducible to a single object or body but is the whole tangled and *in-tensional* web of fluctuating signals and responses.[20]

Vegetal communication is not merely a linear, mechanistic response to stimuli. It is a change across or through a whole network. For example, when one cuts sagebrush, it gives off a volatile chemical that helps wild tobacco resist grasshoppers and cutworms.[21] Each change is a change in the whole. The chemical language of plants does not have a fixed semiotic structure. Plants can use old chemical words in new ways or in new contexts with different meanings.[22] The language of plants is a collective affair in which plants can eavesdrop on one another and use this shared knowledge to prepare themselves ahead of time for attacks.[23] Underground communications travel through vast fungal mycorrhizal networks, affecting numerous bodies in the process.[24]

Recent discoveries have even shown that plant relatives across a wide range of taxa can recognize one another using specific leaf gestures and light signals,[25] and use this to work together cooperatively.[26] Plant language thus always emerges in an embedded context and is not a fixed property of the organism or the chemical compound. Plant language is a material language. It is a performative kinetic act that changes in tension with the world and with or across other plants.[27]

Plants can even use biochemical cues to summon the predators of herbivore insects that endanger them. They can regulate their root volumes in a tensional response to their neighbors. They can share and distribute nutrients among different plants. They can release chemicals into their leaves to ward off predators. Root apices even have brain-like sensitivities to all kinds of minerals and dangers that can affect the whole plant.[28]

All of this trans-plant signaling and communication takes place through a material kinetic movement of volatile gases, chemicals, light, and electrical pulses that touch and move through a tensional network. Material signaling thus requires a tension in which different bodies hold together and apart in a network across which material signals flow.

Cosmos

Trans-cellular vegetal motion also follows a tensional movement of cosmic matter through vegetal bodies. The earth is composed of meteoric minerals accreted from the cosmos, vaporized into the atmosphere and ocean, and then folded up inside out into billions of tiny walled cells in tension with one another. Vegetality takes matter from the cosmos and celluarlizes it into compartments in kinetic tension with one another. Photosynthesizing plants thus not only dig up the dust of fallen stars, they also live off the luminous dust of our dying sun. Plants are star eaters. They live on the death and decay of cosmic bodies.

Vegetal bodies are a connective tissue that brings the cosmos to the earth in a series of pressurized pockets. Through them, mineral, atmospheric, oceanic, and cosmic matter is brought together and held apart in a network of energy-rich tissue. Terrestrial plants thus have their own little worlds—with their own minerality, atmosphere, and cosmos. They are not just "hyperseas," they are "hypercosmoi." They are not just pockets of water on land; they are pockets of sun, pockets of minerals, pockets of air. They are "hyperearths," "hypersuns," "hyperatmospheres." Vegetal life remakes the entire earth, sky, and ocean, cell by cell—tensionally—link by link.

Plants are, therefore, the ultimate material-kinetic tension on the earth. They live in constant tension between the sky and the earth. They grow up to the sky to pull the sun into the earth, and at the same time, they dig deep into the earth to push it up to the sky. Plants are the earth inside out, upside down, and *com-pleatly* (un)folded. We live on a vegetal-origami earth. Since their entropic heat eventually radiates into space, plants are the cosmos made earth and the earth made cosmic. In their way, they continue the deep earth's task of transforming the earth into a kind of sun, radiating heat into space. They are, therefore, the cosmos transformed, remade, and returned to itself.

Below the ground, the dead bodies of plants store liquid sunlight as carbon. In this way, their tensional movement is inverted. Instead of turning the earth into sun, they transform the sun into earth. Vegetality is thus a becoming earth of the sun and a becoming sun of the earth, in the same tensional movement that materially courses through their pressurized bodies. Sugars move down from the sun and the leaves to feed the roots and the fungus, while minerals move up to feed the leaves. The sun and the earth are thus held together and apart in the material body of the plant.

There is no alternation or final mixture of earth, air, ocean, and sun that erases their differences. The plant retains the differences in the mix through the tension of its cells. This is why the rooted plant is a heliotrope, geotrope, atmostrope, and aqueotrope all at the same time, without contradiction or synthesis.[29] Through plants, the earth becomes an astrological body in tension with all other astrological bodies. Plants are heaven on earth—the sky made earth and the earth made sky. They are the ocean made earth and earth made ocean again. They are the ultimate shape-shifters and metamorphic bodies that convert everything into everything else.

These are the three main tensional patterns of motion that define the vegetal kinetics of the Proterozoic Eon. However, we still need to see how each of these patterns emerges historically. This is what we turn to in the next chapter.

10 Proterozoic Earth

DURING THE PROTEROZOIC EON, 2.5 billion to 541 million years ago, the entire life-saturated planet began to fold itself up into a vast knotwork of cellularized tensions. The birth of cellular and complex cellular life was not just the birth of a new type of substance "on" the earth, but a new kinetic relation of the earth to itself.

The great oxygenation event that closed the previous eon, the Archean Eon, also marked the beginning of a new tensional regime in the Proterozoic. With this new oxygen, the earth multiplied bacteria like wildfire, using all its planetary powers of convection and respiration. The earth became an enormous lung filled with porous alveoli breathing in gases, liquids, and minerals and breathing them back out again. The earth reinvented itself by folding its self over itself again and again until it overflowed with a million tiny bubbles of "living water" that could eat the earth and spit it back out again. To live, the earth had to die. To grow, it had to eat itself. The result was a massive planetary transformation of the earth's kinetic and tensional relationship with itself.

In particular, early cyanobacteria gathered together in sunlit shallows near the shore and fed directly on sunlight. These were the first stromatolites. They secreted a mucus that collected grains of sediment and then bound these grains together with calcium carbonate extracted from ocean water. These little shoreline "towers" secured the bacteria against the waves and moved them vertically toward the light. The waste product of their photosynthesis was the oxygen we are still breathing today. Specifically, it led to a gradual increase in the concentration of free oxygen in the atmosphere, from 1% to a record-breaking 21%. This, in turn, led to the death of numerous anaerobic bacteria, the creation of the ozone, and the rise of a new oxygen age of breathing organisms.

The "great oxygenation" at the beginning of the Proterozoic Eon was not just a quantitative increase in the percentage of oxygen. More important than this, it was a change in the kinetic structure of the earth itself. The earth began to break itself apart through the pressurized and tensional cell walls of cyanobacteria. Stromatolites produced consumable energy by using the sun to break apart molecules of carbon dioxide and water. These three different molecules started to gather together in the same pressurized space of bacteria and snap apart to release something new.

In short, the earth began to cellularize itself, folding itself inside itself, to work on and transform itself into something new. The earth became a producer and releaser of tensions between light, water, and air. With the help of the sun, the earth began to weave a new fabric by unraveling itself. *Phytality* is the weaving of sunlight into the folded surface of the earth. It is the sun made earth: the textile-like tension between distinct threads held together and apart in the new vegetal body of the earth.

Phytality is, therefore, not merely life or "animacy" in general, or in contrast to dead minerals and atmosphere. Rather, as the Greek word "*phutō*" indicates, phytality has a specifically kinetic definition, meaning "a vigorous movement that rises or springs *upward*." Phytality is not just life but the process by which something holds itself together and apart using pressurized cells that support a verticality.

Tensional patterns of motion can already be found to a lesser degree in the thermodynamic processes of conduction and convection cells holding each other together and apart spontaneously across energy gradients (Bénard cells) and to a greater degree in the earliest lipid colonies, bacterial cells, and algal strands of cells connected by mucilage. Historically, however, tensional motion rose to the prevailing type of motion on the earth with the emergence of eukaryotic cells around 2.5 billion years ago—thanks to the great oxygenation event.

This chapter argues that the emergence of a prevailing tensional pattern of motion occurred increasingly throughout the Proterozoic Eon. I argue that phytality is the material condition for all subsequent eons up to the present. In this chapter, I will look closely at the kinetic patterns produced by vegetal bodies. The thallus, stem, leaf, root, seed, and flower all contributed to the kinetics of the Proterozoic and early Phanerozoic earth. This chapter argues that a distinctly tensional pattern of motion defines each of these major vegetal phenomena.

THALLUS
The first major tensional structure of vegetality was the thallus. A thallus is the undifferentiated tissue of vegetal bodies such as algae, fungi, liverworts, lichens, and some protist slime molds (*Myxogastria*). Thalli, from the Greek word θαλλός (*thallos*), means "shoot or twig" and is the ur-plant body. It is the basic tubular structure of the tensional cell wall linkage system that defines all vegetal life. The thallus is the earth made into a chain or thread of cellular tubes. It is the simplest kind of linked tension in which cells hold together and apart from one another.

Tubular tension: Algae
The thallus, in its basic form, emerged around 2.5 billion years ago and formed the bodies of threadlike microalgae. From single-celled chains, thalli eventually began to share their DNA and to reproduce one another in a series of linked structures with shared cell walls. The multicellular thallus was thus the first major vegetal structure in which the kinetic tensions of all previous cellular motions were completely integrated and reproducible as a whole organism. By the early Proterozoic, thalli were already being woven from countless cellular threads wrapped around one another in braided cables.

The kinetic pattern of the environment also played a crucial role by physically moving single-cell chains together again and again in eddies of ocean water. At least one of the earliest single-celled thalli must have anchored itself to a mineral base through mucilage.

After this event, we have a basic kinetic tension in all plants: a mineral surface upon which to build and an atmospheric region filled with turbulent flows and energy gradients. Algal cells bind together with other algal cells in a chain, but the chain is bound to the ground in order to stay in the energetic tension between a high energy source (the sun) and a low energy sink (the rock). Early algae and marine plants are thus the physical embodiment of the tensional movements between the sun and the earth.

As floating collections of trestles, thalli were pushed and tangled against one another by waves and currents. These knotworks and tangles must eventually have allowed filaments to weave into one another and share genetic material. The sharing of genetic material allowed them to reproduce as true multicellular bodies and to reproduce the contingent kinetic patterns that folded them together in the first place. In this way, algal strands became increasingly large tubular tapestries. The new woven multicellular thallus was a tube made of tubes—a tensional matrix made of tensional matrices. Each strand was held together and apart by other strands in a single thallic body.

Decentralized tension: Fungi

The thallic body moves and acts in tension with itself because it has no central command center that organizes it. Each cellular strand or tube in the multi-stranded tube grows and moves on its own. However, it always does so in a linked tension with the others. Before each cellular tube has organized itself into a holdfast, stipe, or float, it can become anything. It only becomes what it is with the emergence of the whole.

Fungi, for example, diverged from their protist ancestors as long as 2.4 billion years ago,[1] and are similarly composed of decentralized tubular threads, called hyphae. Fungal hyphae are cellular tubes wrapped around one another. They move and function semi-autonomously, but also contain multicellular DNA capable of producing a fruiting body. In the same way that algal tubes can become rootlike, leaflike, or stemlike, depending on their location on the plant, so too can hyphae. Algae and fungi have no central command center, but rather a decentralized network of tubes that share nutrients, fluids, and electrochemical signals in a fully cellularized body: the thallus.

Every cell can take on any function in the body. The thallus is a true "body without organs."[2] It has no organs or pre-organization in advance of the environmental context and process of its unfolding. The fungal thallus functions not by radial command but by tensional networking. When two hyphae threads meet, they can fuse, forming a vast filamentous body woven together in webs called mycelium.

Some of these webbed colonies can extend over three and a half square miles and live for nearly nine thousand years. When conditions are right for reproduction, the filamentous thalli converge and produce a fruiting tubular body with spores, which can then each begin a new nucleated hyphae strand (see Figure 10.1).

In fungi, the tensional structure of vegetal threads is dramatic. The thallus lacks organs but is not strictly homogeneous. Instead, it is tensionally composite because it is woven together from a multiplicity of hyphae tubes. The tubes merge, branch, and diverge in webworks below ground and converge in the fruiting body. The kinetic structure of verticality is the product of a manifold of tensional tubes holding each other together and apart. Vegetal verticality is, therefore, fundamentally possible only because of tubular tensions.

These are the tensions that made it possible for fungi to grow on land long before plants—as early as 1.3 billion years ago.[3] Fungi could live off the minerals of rock, drink rainwater from the sky, and bury themselves in almost any size crack or crevice. The thallus is a shape-shifter, but it was

FIGURE 10.1 Mushrooms are fruiting bodies made of hyphae

kinetic tension that allowed it to hold together/apart and find energy within any surface. Fungi also produced the first soils and enabled some species of mushroom (prototaxites) to grow into the largest vertical organisms on the earth, at more than 26 feet tall. All this happened while other plants were still wallowing around in the mud.

Symbiotic tension: The lichen

It is precisely these tubular linkages between vegetal threads that made possible an extraordinary symbiotic relationship between plants and fungi, blurring the difference between so-called kingdoms: the lichen.

A lichen is a tensional composite organism that arises from the symbiosis of algae or cyanobacteria and various fungi. Algae produce energy from the sun, and fungi gather energy from the mineral earth. Algae are the sky made earth, and fungi are the earth made sky. The lichen is the symbiotic tension between two organisms living together/apart. It is an incarnation of the cosmic tension between the sky and the earth.

Fungi feed on carbohydrates produced by algae or cyanobacteria, and algae are protected from the environment by the filaments of the fungi. The filaments gather moisture and nutrients from the fungal surface area and anchor it to the ground. "Lichens are fungi that have discovered agriculture."[4] Algae and cyanobacteria were initially only able to live outside the water because fungi had already prepared a tensional webwork that could hold them together and apart in the first place. Fungi thus provided the first hypersea by wrapping up plants and bacteria in a liquid webwork of hyphae. The earliest certain fossil records we have indicate that lichens were some of the first plants on land, around 480 million years ago.[5]

The lichen thallus has several tensional layers of algae or bacteria and fungi—each held together and apart in tension. The top layer has tightly woven fungus filaments, densely merged to create a protective "skin" to protect the algae or bacteria from ultraviolet radiation. Below this, the "photobiont layer" is composed of less tensely woven fungal filaments wrapped around, between, and enclosing the photosynthetic partner cells.

The tensional matrix or mesh layer holds each cell together and apart at just the right ratio so that air and light can freely circulate between them. In some cases, the hyphae, using turgor pressure, penetrate directly into the photosynthetic cells to feed on their energy. Below this mesh level are more loosely woven hyphae designed to hold water like a little hypersea and thus keep the algae productive as long as possible. Finally, there is a lower level of more densely interwoven hyphae to sandwich the whole thallic body together.

In short, the lichen is defined not only by a kinetic tension of each layer woven from pressurized cell filaments but by a tensional sandwiching of these layers together. This high-tension structure allows for a linked movement of all the layers as the lichen fills up with water and dries out. The lichen is not merely like a cellular sponge, filling each vacuole up with water. It has a lamellar structure of woven layers with different densities that allow water to fill all the cells in a tension between two densely woven cortexes.

In contrast to the algae and the fungus, the lichen makes its entire thallus into a giant walled cell. It is thick on the outsides and tensionally woven on the inside. Instead of cytoplasmic strands holding organelles together and apart, hyphae weave together entire plant bodies. This is the symbiotic and tensional precursor to the first living bodies with organs that emerge later.

Terrestrial tension: Early land plants

Between 510 and 630 million years ago, fungi brought algae and other plants on land through a symbiosis. Mosses, hornworts, and liverworts, collectively known as bryophytes, evolved tensional lichen-like thalli that were dry on the outside and wet on the inside. Early terrestrial plants thus evolved from algae in symbiosis with fungi. These plant bodies, however, still lacked organs and retained a mycorrhizal symbiosis with fungi whose hyphae acted like roots. The hyphae extracted phosphorus and nitrogen from the early soil in exchange for the sweet products of photosynthesis.[6]

In addition to the previous vegetal tensions described above, early plants added a global tension inside the plant between dry portions on the exterior of the thallus and wet portions on the interior—something no other plant had yet done. With the emergence of land plants, the thallus began to take on permanent physical attributes based on the givens of the surface (air, rock, water, and sun). Land plants needed fungi to get energy from rock. They required a dry exterior to hold water inside and to keep pressure. They needed chlorophyll to feed on the sun, and they required pores for gas exchange with the air.

In short, the thallus of land plants became a tensional site where all four terrestrial elements had to be held together and apart in the same vegetal body. This new tension between interior and exterior made possible a brilliant new invention.

STEM

The second primary tensional structure of vegetality is the stem. As early plants worked with fungi to move further and further inland, their thallic bodies became increasingly dry and rigid on the outside. The vegetal thalli evolved into spindly, leafless, greenish stems and branches that grew a few inches above the ground.

The oldest known plant to have a stem with vascular tissue is the Cooksonia (433 million years ago). Some Cooksonia also had tiny openings along their stem (stomata) to assist in the transpiration-driven transport of minerals through the vertical tubes running to the top (xylem). Instead of leaves, photosynthesis occurred directly through the stem. Beginning with Cooksonia, however, the stem also started to evolve into increasingly thicker, stronger, and taller stems, producing three new types of vegetal tension.

Vertical tension: Lignin
The first new kinetic tension introduced by stems was a terrestrial verticality made possible by the development of lignin. Lignin is a biochemical that rigidifies the cell walls of plants. Using atmospheric oxygen produced by cyanobacteria and lichens, early plants were able to produce lignin that, when combined with cellulose, gave strength and flexibility to stems. This stability, in turn, led to the emergence of vascular systems that could transport water and minerals upward from underground mycorrhizal networks. The same vascular systems could also feed their mycorrhizal symbionts with carbohydrates produced by photosynthesis.

This new vertical movement between the top and bottom of the plant also produced an increasing tension between the interior and exterior. Lignin allowed for larger, longer, and more rigid bodies. This, in turn, allowed for increased tension between the inside and outside of the plant. The vertical structure also introduced an increasing pressure inside the body of the plant that was required for the transportation and transpiration of fluids. In short, lignin made possible an increasing double tension in the vegetal tubes. It created a tension between the inside and outside of the plant and between the top and the bottom.

As a kinetic structure, the stem increasingly hardened off the permeable membrane of the thallus and thrust it up into the air. The rigidity of the stem helped plants maintain their water and shape and replaced the protective exterior skin (cortex) previously provided by fungi.

Horizontal tension: Competition

The second kinetic structure produced by the stem was a horizontal tension with other photosynthesizing organisms. Where there were groupings of early plants, vertical tensions played out horizontally as well. Some plants can live vertically in the sun only if other vertical plants are not growing nearby, blocking their sun. So, the taller the stem, the more surface area there will be for photosynthesis. Ultimately, the higher up the plant grows, the higher up its spores will be released, and the farther they will spread horizontally. The vertical tension of the stem is thus related to the horizontal tension of reproduction and survival. Thus began the epic vegetal race to the top.

The race produced a kind of tensional feedback loop that was structural (vertical), reproductive (horizontal), and vascular (capillary) all at once. In other words, the emergence of tensional structures led to more increasingly tensional arrangements. As stems grew taller, they had to become wider and more rigid or reproduce more rapidly along the ground to outpace the increasing length of their neighbor's shadows.

The vascularity of the stem is, therefore, a function of the speed along these two axes. At the logical extremes of these two axes, we get the vertical tree and the horizontal rhizome: two different solutions to the same problem of tensional motion. But increasing verticality also requires increasing transport techniques.

Vascular tension: Capillarity

The third innovation of the stem, therefore, was to increase the amount of water that could be pulled upward against gravity. Two tensional structures inside the stems of plants are necessary but not sufficient for the transfer of water up into the air: osmotic pressure from the soil into the rhizoids (and eventually roots) and the pull of water upward through transpiration.

Neither of these, even together, though, is powerful enough to move water from the earth into the sky. Only capillary tension is strong enough to pull water as high as plants and trees can grow.

Capillary pressure is facilitated by lignin, which is hydrophobic, in contrast to the highly hydrophilic components of plant cell walls. The dual action of water absorption and water repulsion, though, helps pull water upward. However, much more critical for transport was the size of the capillary tubes themselves. Since the pores in the stomata at the top of the plant are ten thousand times smaller than the xylem tube that contains them, these tiny tubes can, counterintuitively, pull tons of water kilometers up into the air through capillarity.

Inside the narrow tubes, the molecular bonds in water are stronger than the pull of gravity. Thus water acts like a solid and is physically pulled up molecule by molecule. This capillary action produces an incredible physical tension, ever more so the higher the plant grows. Toward the top of trees, for example, the pressures can be fifteen times that of the atmosphere itself.

Vegetal stems are therefore not just filled with little cellular atmospheres but are capable of producing tensional atmospheric structures many times greater than the atmosphere itself. This was the birth of a whole new kinetic regime, defined not just by centripetal or centrifugal circulation and transrespiration but by a new cellular tension of the earth with and against itself. Stems are the earth pulling itself up into the clouds. The more stomata were exposed, the stronger the capillary tensions. Thus, we come to the next ingenious kinetic invention by plants.

LEAVES

The third major tensional structure of vegetality is the leaf. Leaves were born from stems. Leaves are stems and shoots carried out by other means. They are a web connecting several branches; they are a stem that has split open and unfolded itself to maximize its surface area.

This is clear in some of the earliest leaflike structures of the ancient Baragwanathia. This plant had stubby leaflike appendages, about 4 centimeters long, densely covering their stems. Beginning with these stubby stemlike leaves, plants began to develop wider and larger leaf structures—and diversified them to fit all kinds of functions. It is as if all the tensional movement of the stem redistributed itself into an array of morphological tensions in the leaf.

While the stem remained roughly the same ever after, leaves began to experiment dramatically in the air. By 360 million years ago, leaves had produced the first forests. Leaves, then, were not just physical extensions of stems but extreme intensifications of all previous kinetic tensions.

Capillaries

The most dramatic intensification made by leaves is their radical multiplication of stomata. Early stems had only a few stomata in a couple of locations, but when leaves began to grow out of stems, they filled their porous surfaces with stomata. Each tiny stoma in a leaf connects to the xylem tube that supports capillary tension through the stem. Leaves are thus the interior flow of the stem unfolded into millions and billions of tiny tubes, each increasing the tensional pulling power of the plant and, therefore, the verticality of the plant.

The leaf is the material-kinetic synthesis of the vertical and horizontal tensions of the stem into a single transversal surface, both vertical and horizontal at the same time. The more surface area, the more stomata, the more the earth and sky com-penetrate each other, and the higher the atmospheric pressure in the plant.

Phyllotaxis

Each leaf on a branch or stem unfolds the maximum surface area in tension with the other leaves. This is achieved through phyllotaxis. Phyllotaxis is the measured distance on a branch between leaves. The distance varies in response to the size and shape of the leaves for that plant. Each leaf holds together and apart from the others to maximize light absorption. The leaf thus expresses in the air what the stem expresses on the ground with its heliotropism. The stem is locked into the earth and so follows the sun up and down. However, the leaves are only locked to the stem and so can follow the sun back and forth. Together, a whole tensional dance is orchestrated every day.

Stomata

Not only are stomata multiplied on the leaf surface, the leaf also becomes the site of a maximal tension between the interior and exterior of the plant. As the stem hardens off, leaves become the dominant site of tensional regulation through the opening and closing of billions of stomata. If the plant is the earth turned inside out, the leaf is the plant turned inside out. The leaf is the stem opened to the air, and the air enfolded in the stem.

Stomata open and close using the mechanical tension of turgor pressure. The pressurized gas exchange via stomata regulates the whole life and growth of the plant. Stomata conserve water in strict tensional relation to mineral concentrations and atmospheric pressures.

Only a small percentage of the liquid energy taken from the soil is used directly by most plants and trees. The vast majority transpires into the air. This is not a question of efficiency or waste; it is a question of tensional motion between ground and sky. As stomata reveal, tension is not about isolation; it is about the linked and regulated push and pulls of flows. Tension is about circulation through a series of linked cells.

Metamorphosis
Not only does the leaf continue the stem by other means, but the leaf itself is continued by numerous other means as well. Leaves unfold into waxy surfaces, spiny surfaces, flower petals, thorns, succulent water storage, bulbs, tendrils, sepals, stamens, pistils, cones, and fruit. When the stem raises the leaf into the air, the leaf is free to experiment without immediate physical constraint. The leaf is left to become an airy shape-changer.

If the thallus is the body without organs, then leaves are the organs without a body. Partially freed from the tensional internality of the stem, leaves enter into a new linked tension with the environment, including insects and other plants. With the historical emergence of leaves, the horizontal tension that once took place on the ground began to take place in the air, as leaves struggled with one another for light in midair.

ROOTS

The fourth major tensional structure of vegetality is the root. Roots were latecomers (430 million years ago) because fungi and thalli had already done much of what roots do. Given that fungi already acted like roots, it is likely that the mycorrhizal symbiosis between fungi and plants was where plants learned how to act like fungi by making their own roots. Even though most plants can make their own roots, ninety percent of them still rely on mycorrhizal symbiosis. Roots thus remain fundamentally symbiotic bodies. However, they also introduced several new kinetic patterns distinct from those of mycorrhizal fungi.

Osmotic tension

The first kinetic innovation of roots was osmotic tension, which I discussed in the previous chapter. Lower concentrations of minerals in the soil move water by diffusion into the roots, which have higher concentrations of minerals. This intake of water increases turgor pressure in the roots and helps physically push water up into the tree.

Turgor pressure

The second type of root tension is the use of this turgor pressure inside the root tip to burrow further into the soil. Unlike fungal hyphae, root tips do not excrete acid that digests rock. Instead, roots move by cellular turgor pressure alone to push through the soil. Tension is thus the source of their motion and growth.

Roots are the living architecture of the subsurface. They are the earth folded back into itself to unearth itself. They fill the earth with holes and pores and are responsible for searching for and discovering sources of water and nutrients in symbiotic tension with fungi—all to be transpired into the sky.

With the emergence of roots, clouds began to share a new kinetic affinity as the mirror image of the underworld. In other words, roots are the clouds of the underground—gathering tiny amounts of liquid from the subterranean atmosphere and raining them up into the sky, where they are gathered again by clouds and rained back down to the roots. Roots and clouds are thus inverted fractal structures of one another—the Janus face of vegetal motion.

Affective meshwork

The third root tension occurred as a series of electrochemical signals between roots, stem, leaves, light, nutrients, and other roots. Root movement always occurs in a tensional network of various kinds of signals. They sense gravity and move downward in response. Plant stems can even direct sunlight down directly to the root heads.[7] This allows roots to sense light and respond to a colorful underground world through complex signaling.[8] The illuminated roots also affect changes in seedlings, plant metabolism, physiology, and perhaps even the circadian rhythms of the plants.

Roots can sense and actively seek out moisture and nutrients by communicating with one another via biochemical cues. They can sense pressure, hardness, volume, salt, toxins (before they touch the toxin), and microbes and even intercept signals from other plants. Right before a root encounters a surface that it cannot dig through, or a toxic substance, it will change its course before making contact. Roots can even change their volume in response to neighboring trees to share or compete for energy.

The highly sensitive and networked process of root growth led Charles Darwin to describe roots as the "brains" of plants.[9] This thesis has been picked up again in the last few years and given new empirical support.[10] Root-brains are highly sensitive tensional networks constantly transmitting and receiving signals and creatively responding to a large number of environmental variables simultaneously.

Morphological tension
The fourth tension in root movement is that its morphology responds in linked tension with its context and environment. As the environment changes, roots can grow out of stems, branches, or even leaves. Roots can grow into the air (epiphytes) and begin to extract water and nutrients as if they were underground. Roots, like their hyphae cousins, can also become parasitic and grow into the stems and roots of other plants (mistletoe). A root can emerge from the ground and transform itself into a stem (raspberries) or be used to store nutrients (tubers). Roots are the shape-shifters of the plant world—always changing what they are and how they move in response to the network of signals above and below ground.

Mechanical tension
Finally, roots are like tensional ballasts against physical movements that might pull the plant over, such as wind or the weight of its own body. Roots hold the plant vertically in the earth by weaving the plant into the soil. As the wind blows the plant in one direction, the roots hold the plant in a meshwork of cables, like a net. Roots bind the plant with the earth but also keep it vertically above the earth. The vertical structure of vascular plants relies on a radial kinetic tension pulling in all directions, like chainmail below the surface.

SEED

The first seeds emerged around 370 million years ago from early ferns (*Elkinsia*). While differently sexed tiny spores were blown into the wind and had to produce zygotes in the presence of water, seeds invented something new. True, spores had already developed the internally folded structure found in the seed, but the seed significantly advanced this.

The spore and seed are both ways that vegetal life, including algae, fungi, and protists, merge with the world. The spore and, to a greater degree, the seed, are the kinetic image of this vegetal sex with the world. Seeds remake the world in the image of the plant and the plant in the image of the world. In particular, vegetal spore/seeds invent at least three main kinetic tensions.

Inner tension
The seed is, first of all, produced by an inner tension within the plant. The seed is the self made other; the individual made plural. Most plants have both sex organs and are thus already different from themselves. Fertilization by another plant is, therefore, the internalization of the other as one's self through the atmosphere.

A seed is not just the product of a plant. A seed is a folded-up trace of the plant's own sex with the world. The seed is inside the plant, and so is together and apart from the plant. It is fleeting and detachable.

The invention of the seed is the invention of extreme inner tension. The seed is the world reproduced inside the plant in the form of divergent cellularity incompatible with the plant's own body. The seed is like a tumor at odds with its host. It is the "not-being-at-home" of the plant.

The seed strangely reproduces the structure of the plant's cells, but as layers of rigid walls. It is like a massive palisade that gets its rigidity from pressurized cells.

This makes the seed into an inner tension of the plant with itself. What began as the plant's tension with the world becomes internalized inside the plant as a seed. The seed is a part of a plant that constructs a pigmented zone of 15 to 20 layers around itself inside itself.[11] The seed is a total inversion of the outside word in the plant and a total inversion of the plant inside itself.

Tension with the world
Sexual reproduction in plants, through either spores or seeds, is thoroughly atmospheric. Plant sex occurs in tension with the fluid medium of the world. Vegetal sex is exposed to the world and facilitated through the wind,

water, and animal flows in the world. The mobility of plant sex is the mobility of the world made reproductive.

Low-lying mosses, for example, live in the calm boundary layer close to the ground where the temperature is always higher, the air always moister, and the wind always milder than it is just a few feet or inches above the ground. Just above the boundary layer, air drags against the ground and creates turbulent eddies. When it is time to release their spores, mosses send up a tall shoot with a spore pod into this turbulent air. The curving vortices of air then lift the spores out of the pod and carry them away. Just like the origins of life, vortices facilitate reproduction and mixture.

Sexual reproduction is thus a kind of asexual reproduction by other means: through atmospheric flows. Spores and seeds are flows made into folds—the outside folded up inside the organism. The seed is thus both the outside folded in and the inside expelled out. Sexual reproduction is carried out asexually through an external medium.

No other part of the plant folds in on itself to such an extreme degree as the seed. The seed is a physical extension and transformation of the plant's bodily form. The plant reproduces the vegetal earth in a seed, but the earth also reproduces itself through the plant's seed. The plant is pregnant with a seed much like the earth—with its atmosphere, its water, minerals, and photosynthesized sunlight stored as carbohydrates. Each seed is its own world because there is already a world out there that it re-worlds and with which it is in tension. Plants reproduce one another through the fluid medium of the earth, but the earth also reproduces itself through the vegetal bodies of plants.

Each seed is not only born into the world but births the world at the same time through its breath. Each seed is thus the earth's impregnation and embryogenesis in the plant. The entire plant becomes the seed body of the earth. Each new earth is born from its sex with plants. Vegetal sex makes the whole earth fertile.

The atmosphere reproduces itself by using plants as its propagating medium of transpiration. Minerals expand themselves in the bodies of plants and take on new morphologies and mixtures through sex. Strangely, even minerals reach toward the sky and become vertical through plant bodies.

Instead of seeing the so-called "lower" non-living processes as the mere support for plant sex, we can imagine an inversion in which vegetal sex is merely the medium for the diversification and reproduction of certain mineral and atmospheric mixtures and structures. Asexual mineral reproduction is thus the continuation of vegetal sexual reproduction by other means. Each continues through the other.

Each plant has the earth as its parent and offspring at the same time. Sex is thus the asexual process of the earth. Sexual difference, too, is only possible on the more primary asexual condition of bacteria and atmospheric transport. What would sex or sexual difference be without the kinetic medium through which matter is bifurcated from itself only to reunite again?

Tension with other plants
Seeds allow for a much more extensive spread of reproduction through the winds, waters, and animals of the earth. While spores were constrained to nearby moist areas, seeds made possible vast networks of plants across much larger and diverse regions.

Vascular plants, in particular, use only a small fraction of the water and energy they transpire. The rest of it (90%) dissipates into the atmosphere. This only appears to be a "waste" from the perspective of the isolated individual plant. From a kinetic perspective, this is just the way that vascular plants (re)produce and regulate their atmosphere. They do so through a linked tension with one another. Each plant shares its water with others. Aboveground plants share their fluids in the form of clouds and rain. Below the ground, they share their minerals through subterranean root-clouds.

Above the ground, thalli, stems, branches, and leaves alternate with one another vertically and horizontally to absorb their preferred spectrums of light. The mosses on the darkened forest floor absorb the wavelengths of light that the vascular plants do not.

Seeds with wings, barbs, or other transport mechanisms allow plants to keep a distance from one another as they spread out over an area. The seed is the plant become vehicular. Plants are mobile, but seeds are built for great voyages by air, land, and sea (in the case of floating sea beans). But voyages are always relational. Seeds only grow in tension with other plants, like a game of leapfrog. Each plant reproduces the atmosphere, the surface, and the subsurface of the whole so that others may grow.

FLOWER

The rise of flowering plants (angiosperms) occurred late in the evolution of vegetal life: around the beginning of the Cretaceous (145 million years ago). Flowers emerged as a transformation of leaves into brightly colored, scented, and novel forms. How can sex itself be made into a geokinetic act with the cosmos? The flower is an answer to this question. How can matter be centripetally attracted to the plant but centrifugally sent off at the same time? The flower is, above all, a question of a tensional and reciprocal movement of attraction and repulsion.

Tensional attraction
Flowers are the imagination of plants. Like leaves, flowers are suspended in the air and allowed to experiment with their form. Their topology immediately tells us something about their kinetic pattern: they are almost universally curved and bowl-shaped in order to function as sites of gathering and attraction. The flower is the place where the plant's reproductive organs are gathered, focused, and intensified. Their curvilinear and bowl shapes are sites of condensation and accumulation.

Leaves have a similar kinetic structure of attraction and repulsion, but to a much lesser extent because they are attracting light and repelling water—often without much accumulation. Leaves attract animals to feed but, at a certain point, repel them by introducing foul chemical tastes and smells. Animals come to eat a bit here and there and then leave their manure to fertilize the plant. A little bit of damage to a plant encourages new growth, but too much can kill it. Thus the leaf is a kinetic structure of attraction and repulsion. This is what the flower develops to an extreme.

The flower centripetally attracts the cosmos through its cosmetics. Sun from space illuminates its colors, the air carries its scent, and animals carry its pollen. The plant gathers these together and makes them have sex with one another. The communication between atmosphere, light, minerals, and animality is not representational. So-called "signaling" in plants is not a code or semiotic structure, but rather an immanent performance that changes each attracted participant. The plant gathers light to make nectar, air to make scent, and animals to transport pollen. Plants transform the whole relation of agents such that there is not a distinct "message" transmitted separately from the performance itself.

The flower is the space or place of an interpenetrating circulation of flows. Flows of air, light, and animals (insects, mammals, birds, reptiles, etc.) are not exchanged but circulated. Just like leaves, flowers attract but only to repel. There is nectar in the flower but only a limited amount. The flower exists only for a finite duration.

Flowers are forms of attraction whose variations occur in linked tension with the variations of their environment. The form of the flower is an inversion of the world, like bathwater is "the spouse of the body's form," as the French poet Paul Valéry writes.[12] If plants are the earth turned inside out, the flower is the plant itself turned inside out back to the world. The flower is the image of the world shown again to itself, like an echo or inverted mirror reflection. The flower has become a vessel for the earth. Like the seed, the flower is a little earth, a site for accumulation and circulation of air, fluids, and animals.

What better way to attract the earth than with a mirror: a floral narcissus? The flower is an intersex hermaphroditic earth drawn to the mirror image of itself. As the flower accumulates water and nectar on its petals, it mirrors the earth's bodies of water. The soft curvature of the flower can even become a temporary home for insects, like the mineral caves of the earth.

The attractive kinetics of the flower are made clear in the spiral distribution of petals. The spiral is the kinetic geometry of centripetal accumulation par excellence. It is no coincidence that the shape of so many flowers is vortical. This is the shape of the fluid dynamics that made and sustained life, mutation, sociality, and sex. Where there is a vortex, there is iteration, mixture, generation, and death. It is an attractive image of both life and death. It is the shape of galaxies and thunderstorms. Dandelions and other downy covered seeds reproduce this double vortex pattern of air currents as they move through the sky.[13] The spiral is an image of motion and transport through a fluid medium (see Figure 10.2).

Tensional repulsion
At the same time, as the flower attracts, it immediately repels what it has attracted. The flower has only a finite amount of nectar and pollen, and so once fertilized, most flowers begin to wilt, die, close up, or otherwise become repulsive to what they had attracted. This creates not only a direct tension with insects but an indirect tension with other flowering plants. Once fertilized, there is usually no reason to re-fertilize.

The kinetic structure of sex is to be brought together in a tension that does not resolve into unity. It is to be held apart in a repulsion that does not resolve into plurality. Sex is relational, environmental, atmospheric, and cosmic. In this way, each sex is singular but tensional. The form of the gamete is only one aspect of an entire structure of sexual circulation, without which

FIGURE 10.2 Dandelion vortex

Image from Cathal Cummins et al., "A Separated Vortex Ring Underlies the Flight of the Dandelion," *Nature* 562 (October 2018).

sex would be impossible. Individual sex is a conceptual abstraction of vastly more diverse material conditions of sexuation and sexuality—which are not at all individuals. There is never any single species with a single sex. Species are always kinopoietic and trans-species. They are processes of circulation where sex is the held tension between many heterogeneous elements.

The death of the flower repels pollinators, but the birth of a seeded fruit attracts consumers. The flower's spiral thus holds the promise of a life that will undergo death and be reborn in the seeded fruit—whose decay and defecation will provide nutrients for the seed's life. The flower's second attraction is the fruit, and its second repulsion is the consumption of the fruit and the transport of the seed away from the plant. The plant thus repels the seed away from itself through the animal. The finite and seasonal supply of fruit also leads animals away.

* * *

The tensional movements of bacteria, protists, plants, and fungi dominated the Proterozoic Eon. However, there was also significant historical overlap with the increasing rise of animal movements during this time, such that the Proterozoic and Phanerozoic cannot be identified entirely with either tensional or elastic motion exclusively, but instead shared a range of overlapping dates. More specifically, plants and animals began to coevolve symbiotically from the time of the Cambrian explosion onward. The transition from the age of vegetality to the age of animality is thus a matter of degree.

Therefore, in the next chapters, we move to discuss the historical emergence of our final dominant kinetic pattern of motion: elastic animality.

D. ANIMAL EARTH

11 Elastic Animality

ANIMALITY IS THE FOURTH dominant geokinetic planetary pattern of motion. The rise of animality overlapped with the end of the Proterozoic Eon, as vegetality slowly dovetailed into the Phanerozoic Eon. The Phanerozoic Eon, from 541 million years ago to the present, began with the Cambrian explosion of diverse animal and plant life. This explosion was made possible by increased oxygen in the atmosphere and mineral-rich soils produced by vegetal life across the continents.

The emergence and proliferation of animals on the earth was the source of a radical new regime of elastic motion, characterized by the ability of living matter to expand, contract, stretch, and oscillate back and forth to a degree never before seen on the earth. In the course of turning the earth inside out, vegetal life invented a whole new energy gradient by capturing sunlight as carbohydrates, the earth as nitrogen and phosphorus, and the sky as gases. These new gradients were the material conditions for the emergence of new organisms that could further degrade this energy by eating it. Animals, though, were not just some other organisms on top of the existing ecosystem. Their existence changed the entire bio-geo-chemo-atmo-sphere. The earth, in other words, *became animal*.

Animals, like plants, fungi, and microbes, are geological agents. They cause weathering, erosions, and deposition. Their respiration (re)produces the atmosphere and the clouds that cool the earth. Animals make the ozone and break down the sun and turn it into the sky. Animal bodies become soil. The soil becomes the oceans, which in turn becomes the sky and more living bodies. Thus, once animal bodies emerged, they increasingly transformed the entire earth.

GEOKINETIC ZOOLOGY

Therefore, just as mineral, atmospheric, and vegetal matter did not simply emerge on the earth but instead transformed the entire earth, so too did animal life. Animals are not charismatic megafauna that can be isolated, studied, or treated independently of their deep historical, material, and geokinetic being. There are millions of species of animals—more than all other species of living organisms combined. Humans have described only a fraction of them. Every body (mineral, atmospheric, vegetal, and animal) carries in itself the entire history of the earth that led up to its existence and is pregnant with the matter of future bodies.

The body of the animal, like the earth, is mineral, atmospheric, and vegetal as well. We cannot separate animals from the deep history and specific patterns of motion that came before them. This is why it is not enough to say that animals are merely ecologically or symbiotically connected to the biosphere in general. It is not even enough to say that they are *part* of the geological or chemical cycles of the earth. Instead, they, like all the other patterns of the earth's motion, are connected to the cosmos and the stars. They are connected to the deep history of the earth and of the cosmos more broadly. A theory of animality ought to take the ecological, historical, and cosmological being of animal bodies seriously.

This is why the term "animality," as I am using it, does not refer to "Kingdom *animalia*," nor to an individual "animal," nor even to the concept of "the animal," as philosophers like to put it. Instead, animality, in my view, is a process or pattern of motion. Animality is not a category or concept indicating all animals in general but no animal in particular. Everything is in motion, but it all moves with different mixtures of patterns.

Animality, in my definition, describes a historical pattern of elastic motion, which is not radically unique but is historically prominent on the earth so far. Animality is not a set of particular features captured by a general definition. Animality, like the other patterns of motion described in this book, is a kinetic pattern of singular movements or traces. It is like "rounded motion" in contrast to the general concept of "circle" or a particular body such as a "bowl." Animality is not a substance but the immanent and elastic process of real historical matter. The kinetic theory of animality or "kinozoology" describes the process of becoming elastic of the entire earth and cosmos.

Animality is not about the differences and similarities between "animals in general" versus human animals in particular. Nor is it about recent discoveries of "humanlike" qualities in animals or vice versa. My deep historical approach to animality rejects the retroactive fallacy of extension by which late-coming humans project their features onto other animals. Animality is not about shared parts or substances between kinds. This is why cultural histories of plants and animals both remain anthropocentric despite their best efforts.[1] While such histories are fascinating in their own right, they are also, for the most part, not interested in the histories of animals themselves, as material bodies in motion.[2]

Animals are neither identical to nor categorically different from human animals, but this is true of anything and anything else.[3] Everything is different and singular. But this flat ontological conclusion is at best unhelpful and at worst risks erasing the historical specificities of different patterns of motion and their structural asymmetries.[4]

Animal theory

The question of "geoanimality" or "geozoology" is, like the question of geokinetics, not a question of what types of beings have or do not have agency. The point of kinetic theory is that all matter has agency, because agency is nothing other than motion. The kinetic question then is what regime of agency is at work in specific historical and geographical patterns of motion. Kinetic analysis is not about types or kinds of substance but about emergent and dissipative patterns of matter in motion.

The notion that animality "does not exist" independently of human cultural constructions is absurd. Animality is not a representation. It is the material and historical condition for humanity itself, including all its absurd abstractions. Animality is a kinetic becoming of the entire earth. One can, of course, study the cultural history of how humans have used concepts of animality to define themselves as humans.[5] Humans have invented all kinds of little "anthropological machines" or cultural behaviors to establish clear boundaries between humans and animal others.[6]

But the study and critique of these machines risks becoming a narcissistic discussion of human cultural products, limited to human history. We need to look at the deep history of animality as a structure of the earth and the cosmos itself. It is not enough to merely recognize the singular gaze of an animal looking at a human.[7] We need to look at the structure of animality as a larger geokinetic pattern in which singular animal motions are made possible well before humans, or their pets, ever existed.

The question of kinetic zoology is not whether animals have "worlds" of "inner experience" like humans or if animal worlds are more impoverished.[8] The question is not "to what degree can animals relate to their environment or their death 'as such'?"[9] These questions assume precisely what they set out to explain, namely the a priori existence of "worlds of experience" as such and human worlds in particular—which are projected and compared to animal worlds. But these questions are ahistorical, backward-looking, and anthropocentric.

First, there are real material worlds of matter (cosmic, mineral, atmospheric, vegetal, and animal) in motion. Then, through their historical mixture, humans emerge—not the other way around. Animals are not *like* humans. Humans *are* animals. Understanding animality thus requires us to move beyond cultural and ontological perspectives.[10] We need to pursue a material and deep historical approach. This is what the kinetic theory of animality aims to provide.

ELASTIC MOTION

The animal field of motion follows an elastic motion of expansion and contraction. This pattern of motion, like the previous ones, emerged historically out of prior patterns of motion. Elastic motion did not replace the others but was rather the continuation of their motion by other means. The earth's centripetal and centrifugal respiration and reproduction made possible cellular tensions, and all three were enveloped and stretched out into all manner of shapes by animals.

Animality *is* tensional cellularity, broken and stretched apart. Animals, quite literally, ate and broke apart the cellular tensions of vegetal life and, in this way, released energy and minerals stored in their fibers. These broken-down cells turned into stretchy collagen, elastic fibers, oscillating muscles, and mineralized tissues such as bone. But the bodies of animals also break apart at the cellular level into billions of tiny mobile cells pushing their membranes into radically new topologies and functions.

Animals are the earth stretched out, contracted, and oscillated back and forth. Animal bodies are sacks of saltwater stretched out into various morphologies by internal and external calcium structures. They are the mineral earth mixed with ocean water and turned into a stretchy clay. They are like little monstrous deformed earths. With the emergence of animality, it is as if the earth had a chance to grow again. Each animal is a new earth, a new chance to recreate the earth in a new form.

With vegetal life, the earth became manifold and cellular, but with animal life, it became mani*form* and mutant. Animals had higher mutation rates and produced more species than plants because of the exceptional elasticity of their form. Of course, we must also remember that their elasticity was only possible on the more primary conditions of a centripetal mineralization, centrifugal oxygenation, and tensional vegetality. There is no hierarchy of patterns, but rather a continuation of the same cosmic and terrestrial motion by various means and mixtures.

The earth has a bony mineral core that supports a flexible liquid membrane (the ocean). Animality is the process by which these minerals are liquefied, accumulated, mixed, and sculpted into a range of more or less elastic tissues such as muscle, cartilage, tendon, shell, teeth, and bone. Although we think of bones as solid and brittle, they are woven from flexible fibers of collagen and chains of calcium with a jelly-like marrow on the inside, giving them incredible elasticity. Bones are an architectural masterpiece of elastic strength.[11] Like muscles, bones are woven layers folded into a porous stretchy matrix, which is continually expanding and contracting.

Symthanatosis

Even in death, elastic mineralized animal tissues introduced a whole new kinetic ecology to the earth. Animal bodies sculpted the earth, and when they died, other animals, fungi, bacteria, and plants decomposed them and used their minerals. New animals will reshape the dead, or their bodies will stockpile in limestone graveyards awaiting further bio-mineralization.

Animals are geological agents—living, walking, stretching strata of calcium, phosphorus, and nitrogen. They do not merely affect geochemical process by creating and eroding soil. They are themselves geochemical strata filled with minerals. Animals are the earth's most elastic and mobile geological stratum. In addition to living together (symbiosis) and coevolution, we should also take seriously the "dying together" (symthanatosis) and co-stratification of geokinetic life.

The historical rise of animality has three major kinds of elastic motion: cellular, muscular, and neurological.

CELLULAR ELASTICITY

The first elastic motion occurred in the cellular structure of animal bodies. Animals effectively developed what was already an existing elastic structure inside the walled bodies of vegetal cells. The rigid cell walls that surround most fungi, plants, protists, and bacteria offered protection, stability, and structure. However, they also inhibited the elastic mobility of the cell membrane inside the cell wall. The lack of a rigid cell wall in animal cells made animal cells more porous, more mobile, more elastic, and more mutable in response to the environment. Vegetal cells were also elastic, but to a much lesser degree.

The rigidity of the cell wall in vegetative life ended up shaping the structure of growth into a progressive and linear string, strand, or tube, as in early algae and fungal hyphae. The presence of chlorophyll also structured plant mobility, in particular, around the need to make longer, stronger, and higher vertical tubes directed toward the sun. Thus plants sacrificed their elasticity in favor of linearity and verticality.

Animal cells, on the other hand, sacrificed some walled protection in favor of increased mobility and mutability. However, to retain structure, communication, and function, animal cells had to invent a whole new suite of connective intercellular strands that could support them without restricting them. In contrast to the tensional structure of plant cell junctions, nestled tightly together and apart between the inner walls of two connected cells, animal cells used a vast array of dynamic filaments to stay together.

Anchoring junctions, for example, bind cells together across their membranes and to the actin filaments inside the cells. These have incredible morphological diversity, from bands around cells, to strands through cells, to spots on cells tied to other cells. Actin filaments are contractile strands that expand and contract like muscles inside the cell. By binding to these elastic filaments, "adherens junctions" can wrap around groups of cells and fold them like drawstrings into various morphological patterns such as those in neural tubes.

"Gap junctions" between cells allow cells to communicate through the extracellular matrix surrounding them. This allows large groups of cells to act together elastically in heart-muscle contraction or in coordinated brain firing. "Tight junctions" wrap around cells and act as barriers and filters for water and minerals. Compared with a rigid cellular wall, tight junctions are much more dynamic, diverse, and multifunctional.

MUSCULAR ELASTICITY

The second kinetic feature of animality is the invention of an elastic extracellular matrix composed of collagen and elastic protein-muscle fibers. As animal cells grow, the fluid around the cells becomes a flexible framework through which cells can move freely and reorganize easily. This elasticity, plus the elasticity of the cells themselves, made possible the invention of complex and extremely diverse structures and functions in animal bodies.

Collagen

Collagen is one of the main components of this extracellular tissue and the most abundant protein in mammals. Collagen consists of strands of amino acids woven together around one another in a triple helix of elastic fibers. These fibers are also morphologically dynamic. They can be woven together more or less tightly to create a range of elasticity: tendons, ligaments, skin, etc.

Additionally, collagen tissue can regulate the amount of mineralized calcium in it. This allowed for animal bodies to transform their intercellular tissue into softer tendons, firmer cartilage, or harder bone, as needed.

Muscle tissue contracts when calcium and ATP are released in just the right amount around it. This means that animal cells living in the ocean had to carefully regulate their metabolism of calcium, from high concentrations in the ocean to lower concentrations inside their cell. It also meant that this regulation could expand and contract their elastic muscles and allow the body to move.

The earliest underwater worms and Cambrian sea creatures surely must have moved precisely by controlling their calcium regulation.[12] After animals used this calcium, they secreted it into increasingly larger plates, then shells, then bones. The bio-mineralized calcium structures could then be used as armor or as supports between muscle tissue that increased strength, function, and dynamic form.

A flow of matter entered the organism and was folded up like an elastic clay inside the body to form all the inner tubules, fibers, threads, tendons, tissues, and so on of the stretchable body. The movement of water across the membrane gradient of the cell also produced a transport vortex, discussed in previous chapters, that was actively facilitated by the contraction and oscillation of actin filaments in the cell that worked like a pump to bring in, circulate, and expel a regulated supply of calcium-rich fluid.

Elastic fibers
Another significant component of the extracellular matrix in animals was the elastic fibers, made of "elastin" proteins. These fibers could stretch up to one and a half times their length and then contract to their original extent when relaxed. Elastic fibers are in the skin, lungs, arteries, veins, elastic cartilage, fetal tissue, and other structures of animal bodies (see Figure 11.1).

These are the kinetic structures whose coordinated oscillation between expansion and contraction allowed for cellular and tissue motility—which in turn made large-scale animal locomotion possible. However, the coordinated regulation of calcium and tissue mineralization, muscle contraction, and locomotion also required a nervous system to keep the whole organism in communication with itself and its external environment.

ELASTIC NERVES

This is where the third primary elastic structure of animal life comes in. Animals developed specialized types of cells, called nerve cells, which stretched out and elongated themselves throughout the entire body like a woven net. On one end of each nerve cell, there is a rootlike system of dendrites that listen to, or receive electrochemical signals from, the extracellular matrix and other nerve cells. On the other end, the nerve cell has an elongated stemlike axon terminal that transports the electrochemical flow from the dendrites to other nerves.

Elastic Animality 185

Plasma cell

White fibres

Elastic fibres

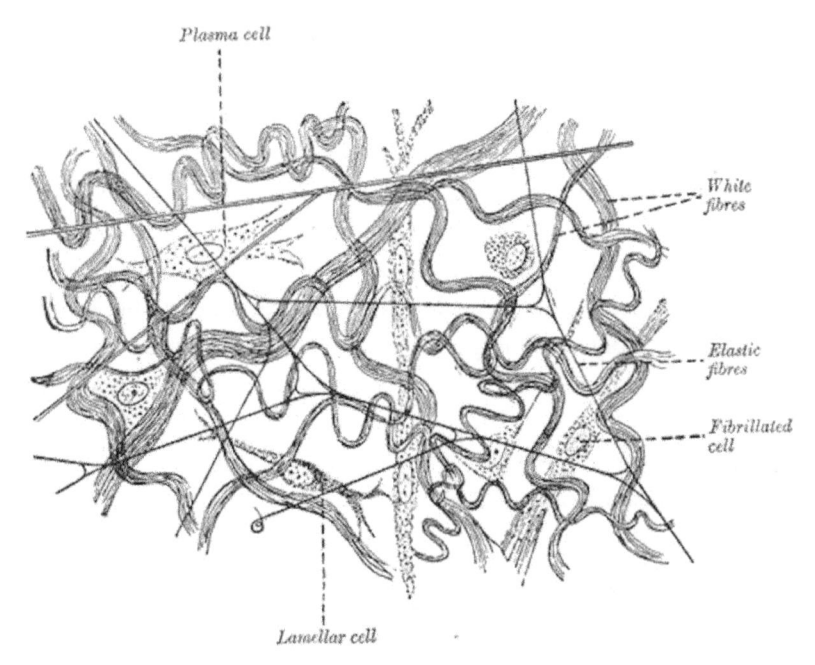

Fibrillated cell

Lamellar cell

FIGURE 11.1 Subcutaneous tissue from a young rabbit, highly magnified

Illustration by Henry Vandyke Carter in Henry Gray, *Anatomy of the Human Body* (Philadelphia and New York: Lea and Febier, 1918).

The nerve cell is one of the most dramatic expressions of the kind of elasticity that a cell membrane is capable of without the restrictive tensions of the cell wall. Without a cell wall, the animal cell was capable of elastically stretching itself into long threads spanning the organism.

The topology of nerve cells is similar to the thallic and vascular structure of vegetal bodies, which also use electrochemical signals to communicate between cells. Some plant cells even communicate as fast as animal cells.

The animal body thus began to make its own body out of plantlike electrochemical communication structures, which evolved prior to animal nervous systems.[13] Complex signal processing and integration, including perception, memory, and decision-making, can already be found in plant "neurobiology."[14]

This is perhaps because plant and animal neurobiologies were both products of a symbiogenesis with bacterial spirochete.[15] Both systems also made use of classical and indeterminate quantum behaviors.[16]

The animal is, therefore, in some ways, like a meta-plant—a plant made of plants. Whereas individual plants are like one big nerve cell, animals are like an assemblage of plants, each differently connected into specific regions of coordinated activity. The animal is like a swarm of plants—like a whole forest ecology filled with electro-communicating roots, stems, and leaves.

Vegetality turned the whole earth into a brain—sharing electrochemical signals through the ground, water, and air. In this sense, the animal is not some discrete being that emerged on the surface of the earth but is rather merely a region of a much larger terrestrial "nervous system." The animal is thus a continuation or extension of the nervous structure already present in the earth itself.

However, this meta-organism of animality introduced a new kinetic structure of elasticity as well. Animal nerve cells are morphologically elastic insofar as they can grow in various sizes depending on their function throughout the body. Nerve cells do not merely communicate signals between tensionally linked cells, as in plants. They elastically stretch their bodies and push signals along through wave action.

In their resting state, nerve cells maintain two kinds of electrically polarized balances inside and outside the cell: one between sodium and potassium concentrations, and the other between calcium and potassium concentrations. When a stimulus such as a sensation or other nerve action affects these concentrations above a certain threshold of chemical concentration or voltage, the cell begins to open up the channels in its membrane and let in calcium or sodium ions. This depolarization changes the action potential in nearby channels, causing them to let in more sodium ions as well. After the initial intake, the sodium is quickly released back into the extracellular fluid (plus a bit more). A feedback loop then ensues, producing an action wave from the dendrites along the axon to the next nerve cell.

Action waves

The whole nerve cell becomes elastic. As the axon fills with sodium, its radius expands; as it releases its sodium, it contracts. As the wave of expansions and contractions along the axon reaches its terminus, the axon begins to shorten.[17] The axon thus transmits electrochemical signals by expanding and contracting itself elastically, like a wave, along its width and length.[18] The elastic deformation of the neuron also produces displacements of the cytoplasmic fluid inside the axon and the extracellular fluid outside in the shape of vortices.[19]

Animal elasticity follows a vortical wave motion continuous with the physical performance and transformation of the neuron. In other words, there is no abstract information that merely passes through the neuron. Instead, there is a collective oscillation, vibration, and vortical elasticity of the neurons. While voltages remain the same in the flow, it is the various frequencies of the oscillations that perform the signal. Neurological communication is, therefore, not merely a translation but a material transformation of the whole vibrating neuronal body.

Neuroelectrical signals are a material and "kinosemiotic" flux. Each pulse is not a binary on or off but rather a continuous flux-wave through the axon. Axons are not like 1's or 0's but rather elastically expanding and contracting bodies whose beelike dance communicates emergent signal patterns in collective relation with millions of others. Fluctuations in the electrochemical composition are constant, but the feedback loops are only triggered when the composition crosses a certain elastic threshold.[20]

Electrical current is a highly stretchable movement of electrons capable of taking on an internally differential voltage or electrical difference between two points in the charge. As a continuously fluctuating kinetic flow of electrons, any given circulation of electricity can thus have a greater or lesser voltage. It all depends on how much is allowed to flow at once—a trickle or a flood.

Electric potential is a flexible elasticity of electrons expanding and contracting as they travel through more or fewer ion channels depending on the strength of the stimulus. Neurological structure and motion is elastic because it continually modulates across thresholds and gradients, expanding and contracting, oscillating in rhythm. In contrast to the rigid tensions and communications among vegetal cells, animal nerve cells are constantly fluctuating and changing shape. They communicate by transforming themselves into waves.

Neuroplasticity
The animal nervous system is also highly plastic. The more elastic electrical waves travel specific pathways between nerve cells, the stronger those connections become. The more frequently stimulations occur, the more likely

they are to trigger an action wave in the neuron or create a new synaptic connection between nerve cells. Conversely, the less they move, the less easily they will move in the future, and the less stimuli will be enough to trigger their action waves.

In this way, the nervous system can change in response to the environment faster than DNA can mutate the organism. Synapse connections between neurons are continually undergoing revision, getting stronger and more elastic in some places and weaker in others. Neural pathways can rapidly expand in some regions of the body and contract in others according to the complex nonlinear dynamics of the whole nervous system.[21]

In short, the animal nervous system is not mechanical, deterministic, or hardwired, but rather elastic, vibratory, and humming with waves of sensation. The difficulty in identifying any linear mechanistic causal connection between a stimulus and a nerve response is that it is impossible to completely isolate a single stimulus from the background electrochemical activity of the body and the environment. The nervous system is continuously transforming itself in response to a non-total set of shifting electrochemical flows inside and outside the body. Because synaptic strength varies continually, it is impossible to say in advance whether and when a single neuron will fire, given a certain external stimulus. It all depends on the pedetic flow history of the neuron and the indeterminacy of quantum and chaotic material fluctuations.[22]

Neuroscientists dream of a master map of every neuron and every synapse. The idea that the nervous system is a linear, deterministic, computer-like system, reducible to a stimulus-response computation, is still dominant in neuroscience. However, an increasing number of experimental findings and theoretical analyses challenge this view, suggesting instead that the nervous system is a complex nonlinear dynamical system,[23] exhibiting highly pedetic movements.[24]

The nervous system is composed of neurons, each with its own electrochemical environment and action potential. When a stimulus exceeds a voltage threshold, the neurons fire. The active nervous system is the collective firing of billions of neurons, in precise patterns, with variable average rates of firing. Interestingly, however, experimental findings show that within any given average firing rate, the individual patterns of particular neurons are completely unpredictable.[25] Although the overall average firing rate remains more or less constant, the individual firing patterns among the neurons that produced this average are entirely stochastic.[26]

Neuron cells receive a variable mixture of excitation and inhibition signals from thousands of synapses. When excitation and inhibition are balanced, small indeterminate fluctuations in synaptic activation can push entire systems across the action potential—having nontrivial systemic effects.

Studies show that individual neuron cells can be removed from the brain and artificially stimulated with largely deterministic results. However, when researchers place the cells into relation with others in a complex environment, they exhibit indeterministic behavior. The likelihood that a neurotransmitter will release in response to a single action potential is thus entirely indeterminate and relational.

Furthermore, since membrane voltages under some circumstances can be affected by indeterminate quantum events such as quantum tunneling, it is also possible that action potentials can be affected by these events.[27] This is the case because the nervous system has complex criticality, in which the whole system exists in a highly sensitive threshold state.[28] Small changes to one area can have global effects that ripple through the system like a pebble dropped into a pond. Nervous systems thus produce highly structured outcomes, but only because an enormous multiplicity of indeterminate material fluctuations in relation has created a structured criticality.[29] Small quantum fluctuations, just like those at the origins of the universe, can become amplified to the degree that they will affect the timing of spikes in neuron firing.[30]

The electric animal

Finally, the animal nervous system is materially continuous with what is outside it. This is, first of all, the case because neurons fire alongside changes in electrochemical gradients in the extracellular fluid. These are, in turn, affected directly by changes outside the body via the sensory system. One of the reasons that patterns in living nervous systems are so challenging to pick out is because they are related not only to one another but to a constantly changing environment. The environment is continually affecting the body through light, heat, pressure, gravity, sound, taste, and so on. Even cosmic rays can affect critically complex nervous systems.

Just as the emergence of life is not an ontologically radical break with non-living processes, neither is the nervous system a break with nonneuronal systems. The animal nervous system has precursors in plants and bacteria. These, in turn, have kinetic precursors in various inorganic structures such as the dendritic structures of mineral crystals, watershed patterns, the atmospheric crystallization of water, lightning strikes, and the distribution of dark matter.

These are not Platonic or geometrical forms. No two crystals or dendrites are precisely alike, because there is no unchanging single eternal form that molds nature. Kinetic patterns like dendrites are matters in motion, or "kinomorphisms." Animal neurology is not a copy of vegetal neurobiology. Animals are the continuation of the material universe by other, more elastic, means.

Animals are electric organisms because the cosmos is already an electric cosmos.[31] Language and communication are not immaterial or arbitrary webs of reference. There is no such thing as a human symbolic language distinct from the electrochemical and material patterns in mineral, atmospheric, vegetal, and animal nature. Webs of linguistic references are part of actual material dendritic webs. In its broadest sense, language is a material affecting of one region of nature by another. The animal body expands and contracts plastically in continuous response to and relation with the cosmos. The cosmos thus becomes part animal through the elastic nervous system.

* * *

These are but the most general material kinetic structures that define animality. The question now is more precisely how this elastic structure developed historically in the Phanerozoic evolution of animals on the earth. This is the topic of the next chapter.

12 Phanerozoic Earth I

Kinomorphology

THE PHANEROZOIC EON (541 million years ago to the present) is our geological eon. It began with the Cambrian explosion of living forms, the greatest number of evolving creatures in a single period in the history of the earth. During the Phanerozoic, the entire planet became increasingly elastic as the proliferation of life forms expanded, contracted, and mutated more rapidly than ever before. The more new organisms emerged, the faster they changed their environment. The more the earth changed, the more this influenced the metamorphosis of species, in a biospheric feedback loop that radically altered everything on the earth.

The great oxygenation event that launched the previous eon, the Proterozoic Eon, increased oxygen slowly over almost 2 billion years to about 10% of the atmosphere. However, with the rapid proliferation of vegetal respiration during the Phanerozoic, oxygen levels grew to 35%, then eventually falling back down to around 20% today. Increased available oxygen made it possible for more, and larger, oxygen-breathing creatures to emerge.[1]

Leading up to, and throughout, the Phanerozoic Eon, the earth invented new and increasingly elastic techniques for moving the flow of energy from the sun out into space. The earth primarily used the waste product of vegetal tensions (oxygen) as a super fuel for destroying (and creating) itself even faster than before. In addition to planetary processes of mineral, atmospheric, and vegetal bodies, the earth began to become an elastic and shape-changing *animal* that metabolized more energy faster than ever.

By proliferating elastic animal cells without walls, stretchy collagen bodies, and plastic nervous systems, the earth increased the diversity of ways and locations of energy dissipation. Increased elasticity of the body meant increased mobility, which meant increased energy consumption for locomotion, which meant an increased capacity to locate new energy gradients to metabolize, all of which involved the material transformation of the entire earth.

Animals can change their body morphology and neurology more quickly than any other form of life. This makes them not only highly mutant organisms, but the ultimate "geomorphs"—transformers of the whole mineral, atmospheric, and vegetal earth. In this sense, animals are perhaps one of the most radical geological processes on the earth. They are like an enormous and diverse living stratum that lives by destroying other strata. They are like a huge vacuum cleaner running over every surface of the earth, spewing energy into the air and eventually into the depths of space. Thus, animality is not just about the historical emergence of animal bodies on the earth but the increasingly elastic expansion and contraction of the earth itself through the metabolism of animal bodies.

Animality is the earth in rapid motion. If vegetality is the earth turned inside out and tensionally linked, animality is the earth continually shape-changing to eat itself—like a snake eating its tail.

Animality, from the Latin word *anima*, means "the state of being filled with breath, quickened, or animated." Animality is a quickened movement made possible by oxygenated breath and increased metabolism. The entire earth thus became a hyper-oxygenated, mobile animal, furiously searching out every crevice in order to devour itself more quickly.

This chapter argues that the emergence of an elastic pattern of motion occurred increasingly throughout the Phanerozoic Eon. In order to do that, we will look closely at the increasingly elastic kinetic structures produced by animal bodies that eventually saturated the late Proterozoic and early Phanerozoic earth: body, head, and tail.

We shall see that each of these significant animal phenomena is defined predominantly by a distinctly elastic pattern of motion. However, before moving on to a historical study of the elasticity of animal morphology, it is crucial to situate it within the broader history of material-kinetic evolution.

EVOLUTION

The evolution of animality is continuous with the evolution of matter more broadly. Evolutionary theory, like much of biology, has privileged life and heredity to the exclusion of many other material kinetic processes and patterns that shape the emergence of forms.

Material evolution

Material evolution is the kinetic process of material transformation, of which biological evolution is a subset. Material evolution is not deterministic or goal-oriented, but it does have directionality. It tends toward the kinetic transformation of higher energies into lower ones. The movement of our cosmos spreads out by the creation of kinetic patterns and forms that increasingly degrade energy. Material kinetic forms, like the energetic dendrites and vortices of the cosmos, storm systems, watersheds, etc., emerge to reduce energy gradients.

Our galaxy evolved into spiraling solar eddies and other dynamic forms to reduce the gradient between its supermassive black hole at the center and its surrounding dark matter at the periphery. Our sun radiates plasma and electromagnetic vortices onto our planet to further reduce the energy gradient between the earth and space. The earth's volcanism further reduces this gradient through vortical convection cycles, thus degrading this energy into space and increasingly evolving new mineralogical forms in the process.[2]

The atmosphere, in turn, also degrades this released mineral heat through vortical storm eddies and ocean currents. It evolves new atmospheric cycling structures (oxygen, nitrogen, and water cycling) to facilitate heat release. These mineralogical and atmospheric vortices produce evolutionary forms of vegetal life and also increase them by eating rocks, air, and sunlight through the vortices of their linked cells. Matter evolves, and the evolution of life is directly affected by the evolution of matter.

Animal life is, therefore, completely continuous with this more massive cosmic degradation of energy. Animals are like the extremely fine capillaries in the leaves of the cosmic evolutionary tree, which branches out in a million tiny directions, actively searching for every tiny energy gradient on the earth. Animals eat and geologically transform minerals, the atmosphere, vegetables, and other animals and then radiate their extremely degraded energy back out into space. Animals, like all matter, give back to the cosmos everything it gave them, but in new and diverse forms (Figure 12.1).

Animal morphology
The material evolution of animal morphology is kinetically structured by increasing elasticity. Kinetic elasticity increased the ability of animals to move and search out untapped energy gradients (mainly by eating other life forms). This initiated a feedback loop in which increased mobility required increased energy to supply this mobility. In turn, new energy gradients needed new and larger morphologies with faster neurologies to take advantage of them, which also required faster metabolic expansions and contractions of the elastic body. Increased metabolism also further increased the degradation of water and atmospheric gases and minerals and required more and larger energy gradients, which required increased elasticity of movement, and so on, as the feedback loop continued.

Animal bodies are like an elastic fluid that can move into every tiny energetic capillary crevice of the earth and consume energy. To maximize the surface area of consumption, two tendencies emerged: increased mobility and increased morphological folding. The more elastic the animal became, the more mobile, and vice versa. The more energy animals required for their increasing mobility and metabolism, the more the body itself began to increase its surface area by folding itself up inside itself.

These folds are the organs: tissues, mouths, heads, brains, intestines, faces, and limbs. It is not by chance that so many parts of the animal body have dendritic patterns that maximize surface area. Lungs, the heart, veins, nerves, and other parts all have a dendritic pattern because they are all trying to optimize energy consumption/distribution inside a finite morphology. The animal body is an elastic body that stretches itself inside itself to produce a maximum surface area for energy dissipation.[3]

Animal evolution
Animals do not evolve through the hereditary transfer of random mutations and toward increasing adaptations to a static environment.[4] Evolution is a vast set of material kinetic processes involving 1) the nonrandom

Phanerozoic Earth: Kinomorphology 195

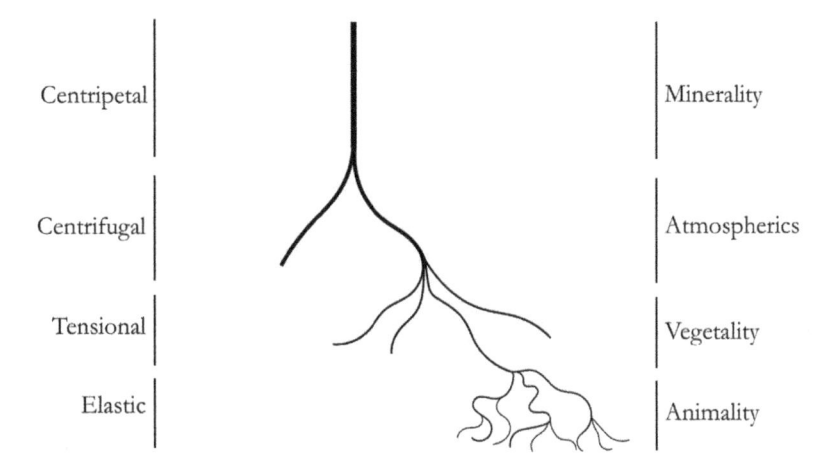

Centripetal		Minerality
Centrifugal		Atmospherics
Tensional		Vegetality
Elastic		Animality

FIGURE 12.1 Dendritic structure of material evolution

kinetic patterns of protein, RNA, and DNA folding,[5] and possible quantum effects,[6] responsible for genetic changes passed on to offspring; 2) epigenetic transformations of the genome based on diet and environment;[7] 3) horizontal gene transfers from viruses and bacteria;[8] 4) endosymbiotic alliances between organisms, often involving horizontal gene transfer;[9] 5) the morphological elasticity of the animal body, including neuroplasticity,[10] behavioral habits,[11] and geometric structures of growth;[12] 6) changes in multispecies ecological/social relations such as nonlinear predator-prey ratios;[13] and 7) transformations in the environment due to changes in animal morphology and planetary and cosmic changes.[14]

All seven of these processes, and perhaps more, are not only distinct factors in animal evolution, but they are all completely entangled with one another in a continuous feedback loop. It is impossible to isolate one from the other in the reality of evolution—if we can even still use the term "evolution" to describe this knotwork of processes. Together, these now well-established seven aspects of evolution deal a crushing blow to orthodox Darwinism and neo-Darwinian evolutionary syntheses. How many "exceptions to the rule" are needed before the rule itself becomes the exception?

Evolution is much more like a "structured criticality" than a linear development defined by heredity and adaptation. Evolution is like a pile of sand—to use an analogy from the Danish physicist Per Bak's study of "self-organized criticality."[15] If grains of sand drop on the top of the pile, one at a time, there will be times when no significant change occurs to the collection. Other times, however, certain regions of the sandpile will produce little avalanches. Sometimes these small avalanches will even create larger ones, and so on. Intervals of relative physical stability are thus punctuated by massive changes, even though the process of minor change is continuous and incremental.

Complex critical systems like evolutionary transformations involve a variety of delicately balanced processes, like neural systems, poised on the edge of critical states, after which the whole system changes. Small changes in any of the above seven factors of evolution may lead to the dramatic transformation of species and the earth. Material evolution pushes in the direction of increased kinetic circulation and the degradation of energy,[16] which leads to an increase in critical states,[17] mutation rates,[18] species diversity,[19] and elasticity in animal morphology.[20]

This is the bigger picture that frames the historical and material emergence of animality on the earth. Now let's take a look at each of the major elastic structures that compose animality.

Body

The first major elastic structure of animality is the body. The animal body is the soft, stretchy sack of water inside all animals. It is the "elastomorphic" bag whose expansions, contractions, extensions, and mineralizations define the shape and function of all animal bodies. The animal body is a sack of sacks; it is the earth made into a ravenous mutagenic slime.

Unlike the vegetal body, the animal body is not woven from threads or chains of rigid cells but heaped into pliable, elastic sheets and folded up inside itself. The body is a surface, a thin sandwich of cellular sheets that began to increasingly pleat themselves into larger and larger areas within a finite volume.

Elasticity produces a manifold, which in turn increases circulation and energy consumption across its pleated area. All other animal features emerge embryonically from the body like crystallizations, extensions, plications, and densifications of this ultra-dynamic hyperkinetic fluid surface.

Elastic circulation: Sponges

Sponges are the oldest known animal fossils. No one knows with certainty what the first animal ancestor was, but its body must have had some features in common with the earliest sponges, whose progeny are still thriving at the bottom of today's oceans. The oldest confirmed sponge fossils are 600 million years old and 1 millimeter in diameter—the size of a very small bead.[21] However, scientists have recently also found biomarker chemicals, that could only have been left behind by sponges, from as far back as 660 million years ago.[22] Sponges were apparently hard at work experimenting with the elastic animal body millions of years before the Cambrian explosion, 541 million years ago.[23]

Gelatinous bodies also emerged from colonial assemblies of small protists with round stretchy bodies, little hairlike collars for gathering food, and flagellated tails for moving around (Choanoflagellates).[24] In Figure 12.2 we can see how several of these bodies have assembled together to create

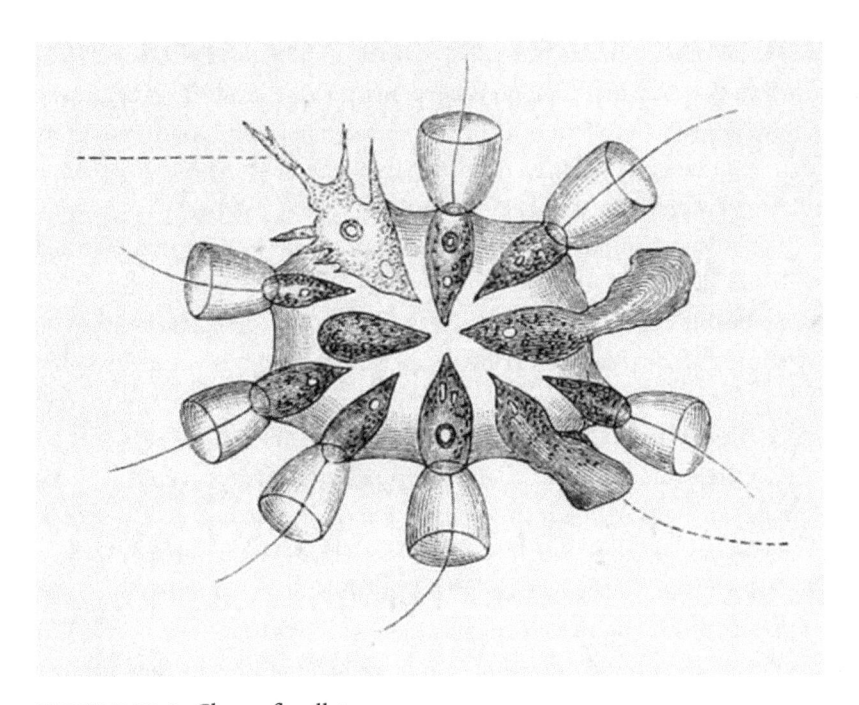

FIGURE 12.2 Choanoflagellates

Illustrated in Iliá Méchnikov, *Embryologische Studien an Medusen. Ein Beitrag zur Genealogie der Primitiv-organe* (1886).

a colony with their tails facing outward. Their flagellated tails whipped in a wave motion to produce a fluid vortex in the nearby water, which pulled nutrients into their collar-like mouths. These are the closest living relatives to animals that we know of before sponges. They were remarkably symbiotic, forming large colonies and sexually/horizontally reproducing within and alongside diverse bacterial groups.

The sponge, though, is the first known organism to have harnessed these basic kinetic features and produced a large-scale, coordinated, and elastic animal body. Sponges thus invented three significant types of material elasticity.

Corporeal jelly

The first significant type of material elasticity invented by sponges was an extended corporeal jelly that could easily change its shape and function much faster than a DNA mutation. Sponges have no tissue, organs, circulatory system, digestive system, or nervous system. As with the jellyfish, but unlike all other animals, the jellylike mass of fluid, sandwiched between two layers of cells, that composes the sponge's body is made up of *non-living* fluids. This jelly functions as a kind of ultra-dynamic fluid zone of transformation into which cells from the outer layers can move in order to combine in new ways and even change function.

What is an animal body? It is a zone of elastic transformation filled with non-living or living fluids and protected by at least two flexible sheets of permeable cells. The animal body is a bounded region of kinomorphic dynamism where life can rapidly become other than it is without waiting for the extended labor of genetic mutation—and without the tensional restrictions of vegetal walls and rigid transcellular strands.

Through this "generalized embryogenesis," the corporeal jelly of the sponge can become firmer in some areas and softer in others. The sponge is not a homogeneous goo. It is rather a kinetically dynamic space where the animal is free to experiment and shape its body without the constraint of organs. Before there were organs, there were vortical flows of collagen fibers that were woven together into new shapes and densities within a changing environment. Function does not precede form, and neither does form precede function: instead, there is a kinetic *entrainment* of matter, as if animal organs were organs of the earth and of the larger cosmos.

We can already see the beginning of this in the sponge. Fluid collagen becomes pathways for tubelike flow gradients of water/energy to pass through. Thicker collagen fibers become muscle-like strands that expand and contract various regions of the body to block or pump water through the body. Even thicker collagen then becomes spongin, which stiffens the jelly into a kind of proto-skeleton. At its thickest, collagen becomes mineralized into calcium or silica fibers—which we find in the fossil record.

The sponge, like all matter, makes itself into an energy-dissipating system. But the grand novelty here is that the sponge does so by taking on a radical corporeal elasticity that can quickly change its morphology and function alongside the changing earth environment.

Kinomorphism

This corporeal jelly zone led to the invention of an elastic kinomorphism, the sponge's second important type of material elasticity. Because the cells of the sponge are not tensionally locked into place by cell walls or rigid cytoplasmic strands, they can all move around freely and change their composition. The jelly zone is also composed of amoeba-like cells that are capable of transformation into any other type of cell (totipotent). Because the two layers of cells holding the jelly together are not bound in place, they are highly elastic and can even remodel the entire morphology of the sponge to take advantage of local water currents.[25]

Most dramatically, sponges fold their bodies up like origami to maximize the surface area, within their limited volume, across which water moves.[26] This pleated structure allows them to take on a variety of shape-shifting forms around numerous ocean surfaces.

Vortex circulation

Sponges also took the vortical kinetics of choanoflagellates to a whole new level. Instead of using individual fluid vortices to capture prey, sponges created a cellular surface covered with choanoflagellate-like bodies (choanocytes) whose collective and coordinated movements produced a massive vortical structure around the entire body of the sponge.[27] As the sponge sucks in water from the periphery and spews it upward, the vortical dynamics of the spiral push away filtered water at the same time as they continually pull freshwater and bacteria back in (see Figure 12.3).

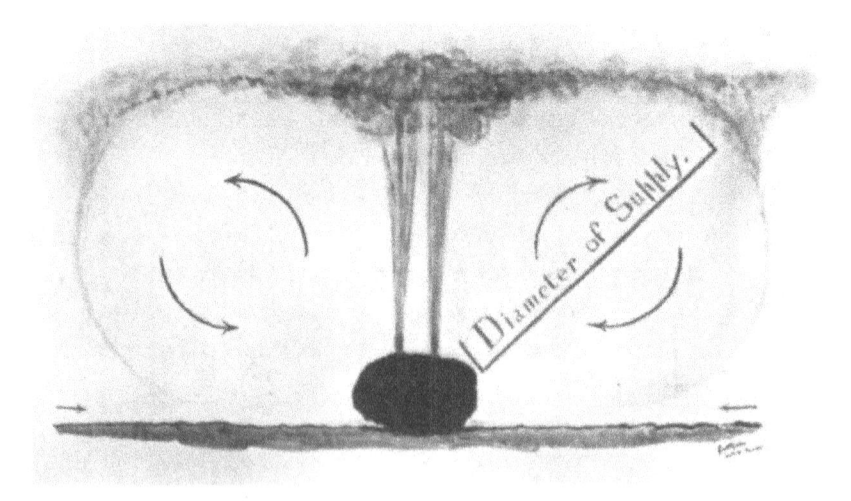

FIGURE 12.3 Sponge vortex. The diameter of supply is the distance water can be recirculated by the sponge.

From G. P. Bidder, "The Relation of the Form of a Sponge to Its Currents," *Journal of Cell Science* (1923): 293–323; 296.

This vortex circulation, the sponge's third important kind of elasticity, is made possible through two related elastic motions: the coordinated elasticity of flagella to produce inflowing waves of water; and the coordinated contraction and expansion of the sponge body itself, even without muscles. When the sponge body expands, the pores are open and allow water to move through and out the opening at the top (osculum). When there is excess sediment in the water or enemies attack, the sponge can contract its body and pores to produce a flexible chainmail-like armor. Sponges growing on one another will even coordinate their expansions and contractions, using neurotransmitter-like signaling.[28]

It is no coincidence that the same vortical motion that gathered life together into proximity and symbiosis was the motion that the sponge adopted to increase its own culture of symbionts. The sponge was not only the first animal but was likely, as sponges continue to be today, a super-symbiotic community. Microbial symbionts cover the outer layer of sponges and can contribute up to 40–50% of the wet sponge mass. Their accumulation is part of the elastic expansion of the sponge body. Some sponges even capture cyanobacteria in their pleated jelly-zone chambers and use bio-mineralized silica to reflect light in to feed them.[29]

Elastic tissue: Sea jellies

The next major elastic animal bodies to emerge were the jellies (comb jellies, jellyfish, sea anemones, corals, and hydras). Cambrian fossil records show that jellies had already diversified significantly by 500 million years ago, evolving four of the five classes still alive today.[30] The two basic body forms developed by jellies are the free-swimming medusa form and the sessile polyp form.

The jelly body further developed the elasticity of the animal body in four new ways: elastic propulsion, the elastic mouth-body, polymorphism, and the elastic nerve net.

Elastic propulsion

The first development occurred in the invention of a new form of elastic motion: the free-swimming medusa. While the sessile jellies had thin two-layer bodies, in other words two layers of cells sandwiching a thin, gelatinous middle section, the medusa thickened the collagen of its middle layer into a springy, rubbery texture. While the sponge maximized its freedom with a loosely elastic but ultra-dynamic middle layer, the medusa increasingly thickened the middle layer into muscle cells. Throughout animal evolution, the animal body generally sacrificed a degree of internal dynamism for a degree of extensive mobility.

Nerve nets and systems in more complex animals are not nearly as dynamic as the gelatinous middle layer (mesoglea) of the sponge, which can completely restructure its entire morphology in ways that nerves cannot. There is, therefore, an inverse relationship between increased mobility and decreasing kinetic dynamism internal to the body. With the invention of tissues (in jellies) and organs (in worms), intensive movement (internal transformation) is thus traded for extensive movement (locomotion). Low energy consumption is thus increasingly replaced by high energy consumption in locomotion.

By producing an increasingly elastic, springy body tissue, the medusa jelly form was able to contract itself with muscle cells and then release this contraction to bounce back into shape. This efficient and straightforward motion produced a double vortical folding of water under its bell-shaped body that helped propel the body forward in water—allowing it to travel 80% farther without any additional energy spent (see Figure 12.4).[31]

For the first time, the animal body took flight on vortical winds by transforming itself into an elastic spring.

Start   Stop

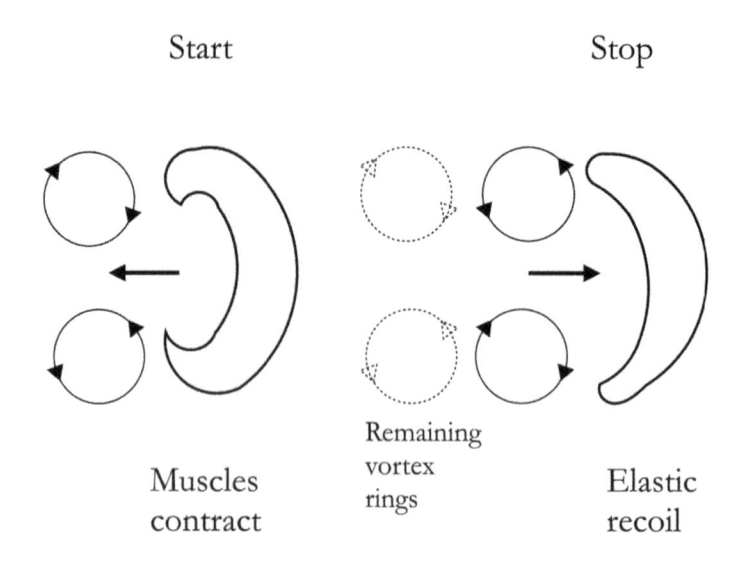

Muscles   Remaining   Elastic
contract   vortex   recoil
              rings

FIGURE 12.4 Jellyfish vortex

Elastic mouth-body

While the medusa form developed into an elastic tissue in order to fly, the polyp forms of jellies filled themselves up with water, like balloons, in order to support themselves on the ground. Hydras and sea anemones contract their mouths when they are not feeding, and digest the nutrients inside the water. Much more dramatically than sponges, their bodies expand and contract elastically with each mouth-body-full.

The entire polyp body became a mouth and stomach. This was made possible by a similarly increased elasticity of the gelatinous middle layer of the body and development of muscle cells. Hydras traded internal dynamism for a more extensive range of locomotion. Hydras and sea anemones also used this increased locomotion to move slowly over rocks or to clumsily crawl or swim by wagging their bases using their new muscle fibers.[32]

Polymorphism

The dual forms of many jellies still retain some of the kinomorphism of the sponge. The same organism can become structurally and functionally two or more types of individuals. Sometimes individual hydras, for example, will change shape and function to participate in colonial structures: some for feeding, some for reproducing sexually, some for reproducing asexually, and some for traveling (such as medusae). They can also elastically regenerate large parts of their bodies.

Elastic nerve net
Sea jellies do not have brains or central nervous systems, but they do have decentralized neural networks that use many of the same neurotransmitters as other animals. This allows many jellies and hydras to sense light, pressure, and motion. Their bodies have radial symmetry without heads or tails—with only one opening for eating and releasing waste. The radical decentralization of sensation and motion thus makes the nerve net incredibly elastic and adaptable throughout the whole body. The elastic jelly body is not disoriented but, rather, ultra-oriented. It has no left or right, head or tail, but a completely global orientation of singular dimensions.

Medusae swim toward a full moon, not as a movement to their left or to their right but as a singular global direction. Without bilateral symmetry, there is no binary, no opposition or interaction of one half to or with another, no negation or balance of one "part" of the body with another. There is only a continuously changing distribution of singular points flowing across an elastic surface. Material kinetic transformations of the world happen at the same time as complete transformations of the whole animal body itself. Movement in this animal body is a decentralized coordination of singularities. Each of them acts and neurologically transmutes the others, on the fly.

With the emergence of the first nervous animal body, coordinated mutation and morphogenesis developed to a degree never before possible. Radial being is not merely "de-"centralized or "dis-"oriented but rather singularly and globally distributed at the same time.

HEADS AND TAILS
As the body became increasingly elastic, it was no longer limited to the single stroke and recoil of the bell-shaped sea jelly. With a few more axes of elasticity and one more folded layer of muscular tissue, the animal body became capable of wave propulsion by alternately rippling parts of its body.

However, the elasticity of wave motion completely changed the topokinetic orientation of the entire body. While the radial body moved along all axes at once (the single jelly bell contraction), wave motion was only possible in a bilateral oscillation between the left and right sides. The increasing elasticity of the body made this possible. Wave motion also introduced a distinct bidirectionality, forward and backward. It is, therefore, the kinetic orientation of wave motion that made possible the emergence of the heads and tails of animality.

What is a head? A head began simply as the direction of wave propulsion. Bilateral equality between sides made possible a front/back equality. However, increased elasticity and mobility also required increased navigational sensitivities in the "front," toward the direction of transport. Thus, bilateral elasticity increasingly introduced an asymmetry between the more sensitive head and the less sensitive tail.

Unlike left and right, which are kinetically equal, front and back are morphologically unequal because they are kinetically unequal. This is due to the directionality of wave transport versus radial propulsion. Sensitive asymmetries like this also entailed energetic asymmetries between the mouth, which emerged closer to the more sensitive head, and the anus, which emerged farther away. Again, the increased elasticity of motion and energy consumption came with a loss of freedom and orientations: from several down to two; from equalities to asymmetries.

The asymmetry between mouth and anus, in turn, resulted in another asymmetry between top and bottom because of the tendency of waste to fall downward. The head and mouth move forward, but waste is what does not move forward. It is what is left behind and sinks. The head receives the motive energy of elasticity, while the tail is what leaves it behind. Evolutionary increases in elasticity thus tend to privilege and diversify the head as a result of motion, but the tail and anus less so. This certainly was the case for the magnificent rotifer. Thus the folding of the head and tail organs into the elastic bilateral body gave birth to the fundamental spatial-kinetic orientation in almost all animal bodies. Let's look at how this happened in a few key animal heads.

Elastic brains: Flatworms

Flatworms were the first organisms to develop bilateral symmetry, a third body layer, and brains—all before the Cambrian explosion.[33] These three coevolutions are all connected to an increased elasticity and mobility of the animal body during this time. A new folded body layer of thick collagen tissue (like a skeleton) bound the two layers of the body tightly together. It allowed the muscles between them to move both sides of the body together, at the same time, in different places along a central axis. The wavelike motion of the flatworm also produced and made use of a series of vortices in the water. These facilitated its movement and efficiently mixed and degraded thermal energy.[34]

A nerve cord centralized the nerve net along the length of the flatworm's body to coordinate this back-and-forth motion. Circulation and respiration dendritically diffused through the elastic movement of the body along this central line. All this coevolved with the emergence of a simple brain whose front part connected with optical nerves and whose hindbrain connected with its swimming movements. The brain's morphology thus already mirrors the coordinated symmetry between left and right and the asymmetry between anterior and posterior. The entire evolution of the animal brain is a morphological and functional distribution built around this basic bilateral orientation, structured by wave motion. Movement structures the body and brain of the animal.

We can also see how bilateralism gets entangled with the kinetic asymmetry of the entire animal. Left and right morphologically mirror each other, but their coordinated motion becomes asymmetrical through elastic wave motion as animals undulate their bodies, just as head and tail are determined by the elastic linear motion of digestion and diffusion from front to back. Motion is, therefore, at the deep historical origin of brain evolution and morphology.

The brain is a highly elastic organ that physically expands and contracts along with the coordinated waves and plasticity of the neurons and elastic muscle fibers. The brain is not a static distributor of information; it is the elastically vibrating organ that coordinates the other vibrations of the body through its motions. The brain is not reducible to an active, passive, or even an active *and* passive organ. Rather, it is a completely entangled organ that is indeterminately related to the rest of the body and the environment. The evolution of the brain is directly related to bilateral kinetic morphology in water.[35]

The kinetic function of the brain is thus continuous with the more extensive use of electrochemical signaling in jellies, sponges, and plants as well as the coordination/diffusion of affective movements throughout the rest of nature.[36]

Elastic tubes: Roundworms

Nematode populations are an excellent visual image of precisely how animal bodies fill in the tiny thermodynamic capillaries of energy consumption. Roundworms, or nematodes, emerged around 470 million years ago and are still around today. If we could put on a pair of glasses that allowed us to see only the nematodes on the earth, and nothing else, we would still be left with a discernible outline of the surface of the earth, its plants, animals, people, and many built structures. Nematodes are tiny organisms that account for about 80% of all individual animals on the earth.[37] They inhabit every region on the earth and are crucial geological agents that decompose organic matter, eat bacteria, and fix nitrogen in the soil.

Kinetically, roundworms introduced a brilliant elastic function to the animal body that connected the head to the tail, the mouth to the anus: namely, the hollowed-out intestine. The roundworm intestine is essentially a duplication of the tubular animal body inside the body itself. Just as the brain is like the dendrite-filled animal body, folded up, so the intestine is the elastic organ that brilliantly exposes the pure surface of the animal body to the outside. The roundworm flips the elastic surface of the body inside out. Now, for the first time, its inside and outside are entirely continuous surfaces. The intestine is, therefore, a radical move in history to secure and reproduce the difference-in-unity of the head/tail asymmetry. The mouth-anus is a kinetic invention that streamlines the movement of water/food through the entire body—fully entraining the whole movement of digestion with the movement of the outside world.

In nematode morphology, animal heads and tails share a common opening to the world, where the inside opens to the outside and vice versa. It makes the animal body and world a single continuous skin. With the hollowed-out body cavity, the animal body was no longer turned in on itself as an enclosure from the world but instead became like a straw that the world moved through. The intestine is like another animal folded up inside the animal. It is like an endosymbiont that eats and shits but shares a portion of itself with the other animal around it, in exchange for protection. All this is possible because of the high elasticity of the stretchy and twisted intestine that moves and folds with the elastic motion of the mobile animal body. This double inversion is radical: the world moves through the tube at the same time as the tube moves through the world.

The intestine is not like the rigid plant tensions of xylem and phloem in the stem, but is completely hollowed out by the hydrostatic skeleton of water pressure. Even the porous sponge is not genuinely hollow, but merely pulls water through its gelatinous membrane body.

Mineral, atmosphere, vegetal, and animal all pass through the anus of worms. The earth becomes worm shit. Radiation from the sun and cosmos accumulates into organic matter and journeys from mouth to anus—becoming asymmetrically bilateral, becoming-animal, becoming-shit. The axis mundi of the animal world is the mouth-anus axis. Worms hollow out the world by digging through it, and the world, in turn, hollows out the worm by digging through its intestine. Each one digs through the other and shits the other out. Worms do not just invert the earth—they digest it, transform it, and give birth to it.

Elastic mouths: Rotifers

The asymmetrical evolution of the head and tail also resulted in the astounding new elasticity of the mouth. We can see this in the magnificent corona of the rotifer. Rotifers were and still are incredibly tiny wormlike organisms with brains, organs, intestines, stomachs, and nervous systems. What is so incredible about them is the degree to which their heads and mouths move and fold so elastically.

The rotifer made its mouth into an enormous double crown structure protruding from its body and covered with synchronized cilia. These cilia move together to produce a large vortex structure in the water that sucks dead organic matter, algae, bacteria, and protists into its mouth. It then tears them apart with its powerful muscular mouth and tiny, calcified, jawlike structures.

* * *

The increasing elasticity of the head sets the historical stage for the next major transformation in animal morphology: the concentration and development of the senses. Increased physical elasticity in animals led to bilateralism, which in turn led to a kinetic asymmetry between the head and the tail, which led to the further development of the head and mouth, and eventually to sense organs, spines, and limbs. The development of these later organs and their power to transform the earth is the topic of the next chapter.

13   Phanerozoic Earth II

Terrestrialization

THE THIRD MAJOR historical-morphological event of the Phanerozoic Eon was the explosion of elastic sensory organs and limbs in the animal body. With the evolution of mollusks, arthropods, and vertebrates, an enormous transformation occurred as animal life in the seas spread to the land and the skies. The process of terrestrial animalization saturated the untapped energy of these new regions—completing the transformation of the earth into an animal.

Especially with the dramatic increase in highly sensitive and limbed animals, we see the most widely diverse and elastic morphologies—as each transformed a more extensive range of environmental conditions and changed itself to fit them. Increased mobility, increased elasticity, and increased sensory organs all coevolved together under the conditions of terrestrial animality.

SENSE ORGANS

The animal body's skin and surface folded back over inside itself creates sensation. Sense organs are the pores and openings that allow for the self-invagination of the body. The senses are the body's skin pulled inside, like a thousand tiny mouths or holes that carve out the body like a sponge.

Animal sensation initially began as the movement of the skin to open itself up to the outside. In this way, it transmitted the motions and vibrations of the world into the body. Nerve nets did not begin as centralized brains but rather as decentralized networks, rendering the animal body into something increasingly porous and increasingly affected by the world inside itself. The nervous network of animals thus makes itself into the shape of the world by letting the world move through it, hollow it out, and open it up.

The eventual concentration of nerves into the brain was thus the product of bilateralism, which created one giant bodily hole in the head: the mouth. The brain is the by-product of the fact that the bilateral worm mouth is such a large sensory pore. The brain is a concentration of neurons because the mouth touches the largest flow of matter and energy through the body. This movement of matter from the mouth through the body to the anus is what carves out the dendritic neuronal patterns of the body and reshapes the animal's form.

This is also why the brain began as a mechanism for simply coordinating the bilateral motions of the body back and forth, as we saw in the previous chapter. Only after this kinetic coordination do we see the emergence of more large sensory pores (eyes, mouth, ears, and nose) and, consequently, of the sensory-motor brain that coordinates the sensory flows with bilateral movement. The brain was the result of a concentration of sensuous pores in the head, not the other way around. The brain did not emerge so that animals could move but as a result of the movement of matter through the body.

The brain evolved so that the body could feel the world and feel itself feeling the world. The animal body increased its plasticity by increasing its exposure to the outside. The more the body was vascularized by the flow patterns of sound, light, touch, taste, etc. moving through its hollowed-out body, the more it was able to respond plastically to these movements. This is true not just at the genetic level, as the ways and intensity with which flows affect and touch the body are increased, but also at the neurological level, as the body rewires its flow pattern of electrical waves to move differently depending on how the world touches it.

It is as if the animal body wanted to completely hollow itself out from mouth to anus and from pore to pore. The more pores it creates, the more it opens itself up; the more it hollows itself out and the more flows of matter sculpt the inside of the body, the more the body can respond in the most subtle ways to slight changes in the world.

Multiplying sensitive pores, however, also increased the energy needs and metabolism of the animal, which also increased its motion. This is the sensory-kinetic feedback loop, where motion, metabolism, and sensation enter into a feedback loop of increasing reciprocal transformation. The animal body expends energy, but then consumes even more energy in order to sense or feel its expenditure of energy through nervous sensation.

To become the world, the animal must destroy it. The animal earth thus senses itself more carefully and minutely, not to know itself but to destroy and transform itself into metabolic waste: to degrade itself. Animal sensation is a way to "eat" light, sound, and other vibrations. The body takes them in and expends increasingly large amounts of energy in order to *feel* them through its body.

In this way, all sensory openings are like mouths that eat the vibrations of the world and digest them with their neurons—expelling the waste as motion. Sensation is a double consumption, or double expenditure. To eat light requires that the animal eat even more nutrients in order to sustain the digestive act of sensation. The more it senses, the more it eats; the more it eats, the more it moves; the more it moves, the more it senses; the more it senses, the more it adapts elastically, which allows it to find more food and develop its sensation. It is a feedback loop of energy degradation, expenditure, and increasing elastic motion.

Sensation and affection are active throughout all matter to one degree or another. The animal is the body with the most internally varied system of sensation. It senses many different kinds of vibrations and most quickly modifies its body's elastic form in response. Increased sensation and movement are reciprocally related: animals do not merely invent sensation in order to move better. Moving better already presupposes the increased senses of predators, who compel the movements of their prey.

The kinetic heart of the matter is that the elasticity of the animal body makes the feedback loop between sensation and motion possible in the first place.

Elastic tongues: Mollusks

Sensation increased in extent and intensity throughout animal evolution: from quorum sensing in bacteria and light sensing in eukaryotes to increasing neurons in nematodes (302), jellyfish (7,000), sea snails (11,000), and octopus (500,000,000).[1] With their increasing number of nerves, mollusks increased the range and sensitivity of their bodies with new eyes, antennae, and sensors of all kinds to detect chemicals, feel vibrations, balance themselves, and touch. Mollusks emerged in the early Cambrian period, 541 million years ago. Most uniquely, however, mollusks invented radulae: tongue-like appendages with rows of teeth for scraping bacteria off rocks. Their highly muscular mouths also secreted a mucus that was sucked in by beating cilia in a "food string," flowing through the body from mouth to anus.

Here we can see not only the connections between increasingly complex oral sensory organs but an increasingly sensitive body corresponding to the increased flow of movement and energy through the body via the mouth. The mouth is the gateway to sensation. It is the largest sensory hole, through which all other sensory nerves dendritically receive their nutrients. If the inside of the nervous animal body is shaped like a tree, the mouth is its trunk. Accordingly, we also see an increasingly sensitive and complex brain wrapped around the esophagus. It is as if the animal body were a series of symbiotic bodies, each eating the waste from the others (see Figure 13.1).

Elastic eyes: Cephalopods

The oldest cephalopod fossils come from the late Cambrian. Cephalopods evolved from early mollusks when they began to detach themselves from the ocean floor and internalize their shells inside the body or lose them altogether. Cephalopods thus gave up the security of their shell on the ocean floor to become increasingly mobile flying predators. This exposed their sensitive bodies to the world and resulted in an incredible development of complex sensory organs.

Here we can see precisely the wild feedback loop between increasing bodily elasticity, sensation, mobility, and energy expenditure. The more the cephalopod body became elastic and freed from its shell, the more mobile it became; the more mobile it became, the more sensitive it also had to become; and the more sensitive it became, the more energy it needed and also found through predation. However, the more sensitive its soft surface became, the less it was protected and therefore the more mobility and sensitive organs it required in order escape predation itself.

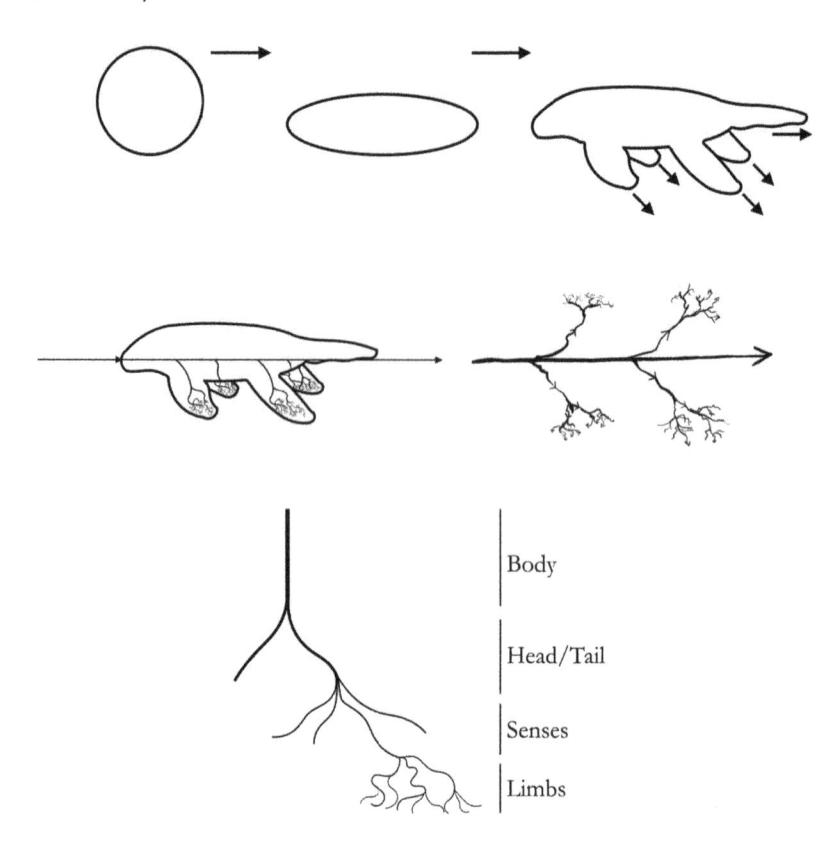

FIGURE 13.1 Animal dendrite body

    These are the two sides of an incredible Cambrian feedback loop of animal bodies caught in a kinetic cycle of predator/prey relations. The faster each animal increased sensation, mobility, energy expenditure, and elasticity, the quicker the whole chain of predation did. One brilliant solution was to turn the mollusk body inside out, keeping the shell as a mouth/beak and turning the entire bodily surface into a sense organ. By making the entire body into a highly elastic and neurologically plastic surface, cephalopods leaped ahead as one of the best predators and most elusive prey at the same time.

    The key to this success was the cephalopod arm. Unlike other organisms, octopuses, in particular, distributed a concentration of neurons throughout each of their arms. The octopus's brain is distributed through its arms, allowing each arm to act independently but in coordination with the others. The brain in the head was thus only one of several regions of distributed sensation. The increasing density of nerves in its arms gives an intelligence, sensitivity, and elasticity to the octopus's motion that almost no other animal has. Cephalopod arms can even regenerate because they are filled with and draw on the plasticity of the nervous system to rebuild itself. This is not like a lizard losing a tail. Cephalopods regrow their brains, which are also in their tentacles/feet, hence their name, from the Greek words "*kephale*" (head) + "*podós*" (foot).

Cephalopods also have advanced vision, can detect gravity, and have a variety of chemical sense organs. For example, their highly sensitive arms allow for depth perception, while the suckers on the arms allow them to taste and smell what they are touching. Octopuses are thus the original synesthetes. They are extreme cases of the kinetic feedback loop between increasing movement, sensation, and elasticity. They are the first great evolutionary experiment of the sense organs, in which the entire body became one big distributed and highly sensible organ.

The cephalopods were also the first to develop complex camera-like eyes that used a lens to focus an image on a retina. These evolved separately, as eyes in general did, across various animal phyla. Unlike the vertebrate camera eye, which was an outgrowth of the brain, the cephalopods' eye emerged as an invagination of the body's surface. The cephalopod eye focused by expanding and contracting, like the lens of a camera.

The cephalopod thus harnessed the basic kinetic pattern of all sense organs: the elastic expansion and contraction of a folded region of the body's surface. Vegetal life had already been using this technique to some degree with eyespots and leaf pores. However, cephalopods took this kinetic technique and multiplied, complicated, and diversified it all over their bodies. The eyeball is only one of many sensory "eyes" on the cephalopod. Each sucker on a cephalopod's arm is like an elastic fold of its body that can open and close independently and connects directly to the nervous system.

Cephalopods even transformed the entire surface of their skin into one enormous elastic sense organ filled with expanding and contracting pores. They turned their bodies into fabrics of little pores that expand and contract in response to light, like little "eyes"—but independent of their eyes. These chromophores directly sense changes in light and can produce camouflage, entice prey, startle predators, and impress mates and otherwise communicate with other cephalopods.

What is remarkable about this moment in evolutionary history is that it demonstrates the basic kinetic structure of animal sensation: the increasingly elastic transformation of the surface of the body into a sensitive "eye-surface." Animal sense organs became increasingly porous as the body filled with neuronal tubes that connected the sensory pores (eyes, mouths, ears, dermal pores, etc.) throughout the body.

The more sensitive the animal body became, the more quickly and adeptly it could change, both neuroplastically and genetically. Sensation and sense organs increasingly became kinetic structures to increase and diffuse matter and energy through the living body, like channels through a watershed. Animal sense organs have evolved to become increasingly plastic and to improve the plasticity of the entire body.

Therefore, the animal cannot be understood independently of the material kinetic structure of elasticity that defines it. The emergence, diversification, and multiplication of elastic pores, eyes, and mouths is something that reached a dramatically new level with animal sense organs. Cephalopods and vertebrates, in particular, made enormous strides in this regard.

Sense organs are not just about sensing the outside world. They evolved alongside the material evolution of the world and its tendency to increase the expenditure of energy. To feel is to consume, and thus all sensation is like a mouth, feeding off the world—not merely observing it, but eating it. What the increasingly sensitive animal body feels is only a by-product of the material kinetic expenditure of the energy it consumes. Sensation is destruction with the aftereffect of feeling.

LIMBS

Animal limbs are the dendritic extrusions of the sensuous invaginated body. Once the body became a hypersensitive surface saturated with pores, openings, and tubes running dendritically throughout the body and hollowing it out, these sensitive inner channels continued to spread outward into limbs and physical protrusions of all sorts. It is as if the nervous system were a kinetic virus. First, it devoured as much of the inside of the animal as it could, and then it began to overflow the body, leaking outside it. Limbs are first and foremost formed by the kinetic flow that rippled through the animal body and poured out through its sensitive holes as heat loss.

To the triple feedback of increasing sensation, elasticity, and mobility, we can now add kinetic extrusions. Extrusions from the body (limbs) and the skin (bristles) increased and extended sensation, which in turn increased the animal's capacity to respond neuroplastically and genetically to changes in the world, which in turn required more energy, which required more mobility, and so on. Each of these four systems began to feed back into each of the others. The Cambrian predator/prey feedback is only one aspect of a much larger kinetic feedback system within and outside the animal.

Limb extrusions (arms, legs, and bristles) increased sensation by extending the nervous system out into the world. They also increased the mobility of the animal. They required increasingly elastic structures that could support the rapid mutation of the animal during its lifetime (molting, scales, fur, skin, etc.). Limbs increased movement, but increased movement comes with increased exposure to diverse aspects of the world, prompting increasing evolutionary adaptation, and thus the morphogenetic elasticity of the body and the limbs themselves. Limbs are elastic, stretching, and moving extensions of the animal. Limbs are like animals growing dendritically out of other animals.

For example, the animal segmentation found in limbed arthropods and vertebrates is an inner iteration of the animal body back onto itself. Animal segmentation is the limb-ification of the entire body. The body multiplies and adds to itself to extend itself—each segment becomes a limb for the others. The segmentation of the nervous system into clusters (ganglia) distributed throughout animal bodies reveals the segmental distribution of sensation, like "microbrains" of the iterated animal body: a ganglion brain for each leg pair.

Crucially, it is the dendritic flow of matter-energy through the body that carves it out and drives the eventual externalization of motion in the form of limbs, bristles, and toes—like the energetic capillaries of the animal body tree.

In other words, evolutionary morphology is a kinomorphology driven by the movement of matter through the animal. The form and growth of animals is the form and growth of kinetic energy degradation and expenditure.

The animal body leaks energy, and the limbs emerge to follow it out into the extremities—arms, legs, tails, fur, etc.—like kinetic capillaries. The inner organs of animals are typically the warmest parts of the body and the limbs the coolest. Changes in the environment, epigenetics, and the seven aspects of evolution noted in the previous chapter also contribute to evolutionary morphology. However, they do so only on the condition that morphology does not vary significantly from the allometric scalar ratio of the dendritic movement of matter-energy dissipation through the body.[2]

For example, the surface area of animal bodies tends to scale with overall mass to the power of 2/3. Their lengths also tend to vary with their mass, to the power of 1/3. In other words, small animals tend to have higher surface areas compared to their mass, and large animals tend to have a lower ratio of surface area to their inner mass. Limbs are thus a way for animals to increase their surface area. If I am right about this, it is precisely why the smallest animals (arthropods) tend to have the most and longest limbs relative to their bodies.

Elastic skin: Arthropods

Forty million years before vertebrates set foot on land, arthropods were already well established there. Arthropods such as trilobites roamed the oceans as early as 555 million years ago, and others like arachnids came on land around 420 million years ago. In comparison, insects emerged on land around 400 million years ago. Arthropods were the first and most diverse of all animal species. Today there are over a million described species, making up more than 80 percent of all described living animal species.

Exoskeletons

Arthropods are extremely elastic animals because their sensitive limbs and exoskeletons are made of stretchy chitin and resilin. Arthropod exoskeletons can thus handle a high degree of deformation. The outer cuticle will return to its original shape when no longer stressed. This is because chitin is an extremely pliable and resilient material that can mix with other materials like calcium carbonate to alter its degree of flexibility and strength.

Resilin, the other key component in arthropod exoskeletons, is an elastomeric protein that provides a soft, rubbery elasticity to mechanically active organs and tissue. For example, resilin is what allows winged insects to pivot their wings, flex their legs, and jump many times their body length. In the exoskeleton, the chitin functions as a firmer structural component while the resilin provides a more dynamic elasticity. Resilin plays a central role in the movement and mobility of arthropods, allowing their wing hinges to recover from deformation and controlling the aerodynamic forces on the wing.

Resilin also supports the ambulatory mechanics of cockroaches and flies by allowing their limbs to bend freely without breaking. Resilin can also store kinetic energy with incredible efficiency and release it to produce incredible strength and bursts of mobility. It is even found in the abdomen regions of ants and bees and allows their bodies to expand and swell during feeding and reproduction. Arthropod limbs are an extension of this superelastic exoskeleton.

Arthropod exoskeletons are one of the most brilliant inventions of elastic animality ever developed because they allow for (a) increased terrestrial mobility; (b) increased storage of kinetic energy in the body; and (c) increased sensitivity to the world by protecting increasing complex nervous systems beneath their armor. Resilin-based exoskeletons transformed the entire animal body into a coiled spring of kinetic energy. This increased mobility allowed arthropods to explore and exploit a whole new world of terrestrial energies, previously untapped.

In particular, their highly porous exoskeletons allowed for high levels of oxygen consumption and an increased hollowing out of their bodies through dendritic tracheae tubes. These tubes bathed their entire interior with high levels of oxygen, especially during the Carboniferous, where oxygen levels were the highest in the earth's history.

Bristles (setae)

The second major elastic innovation of the arthropods was the protrusion of sensory hairs, or sensilla, from the pores of their exoskeletons. These bristles (setae) had a variety of functions, as mechanical and chemical sensors, mouthparts, and grooming tools, and in some cases were modified into scales and appendages. The bristles were ultra-elastic, highly sensitive, and mobile. Inside or at the base of each bristle, they were attached to neurons. Here, they effectively extruded the inside of the body into their ambient environment. The animal body not only expanded its surface area through this extrusion but increased its surface of sensation and energy consumption.

Insects used these setae as sensors to detect air or water currents or feel objects. Aquatic arthropods, for example, used feather-like setae to increase the surface area of swimming appendages and to filter food particles out of the water. Aquatic insects used thick, feltlike coats of setae to trap air and extend the time they could spend underwater. Some arthropods even rigidified their bristles into heavy, defensive spines.

This was a brilliant way to increase bodily surface area. The internal folding process that had hollowed out the animal body with respiratory, nervous, and circulatory tubes eventually began to leak through the dermal layer and extend into the sky in the form of a million tiny waving limbs. Each hair was a limb that reached out to touch the world and was touched back by it.

This is the material meaning of the limb: (a) to protrude or unfold the body into the world, and (b) to maximize the surface of the body and thus maximize the becoming-world of the animal. With the invention of bristles, the animal body became increasingly continuous with the most subtle changes in air, temperature, chemical flows, and so on. Bristles are like legs that walk in the air, or tongues that lick the sky.

Mutation

Each hair allowed the animal body to become an elastomeric experiment in neuroplasticity and genetic mutation. Arthropod bristles are like the Swiss army knives of the animal body.[3] The segments of arthropod limbs were modified to form gills, mouthparts, antennae, claws, and other specialized body parts.

The combination of the elasticity of the exoskeleton and the protrusion of elastic bristles allowed for something amazing. Together, they produced an extremely flexible form of life that could rapidly adapt itself, neurologically and genetically, to an enormous variety of changing situations. That there are so many more species of arthropods than of any other animal is a profound indication that they have maximized their sensuous connection to the world through an incredible use of elastic skin and limbs.

The arthropod body was the first animalization of the "hypersea." Life on land is, kinetically speaking, the movement of sacks of water—a *kinosea*.

Plants are living pools, rivers, and flows pouring into the sky through transpiration. Life on land has its own hydrokinetic topology. Animal bodies, too, are little eddies in these flows of water. They drink from plants and from one another, recirculate these fluids, and ultimately radiate them into space—fulfilling their cosmic expenditure. Arthropods, in particular, have an open circulatory system that bathes their organs in one big pocket of water.

A kinetic perspective helps us to see something extraordinary and new about the evolutionary morphology of arthropods. Scientists typically think that the exoskeleton evolved for defensive[4] or offensive purposes.[5] However, this is a backward-looking, teleological, and functional understanding that assumes the product before the process. Evolution does not tend toward this or that function or survival but, rather, toward the increased efficiency of *further evolution*. Material evolution is an end in itself, without any higher purpose. Mutations are neither random nor determined but responsive, in such a way as to increase *responsiveness*—like capillary tributaries saturating a river delta. Nature is not instrumental but tends increasingly and more efficiently toward its own energetic dissolution.

The more fundamental and kinetic question is: "What are the material and kinetic conditions for these adaptations in the first place?" The answer, I believe, is that the invention of bio-mineralization in arthropods is continuous with the larger process of elastic animalization. It is part of a broad tendency to increase nervous sensation, surface area, mobility, mutability, and energy expenditure. Defense and offense are thus secondary effects, emerging only after the fact, of a more primary process of elastic experimentation and expenditure.

In other words, elasticity has many side effects, but the most important and only primary effect is the increase in elasticity itself. This is why the study of kinetic patterns is so important, if not primary, in the study of nature. Increasing morphological elasticity further increases elasticity, for no higher functional end but the increasing expenditure of energy through sensitive mobility.

Elastic bones: Vertebrates
The first vertebrates also emerged around the time of the Cambrian explosion. From the first bony fish to amphibians, to reptiles, to birds and, eventually, to mammals, animal bodies began to produce bony skeletons *inside* their bodies. Once again, animalization turned the body inside out. The

emergence of ossification in animals entailed three major elastic transformations of the animal body.

Bone remodeling

The first major transformation was an increase in the elasticity of the skeletal structure itself. Vertebrates, instead of having a skeleton on their exterior, dangerously molting it off whenever they grew, invented a new, collagen-based, elastic body *inside* their body that they wove, unwove, and rewove alongside the rest of the body's metabolic processes.

Bone is not a solid substance but is composed of about one-third elastic matrix and two-thirds bound minerals, which are always being woven and remodeled by bone cells. The elastic matrix is flexible because it is composed almost entirely of collagen fibers (ossein), bathed in a clear fluid medium called "ground substance" that facilitates internal movement, development, transport, and lubrication.[6]

The elasticity of the collagen improves the fracture resistance of bones. It tends to accumulate near the ends of bones where they meet with other bones, move, and experience stress, including along the vertebrae. As bones develop, they begin first as thick collagen or cartilage structures whose woven, porous, honeycomb-like nets provide the basis of the laminar columns of hardened calcium phosphate that will cover them. These laminar sheets of calcified bone provide vertical rigidity, while their elastic matrices offer flexibility.

While the laminar cortical bone provides rigidity, the cancellous bone provides a more open and porous woven space, with a high ratio of surface area to volume, where metabolic activity is free to move, reproduce, and alter the bone. Laminar bone is thus an emergent property of a more primary woven, folded, and kinetic metabolic activity of cancellous ossification. These two kinds of bone are crucial not just for making bones more elastic and impact-resistant but for further increasing a sensitive feedback mechanism between the animal and the world. Bones do not rigidify the animal but, rather, increase its elasticity, by allowing it to reweave, during its lifetime, its hardest and most durable structure. Just as neuroplasticity allowed for a rewiring of the moving animal's brain, osteoplasticity allowed for the restructuring of its body.

Bones are therefore not only materially but also developmentally elastic, insofar as they are continually undergoing microfractures, which they regularly repair. Their "weakness" is thus precisely the strength that allows them to respond directly to their unique conditions of motion in a particular animal body and environment. Bone cells both reabsorb hardened bone, through little pits or lacunae, back into its elastic matrix, and also create new bone tissue in its place.

Like weaving, each iteration is different and responsive to its relational context. We tend to think of bones as permanent and unchanging, but they are precisely the opposite. The animal endoskeleton is like an inner arthropod that is continually molting into and out of itself. It does so in a sensitive response to a whole range of biomechanical factors. Bone remodeling is not the carrying out of a fixed blueprint but rather a pedetic and kinetic process of ossification and reossification with the world. This process of continual material circulation is also what makes bone a reservoir for minerals like calcium and phosphorus.

Dermal regeneration

Animal bones also made possible a radical intensification of sensation by exposing the animal's highly sensitive nerves to the world through the use of skin. Arthropods had bristles that could poke through their exoskeleton here and there, but by putting their bones on the inside, vertebrates were free to multiply their sensitive pores, hairs, scales, and feathers on the outside without compromising an exterior skeleton. This simple change contributed to a radical increase in neurons, from the hundreds of *thousands* in arthropods, to the tens and hundreds of *millions* in fish, amphibians, reptiles, and birds, and the *billions* in large mammals.

The more the nervous body is sensitive to the world, the more it can respond elastically to that world. The more energy it needs to respond, the more it moves and eats; the more it moves and eats, the more sensitive it needs to be to the world-on-the-move as mobile predator or prey; the more sensitive it is to the world, the more elastically it responds to that world, and so on. The endoskeleton made it possible for the first time to expose the largest sensitive organ on the animal body: the skin.

It is no surprise, then, that animal skin followed an elastic model similar to that of animal bones. Bones are not permanent, and neither is skin. Both are woven works in progress. Animal skins are naturally broken down, recycled, and regenerated in direct response to the context of their usage, just like bones. Skin protects the body, regulates metabolic activity, and has all kinds of highly sensitive functions including touch, temperature sensitivity, and pressure sensitivity. Skin has many pores for releasing increasingly large amounts of heat from the increased energy expenditure of the elastic animal bodies and limbs.

Not only is skin regenerated, but the dermis has highly elastic collagen fibers woven together in crosshatched layers for durability, elasticity, and circulation. The skin is so elastic and porous that its surface area is much larger than it appears. If we consider that each pore opens up the body to release heat, then the bony vertebrate is essentially the most exposed animal. The inside of the body becomes nothing but the outside folded into a million passageways for blood, oxygen, nerve waves, sweat, and other fluids cycling through the body like one big surface—for maximizing the dissipation of energy into the air. Just as water, for example, is more rapidly evaporated from a larger area than a smaller area, so by increasing the porosity and exposure of its skin, the vertebrate animal can release more energy faster than any other class of animal.

Animal muscle

The third interrelated consequence of ossification in vertebrates is the dramatic increase in muscular mobility. Bony endoskeletons and limbs did not just support the animal body so that it could walk on land; they allowed for the storage of energy in the elastic material of muscles.

Most animals have some elastic, collagenous muscle fibers, but such fibers become unusually large and robust in vertebrates precisely because of their bones. Muscles are highly elastic collagen fibers that can store and rapidly release energy. Vertebrate backbones and endoskeletons are the relatively rigid structures inside the body that can pull muscles apart and together and thus allow those muscles to acquire, release, and transfer motion through the body and limbs.

Muscles, tendons, ligaments, and joints are like the rubber bands of the animal body. Just like bones and skin and much of the elastic animal body, muscles are continually unwoven and rewoven in response to their contextual use. Muscle, like bone, strengthens itself through micro-tearing. Continual self-destruction/regeneration is the price for maximum elasticity, responsiveness, and sensitivity. Muscles must die to live, must be torn apart by bones to build new and stronger tissues. This is the lesson of the elastic vertebrate.

Animal limbs do not need to wait for generations of evolution to change their muscular structure but can modify it easily throughout their life. Bone, skin, and muscle together transform the limbed, bony, animal body into an utterly elastic surface. They increase the animal's direct nervous connection with the world. By continually reweaving bone, muscle, skin, and nerves, the animal ensures that the body can easily adapt to any changes in the world. The body can then rewrite, restructure, and retrain itself toward the most subtle, fleeting, and tiny opportunities for energy consumption.

The energetic cost of constantly rebuilding, expanding, and contracting muscle is comparatively quite large. The more elastic the animal body became, the more rapidly adaptable it was to changing circumstances, and the more mobility was required to secure the food to make that responsiveness possible. It was a brilliant but volatile evolutionary experiment, compared with mineral or vegetal kinetics. At the capillaries of evolutionary history, limbed vertebrate bodies were balanced on a knife edge, innovating widely and quickly to stay alive.

* * *

The material evolution of animal morphology is also a kinetic evolution toward the increased elasticity, mobility, sensitivity, and energy expenditure of the earth more broadly. Animals are not *on* the earth but are aspects of the earth itself—the earth's own becoming-animal and becoming-elastic.

However, the emergence and flourishing of one particular kind of vertebrate (homo sapiens) has had an increasingly transformative effect on the four kinetic patterns of the earth. This tiny region of the earth—humans occupy a mere .01% of all animal bodies—thus requires its own contemporary kinetic analysis in order for us to understand what is happening to the earth today. This is the unique challenge of the final part of this book.

PART III
THE KINOCENE: A DYING EARTH

14 Kinocene Earth

TODAY, THE EARTH is in increasingly unstable motion. The earth has always been in motion, but today these four major patterns of geological motion—mineral, atmospheric, vegetal, and animal—have become increasingly disrupted due to the coordinated efforts of particular human groups.

What I am calling the "Kinocene" is a new geological period not because motion is new to the earth but because of the increasing mobility of the earth's geological strata. However, we are also witnessing a significant *reduction* in the net kinetic expenditure of the whole planet. This is the strange paradox of the Kinocene: as mineral, atmospheric, vegetal, and animal strata are becoming increasingly mobile, the earth is also expending *less* energy. How is this possible? And what should we do about it if we want to survive? This chapter and the next aim to answer these questions.

The contemporary state of the earth is one in which each kinetic pattern described in Part II has today become increasingly entwined with the others. With the historical addition of each new metastable pattern to the previous ones, the earth has slowly increased its capacity to more quickly and thoroughly break down energy from the sun. Today, all four patterns of planetary motion are integrated systems that work together to maximize one another's dissipation of energy into back into the cosmos.

The geokinetic theory and history of Parts I and II of this book have been essential for us to see the full reciprocal integration of various planetary systems of motion. Each subsequent pattern of motion was not determined by the prior, but emerged from it and transformed the whole earth. The idea that the earth is relatively static or stable is false. The earth is not a passive object of causal laws, nor is it a vital subject here to support life. The earth is an unpredictable, unconstructible, kinetic process radiating itself into space. All the patterns of motion on the earth, including ourselves, are experiments toward increasing this energetic dissipation.

After the discovery of the four significant patterns of motion covered in this book, we are finally in a position to assess the full geokinetic history of our present epoch, where all the patterns mix. Part III of this book draws on Parts I and II to show what this mixture looks like today and where the human animal, in particular, fits into it.

This effort takes place over two chapters. Chapter 14 shows how the kinetic patterns of nature described in Part II all work together to increase cosmic and planetary kinetic expenditure. This is the deeper historical frame within which the earth and human beings emerged. Chapter 15 then looks at how the history of certain human animals has slowed this dissipative process down—stunting numerous earth processes and driving various species toward extinction. Chapter 15 also looks at the consequences of these historical actions and proposes an ethics based on the movement-oriented perspective developed in this book.

The project of modern anthropocentric and biocentric civilization has come to an end. Fossil fuels are finite, and the earth is in unpredictable motion. Human domination is today exposed as the pipe dream that it always was. The future of this planet will not be a return to a stable, static, conquerable earth (that never existed). It will have to be a new metastable formation.

THE KINETIC EXPENDITURE OF NATURE

Scientists who study energy tend to count only the movements that are useful to humans. They count how many calories there are in our food (1 calorie = 4.184 joules) and how much energy it takes to run a lightbulb (1 joule to light a 1-watt LED for 1 second). If humans are not eating it or using it to power their technologies, energy scientists tend not to think much about the energy produced and consumed by the rest of nature.

Nature's expenditure is all just energy that is effectively useless, unharnessable, and unconstructible—including our physical, energetic waste. Energy scientists even tend to define energy itself as "the ability to do work," which is frequently interpreted as "work that is useful *to us*"—although this is not the precise meaning of the term "energy" in physics.

Physicists also distinguish between two types of energy, only one of which actually exists, in my view. "Kinetic energy" is the energy released through motion, and "potential energy" is the energy that an object will potentially be able to release if and when it is in motion. In other words, only the former is genuinely energetic, while the latter is merely a formalized abstraction of kinetic energy: a theoretical possibility. Strictly speaking though, nothing in the universe is completely static. So nothing can have a pure "potential" energy. The notion of potential energy is just a useful heuristic masquerading as a metaphysical type of energy, historically and unreflectively lifted from Aristotle.[1] In the end, in my view, energy is nothing other than matter in motion.

To the best of our knowledge, all matter in the universe is in motion. Matter is in movement, as energy, and all movement is material. There is no static *substance* in nature, only material kinetic *processes*.[2]

PEDETIC EXPENDITURE VERSUS ENTROPY

Matter is neither created nor destroyed but, rather, redistributes itself. What I call *pedetic* or *kinetic* expenditure in this chapter is the historical tendency for movement to increase its rate of redistribution from hot to cold through various patterns of motion.[3] This is not the same as entropy, as currently defined in physics, namely the random and probabilistic movement of discrete particles toward equilibrium. I do not accept this definition of entropy and have criticized its metaphysical assumptions at length elsewhere.[4]

Entropy, for me, stripped of its metaphysical assumptions, is simply the historical tendency for matter to spread out from hot to cold in our universe. Everything else added to the essence of this concept (randomness, universality, discreteness, probability, equilibrium, etc.) is metaphysics. Entropy is what I call a "kinetic expenditure," or the tendency for matter to increase the rate at which it moves through self-organized patterns of motion.

The cosmic dissipation of matter from more concentrated to less concentrated can happen more or less quickly. All material structures have a rate of expenditure below which their pattern of motion will not be able to keep up with heat degradation and will eventually fall apart, ultimately lowering the total rate of expenditure. Material structures also have a rate of expenditure above which they can increase their total energy degradation, but only for a short period, before their pattern of motion falls apart, thus also lowering total energy dissipation and the rate of expenditure. For example, a whirlpool will form as you drain your bathtub, but only if enough water is moving fast enough, but not too fast, to sustain it.

Imagine a sedentary person whose life span declines due to inactivity; then think of an ultra-active athlete whose life span also decreases, but in this case due to overactivity. In between these two poles is an optimal range of kinetic expenditure, where short-term activity is less than that for the athlete, but where total long-term activity is much higher because the person ended up living longer.[5]

Entropy, in my strictly kinetic definition, is not an absolute ontological law forever and for all time, but an emergent property or tendency of our universe so far. Because our universe emerged from an indeterminate quantum process, as detailed in chapter 1, entropy, along with all other laws and fundamental fields, is strictly an emergent historical product of matter and motion. The ultra-fast motion of the post-indeterminate epoch (low entropy) tended to spread out and cool down (higher entropy) through distinct patterns of motion. If or when the universe completely dissipates, quantum fluctuations of gravity are capable of reversing this process and thus reversing entropy itself.[6] So cosmic entropy is both emergent and possibly reversible.

Accordingly, the historical existence of expenditure or entropy (in my definition) does not assume a closed totality of the universe, a series of "possible" states of that totality, or a supreme law given in advance of the universe. The universe is an open system that is far from equilibrium.

The spreading out of motion/matter can occur faster or slower, depending on its distribution. Some patterns of motion tend to increase the speed of kinetic dissipation, and others tend to slow it down. However, because the cosmos began with indeterminacy and is not a closed system, there is no pre-given path to its dissipation. This means that the cosmos is indeterminate, relational, and experimental in all the ways it tries to dissipate and move into the cool.

The cosmos is like a lightning strike that experiments with an eye toward the most entropic ways to get to its ground. This means that nature tends toward self-destruction, but that it also makes mistakes (slower patterns of motion) and ends up redirecting itself in response to emergent problems of dissipation (to move more quickly). This also means that the cosmos is not equally "efficient" in all its regions and directions as it spreads out.

This then means that entropy, like the path of a lightning strike or growth of a tree branch, does not predetermine the precise kinetic and topological structure of the universe. The universe makes its path by walking. The cosmos has to experiment with itself pedetically and relationally as it goes. We could say that expenditure is the way that the cosmos comes to relate to itself and know itself. It tries and learns faster and slower paths to its death and rebirth.

Historically, the portion of the cosmos that we know thus far has learned some pretty efficient ways to dissipate itself: dendrites and vortices are two of the most common. Trees, for example, have developed a material knowledge of how to live and die well that responds dendritically to their environmental conditions. This knowledge is not guaranteed, because trees can make mistakes in which direction they grow, or they can grow so large that they deplete their resources. Even if most trees grow in a dendritic pattern, the fine details of that pattern are different each time and change as the tree learns to optimize its expenditure of motion.

Nature accumulates itself not to avoid dissipation but to hasten it. The more quickly material structures can dissipate motion, the longer their form tends to stick around and multiply. This is why black holes have been so successful. Their vortical patterns tend to accumulate and concentrate enormous amounts of energy and then release it in its smallest and most degraded form.

In other words, material accumulations and evolution tend not toward conservation but toward dissipation. There is thus a kind of anti-telos to our cosmos: to render itself useless. It uses itself to destroy itself; its use is to abolish utility. Eventually, the universe will unweave and reweave itself through entropy and negentropy. The question is *how* (using what patterns?) and *how quickly* (at what rate of expenditure)? This is a fundamentally relational, pedetic, and open question. Nature is capable of genuinely radical novelty without a pre-given set of possibilities.

THE EARTH

The earth is one region of this much larger cosmic "autophagy" (self-devouring). Each epoch in the earth's history has offered further experiments and refinements in the fastest ways to dissipate the most movement. The following sections demonstrate two things: first, that the total amount of energy expended (in joules) decreases from older to more recent geohistorical patterns of motion, and second, that the *rate* of energy expenditure (measured in watts) per gram increases over the same history.

In other words, there is an inverse relationship between total energy expenditure and the rate of energy dissipation per gram. Nature abhors a gradient and tends to develop new kinetic structures in order to expend it faster and faster. The history of the earth is the history of these interlocking patterns of dissipative motion.

Solar expenditure

We begin with the sun, whose total kinetic output is absolutely enormous, although relative to its mass, the sun is in fact extremely inefficient at expending its energy. While the total energy of the sun is gigantic (1.74×10^{47} joules), as is its rate of solar luminosity (390×10^{24} watts), its rate of expenditure per gram is relatively low. For example, the metabolism of a schoolchild makes the child 15,000 times more dissipative than the sun, measured per gram. Some bacteria, such as Azotobacter, even reach energy conversion rates of up to 500 million times greater than the sun's![7]

In other words, the sun expends energy relatively inefficiently, and does so mostly into empty space. The earth intercepts only a small fraction (4.5×10^{-10}) of this kinetic energy, and yet the total annual aggregate of this intercepted energy is still incredible (5.495×10^{24} joules). By contrast, at the beginning of the twenty-first century, the global consumption of all fossil fuels was equivalent to only about 0.006% of this solar irradiance. Even the total estimated energy contained in all fossil fuels on the earth is equal to a mere 13 days of the sunlight intercepted by the earth.[8]

The point is this: the sun is a large but inefficient dissipative system whose kinetic gift to the earth dwarfs all energy dissipation systems on the planet in size, but not in efficiency of expenditure *relative to size*. The earth and all of its processes are much faster at expending energy. This defines the kinetic relationship of the earth to the sun.

The earth helps the sun's flows to die faster, better, and more thoroughly. The earth is one big gradient reducer between a high concentration of energy (the sun) and a low concentration (space). Nature abhors a gradient and experiments with the fastest way to break that gradient down. The earth and its history are thus experiments in cosmic gradient reduction. The earth is not an isolated ecosystem but is kinetically connected to a vast "uranosystem," within which it creates and invents new, faster, and more diverse ways to spread out.

The earth is thus a kind of ethical agent in the cosmic drama, trying to figure out how to degrade the sun and itself into space most rapidly. Our planet is one big attempt to reciprocate what the cosmos has given it: the gift of death.

Mineralogical expenditure

The mineral body of the earth is not merely a passive surface for solar absorption, reflection, and transformation—it is itself an active expender of motion. Early Hadean and Archean volcanism and meteor impacts were constant and likely expended an enormous amount of all kinds of energy flows, including nuclear explosions. Any estimate of these enormous energies, however, would be highly speculative at best.

These prolonged high-energy expenditures are what produced the atmosphere and the oceans. However, after the earth cooled (in part due to the new ocean and atmosphere), the earth also continued to steadily release geothermal energy as radiation and heat. Recent (2005) estimates put this heat release at an annual average rate of 44 terawatts—70% of that released through the oceans and 30% through the continents.[9]

Additionally, global volcanism adds something to this punctuated release of mineralogical energy. Volcanism is energetically and geographically diverse, but recent estimates claim the annual average total release over the twentieth century to have been about 36×10^{15} joules.[10] The nearly constant average of thermal and kinetic energy released by earthquakes adds another 800×10^{15} joules to this annual total.[11]

But even this enormous total (836×10^{15} joules a year) pales in comparison with the total gravitational "potential" energy of the planetary body itself (-2.6×10^{33} joules). A negative potential energy here means that work must be done to extract this planetary energy against the force of gravity. In other words, the mineralogical earth, like the sun, has an incredible amount of energy but is not very efficient at expending it.

However, the mineralogical earth also produced the immanent material conditions for the emergence of even more rapid expenditures of energy in the atmosphere, oceans, and biosphere. Without tectonic energies to raise the crust above the ocean, terrestrial life would not exist. Mineralogical expenditure is also a significant factor in global oceanic and atmospheric circulation and convection. High mountains cool the planet, and warm Gulf Streams heat it. Volcanism is also the most important source of carbon dioxide, which has provided the stuff of life and the long-term balance of biospheric carbon. Volcanism also contributes by far the largest source of aerosols injected into the stratosphere, which create global cooling.[12]

In short, mineralogical expenditure provides the immanent material support for all other planetary processes. The mineralogical earth dies so that others may die better and faster together. Mineralogical expenditure responds to solar expenditure by facilitating its transformation and decomposition into the atmosphere, oceans, and life. Just as the sun is trying to reduce the gradient between itself and cold space, so too is the earth. The sun and the earth combine their expenditures to increase their total rate of release into space—toward their cosmic heat dissipation. Atmosphere, ocean, and life are techniques to speed things along.

Atmospheric expenditure

The earth and sun together invented a new kinetic pattern of motion with an enormous ability to expend motion: the atmosphere. The atmosphere is such an immense and rapid process of kinetic expenditure because it is continuously fueled by the mineral earth and by the sun. The ongoing alternations of hot and cold connected to the earth's rotation (day and night), the earth's tilt (the seasons), and the earth's internal convection and kinetic release cycles (thermal release, volcanoes, and earthquakes) all drive an extensive vortical and circulating atmospheric system with massive energy accumulation and rapid dissipation.

There are, for example, about 1.4 billion lightning flashes per year on the earth.[13] Each one releases an average of 1 billion joules of energy. That means that lightning alone expends 1.4×10^{18} joules on average each year. This is several more times than all volcanoes and earthquakes combined over the same period—and has a much higher energy dissipation rate per area. Furthermore, lightning is only a minuscule portion (4×10^{-7}) of the atmosphere's total convective energy.[14]

Other ways that atmospheric energy is released include thunderstorms, cyclones, atmospheric pressure changes, and water movements. For example, average-sized thunderstorms that leave behind just 1 centimeter of rain release 2.5 to 5×10^{15} joules of energy through evaporation. That is an impressive 10 to 100 times more than their total kinetic energy of 30 to 300×10^{12} joules. Large cyclones discharge less than .5% of the planetary heat flux, with rates for individual events mostly between 10 and 50×10^{18} joules (at a rate of 120 to 580 terawatts). Even ambient atmospheric pressure, heat, and air currents add up, at the global level, to about 300×10^{18} joules a year.[15]

The larger the atmospheric pressure gradient between inside and outside the storm system, the more massive and faster the hurricanes. Interestingly, hurricanes follow a kind of allometric scaling of size and "metabolic" speed similar to biological metabolic scaling. The larger they are, the faster they break down the energy gradient. In this way, self-organized systems like vortices in the atmosphere and ocean speed up the degradation of heat much more rapidly than does the linear conduction of heat from the earth's surface into space.[16]

The movement of water across the earth also expends an incredible amount of energy. Surface water runoff, for example, excluding the Antarctic ice flow but including runoff to continental interiors, releases an average potential energy of 367×10^{18} joules a year. The total power of ocean currents has been estimated at 100×10^{9} watts.[17] The global water cycle evaporates about 430 gigatons of water but expends only a small but efficient amount of kinetic energy at 360 joules per square meter.[18] Average-sized tornados dissipate a total of around 275×10^{9} joules, and the largest dissipate over 150×10^{12} joules.

Since water can hold much more heat per volume than air can, the oceans are disproportionately responsible for global energy transfer. The Gulf Streams and the Kuroshio Current, for example, mix warm water from the equatorial Pacific to the poles. Satellite views from space show

Gulf currents curling into a thousand tiny vortices of energy degradation. Dense, cold, oxygen-rich saltwater from the poles sinks to the deepest parts of the oceans and is then heated by the earth to rise and cool, then fall again.

Ultimately, air and water movements on the earth have less total energy than the sun or the earth themselves. However, what they lack in total joules they make up for in their energetic "intensity," or expenditure per unit area. Lightning, for example, releases up to 2000 joules per meter and even more per gram.[19] It is thus a much more efficient expender of energy than is the sun or the mineral earth. Air and ocean convection, along with the water cycle, release an incredible amount of energy in relation to the relatively small mass of the atmosphere (5.14×10^{18} kg) as compared to the mass of the mineral earth (5.972×10^{24} kg).

Although the atmosphere has less total energy and mass, it speeds up the process of energy dissipation through the development of dendritic and vortical patterns of motion (dendritic river basin flow paths, vortical hurricanes, tornados, and so on).[20]

Vegetal expenditure

Life, though, is perhaps the most advanced instrument yet evolved for degrading energy. In particular, the transformation of sunlight into thermal and chemical energy in chloroplasts is the most significant and most efficient process of energy conversion on the earth. Organisms are not just bodies or populations. They are liquid eddies and pools of energy organized into enormous metastable patterns of mutual energetic degradation.

All life takes energy from the environment and transforms it through metabolism into adenosine triphosphate (ATP). This is the energy currency of life and powers all living cells. When its bonds break down, it releases an incredible amount of energy compared to other molecules (31 kilojoules per molecule). These phosphates were not selected by accident but through a long material evolutionary process.[21] This may sound like a small amount of energy, and it is, but not relative to the size of the organism. Per molecule, it is some of the highest metabolic expenditure of energy on the planet.

From microscopic bacteria to redwood trees, life on the earth evolved toward increasing capture and efficient dissipation of energy. The organisms that can release the most energy in the form of offspring are the least likely to be eliminated in the evolutionary process. Natural selection thus tends towards organisms and ecosystems that waste, expend, or share their energy with others in the form of degraded heat. Historically, the most effective way to increase overall metabolic expenditure is by maximizing biodiversity to achieve an optimal energy flow. The more biodiversity there is, the more ways energy optimally breaks down.[22]

The evolution of life on the earth exhibits a clear historical tendency for organisms to increase their metabolic rate of energy expenditure per body mass. In other words, as organisms evolve into larger and larger complex composites (body mass), there is a general (although not universally predictive) allometric scaling ratio. Mass changes proportionally with the increasing rate at which organisms metabolize energy. Larger organisms not only expend more energy to survive, but they dissipate this energy more rapidly relative to their body mass.[23]

This is true not only for the evolutionary emergence of larger organisms, but even relative to similar kinds of organisms over time as well. For example, early "reducing bacteria" have slower metabolic rates than later, more evolved aerobic bacteria. Over the past 600 million years of evolution, macroorganisms, in particular, have increased their metabolic rate fourfold over their Cambrian counterparts.[24]

Life, it seems, has always been headed full speed into the cool—driven by the optimization of energy expenditure and increasing biodiversity.[25]

But allometric scaling, entropic expenditure, and biodiversity are not deterministic laws. Nature is not a closed system and does not know in advance the best way to optimize expenditure. It has to experiment. Meteors strike; predators devour all their prey and die off; the earth fluctuates between baking hot temperatures and freezing cold ones.

There have already been five mass extinctions on the earth that had nothing to do with how well a certain organism metabolized or failed to metabolize energy. Sometimes global feedback effects overwhelm relatively successful organisms. Incredibly, 99% of all species that have ever lived are now extinct, for countless reasons not directly related to their metabolic expenditure.

The earth's phylogenetic tree of life is like a giant lung that occasionally fills up with energy and distributes it intensely in vast explosions of biodiversity. Life evolves to expend it as quickly and diffusely as possible, down to its most diverse bio-capillaries. Then, as rapidly as life enters in, huge portions of it die out: 444 million years ago, the first great extinction killed off 86% of all living species; just over 100 million years later, the next one killed off 75%; the next one killed 96%; then 80%; then 76%.[26]

Of all living organisms, plants are the most advanced degraders of solar radiation.[27] The transpiration of water through trees, for example, requires about 580 calories per gram of water transpired. A single tree transpiring 100 kilograms of water a day will use nearly 60 million (5.8×10^7) calories.[28] If the average tree evaporates 151,600 kilograms of water a year, that is nearly 88 billion calories per tree per year (368 billion joules); multiplied by the 3 trillion trees currently on the planet, that results in an incredible 1.1×10^{24} joules a year.

Trees are powerful dissipative systems that degrade solar energy into low-grade heat energy, water, and oxygen. However, only about 1% of most plants' energies go into making new biomass. Vegetal structure and accumulation are, therefore, secondary to dissipation. Vegetal life is, first and foremost, a massive energy-degrading process of radical expenditure and waste. Even when new biomass growth occurs in root structures, branches, and phyllotactic patterns in leaves, the tendency is to seek increasingly efficient ways, experimentally, to expend more energy faster. Plant growth is both thermodynamic and evolutionary at the same time in its flow optimization.

Of the 800 watts per square meter of solar radiation that falls on plants, they convert 18% into sensible heat, reflect 15%, transpire 66%, and convert 1% into biomass—most of which they feed to heterotrophs such as fungus and herbivores. The waste products of herbivores are even further broken down by bacteria. Plants eventually expel all energy as low-grade heat into space, but before that happens, it is shared widely with the rest of the earth and trophic levels.[29]

Estimates of the total plant biomass currently produced on the earth vary historically. Recent estimates are around 121×10^{15} grams (121 trillion metric tons) of carbon per year—only about half of which remains after plant respiration, metabolism, and heterotrophic consumption.[30]

When organisms work together in biologically diverse ecosystems, they increase the total amount and rate by which they degrade solar radiation into lower quality non-radiative processes. A clear-cut forest, for example, degrades only 65% of its incoming energy, while a 400-year-old old-growth Douglas fir forest degrades 90% of incoming solar energy.[31]

Plants transform solar heat into airborne oceans of water vapor that cool off the planet and are rained back down by thunderstorms, to be degraded again and again in an enormous recycling process of expenditure.

The biosphere absorbs solar radiation and increasingly tends toward reducing that energy to the cold background heat of the universe. Dense old-growth forests, for example, are organized such that every little beam of sunlight gets captured on its way down to the forest floor. Each trophic layer then lives by eating and expending the waste of the prior stage of kinetic extraction. Ecological diversity, therefore, results in increased energy gradient reduction.[32]

Animal expenditure

Although its total expenditure is not quite as massive as the total energies expended by vegetal life, animal life is by far the most intensive expender of energy on the earth. Within the broader trend of all life's optimization of metabolic expenditure, animals occupy the upper reaches of the allometric scale. Even for animal life itself, we can see a clear and dramatic evolutionary tendency not just toward larger animals but toward animals with higher rates of metabolic respiration and thus kinetic expenditure.[33]

Animals live on the energetic coattails of vegetal life—which are themselves riding the flows of solar and geokinetic waves. Plants transform only about 1% of the energy they consume into biomass that can be eaten by herbivores, which can be, in turn, eaten by carnivores. At each trophic level, 80–90% of the energy consumed by animals is dissipated as heat. Animals thus continue the biospheric trend toward increasing not only metabolic intensity but also metabolic waste per area.[34]

Animals are expenders of expenditure, recyclers of recycling. They are less than 10% efficient at converting energy into "useful" biomass. Just as solar expenditure is greater than mineralogical expenditure, which is greater than atmospheric expenditure, vegetal expenditure is greater than animal expenditure. However, each further pattern of organized expenditure is more concentrated and intense per unit mass than the previous.

Estimates of total invertebrate and vertebrate animal biomass (including fish but excluding domesticated animals) put it at close to 10 trillion metric tons. Wild mammals contribute only a tiny 10 megatons to this figure. This is significantly less than the 121 trillion tons of plant biomass, but considerably more intensive.[35]

There are so few large animals because they regularly require enormous quantities of energy, which require exponentially larger stored biomass energy for them to consume.[36] In other words, energy expenditure in animals is no less impressive in its intensity (thousands of times more intense than the sun) than the sun is in its overall magnitude.

At the tiny top of the trophic pyramid are humans—occupying a small fraction of all life on the earth. As large mammals, humans need to consume a relatively large amount of concentrated biomass, but also, like most mammals, have extremely high metabolic rates. Just for the human body to continue to power all its vital functions (basal metabolic rate) it needs to expend a relatively large amount of energy compared to other animals.

For example, the basal metabolic rate of an adult woman weighing 130 pounds is about 5.5 megajoules (about 1400 calories), and for a man weighing 150 pounds it is about 7.5 megajoules a day (about 1800 calories). If they do hard work and increase their mobility, then around another 2 megajoules are required.[37] Without even breaking a sweat, we expend a full 60% of this energy through respiration and skin diffusion at the rate of about 12 watts per square meter. Another 20–25% of this energy is consumed by the brain, compared to 8–10% in other primates and just 3–5% in other mammals.[38] By sweating, a human can even increase its energy expenditure up to 500 grams per square meter, compared to a horse, which can lose 100 every hour, or a camel, which can lose up to 250.[39] Humans, it turns out, are designed to rapidly expend enormous quantities of energy per square meter—exceeding the rate of most other organisms and natural processes.

After energy inputs and respiration, human animals can directly expend around 16–20% of consumed energy through kinetic energy (moving around), growing and repairing the body, nerve functioning and plasticity, cellular metabolism, and storage (in fat and protein, for example).[40]

Physiologists love to tout this percentage of energy available for our direct use as an efficiency of the human body, comparing it to the 1% in plants and less than 20% in other mammals. However, we should keep in mind that humans are still, far and away, the most efficient *expenders*, or squanderers, of energy per area of that other 80% of energy. This means that we still expend more energy per area than do trees and most other animals.

Furthermore, even the so-called "useful" 16–20% of energy that humans "conserve" is almost entirely consumed by various parts of the body, which also release metabolic wastes. Even when we temporarily store energy in our liver or muscle cells, we eventually use it. In other words, even most of our "efficiently used energy" is still precisely that: *used*, expended on our bodies. "I waste, therefore I am," is true not only of humans but of all of nature.

★ ★ ★

Material evolution tends toward optimal kinetic expenditure at every level, from galaxies to microbes. The Kinocene is an age of motion in which all of the earth's historical patterns of motion mix and rapidly expend enormous amounts of energy. This chapter has aimed to show the contemporary integration of all the earth's patterns of motion in what I am calling the Kinocene.

However, the historical tendency of the cosmos to optimize flow patterns toward dissipation is not always successful. Nature experiments, produces waves of extinction, and then begins again. The activity of certain human groups, in particular, has increased specific mobilities at the cost of *decreasing* planetary movement more generally.

Burning fossil fuels temporarily increases one kind of expenditure, but destroys the biosphere and also replaces human labor with machine labor, which has slowed down planetary kinetic expenditure as a whole. Material evolution tends to select for optimal gradient reduction and has feedback loops that tend to hasten the destruction of processes that do not optimize expenditure. This is what we face today. If humans have any hope of surviving, we need to understand what is happening on a deep historical and kinetic level and find a different place for ourselves in it.

15 Kinocene Ethics

THIS BOOK HAS SO FAR BEEN a long preparation for the introduction of an important shift in perspective. We have moved from a theory of a static or stable earth, viewed through the narrow lens of human history in the Holocene, to a much more mobile and experimental earth, tending toward patterns of optimal kinetic dissipation. The earth is not a passive object or stage for human activity, nor is it a benevolent subjective agent of life (Gaia). If anything, the earth is a process of expenditure and death—more like Python than Gaia.

Human animals do not live *on* the earth. They *are* the earth. Technically, even the earth itself is not *of* the earth. It is a branch of a massive cosmic dendritic flow, of which humans are one tiny experimental capillary seeking out optimal kinetic expenditure into the cool darkness of space. The duration of our life on this planet is related to how well we help our planet and the cosmos expend itself. The biosphere has tried five different capillary paths so far, each ending in mass extinction—like river tributaries that dried out in the sun before making it to the ocean, or the tentatively stepped leaders of a lightning strike abandoned for another direction.

There is a double meaning to the Kinocene. On the one hand, it is a planetary epoch defined by the increasing geological mobility of various strata in broader circulation than ever before. However, despite this burst of temporary mobility for some parts of the planet, the Kinocene has led to a net decrease in total planetary expenditure and mobility. This is because it has destroyed much of the most efficient dissipative system on the planet: the biosphere. I call this tendency "Kinocene extinction."

On the other hand, the Kinocene is also the geological epoch in which climate change has forced us to confront our entanglement in cosmic energy expenditure. The Kinocene could, therefore, be the epoch in which global humanity actually rediscovers its place in the energetic nature of things.

Human animals are not at the top of a cosmic hierarchy, but at the bottom of a river delta. We are not unique in our ability to help or hinder the optimal expenditure of matter in motion, but we are distinct in our capacity, at the current historical juncture, to correct an error in our own practical relation to the earth and cosmos. Instead of struggling against death, expenditure, and the motion of the cosmos, we can embrace it and learn to share generously and reciprocally with the rest of nature.

However small our contribution may seem on the grand scale, humans, at this point, can either increase or decrease the metabolic dissipation of motion in the cosmos. We can hasten our extinction by trying to preserve ourselves, or we can survive by dying well, along with the rest of matter on its journey into the cool (*psychros*).[1] I call this change in behavior "Kinocene expenditure."

In this chapter, I look at each side of the Kinocene in turn: first, looking at the history of human ecocide and its net effect on planetary expenditure; and second, looking in detail at the ethical practices that tend to increase planetary expenditure.

KINOCENE EXTINCTION

Humans currently make up .01% of global biomass, and yet since the rise of civilization, certain groups of humans have been responsible for destroying a full half of planetary biomass. The vast majority of this destroyed biomass has been plants and trees, which make up 80% of all biomass on the earth. The primary historical reasons for this have been deforestation and agriculture. Crops only comprise 2% of plant biomass, but are responsible for an enormous loss of biodiversity and biomass. Various human groups throughout history have been responsible for the loss of 85% of wild land animal biomass, 80% of marine mammal biomass, 14% of fish biomass, and the decline of 41% of all insect species.[2]

All life is knotted together into a trophic feedback loop in which pollinating animals and insects are needed in order for plants to reproduce. Plants, in turn, are needed in order for animals to eat and reproduce. Plants are also needed to regulate the carbon cycle and global weather and temperature. These, in turn, are required for most of the biosphere. Although we do not yet have accurate statistics for bacterial and fungal biomass decline, they currently make up a vast and crucial portion of global biomass, at 70 gigatons of carbon for bacteria and 12 gigatons for fungus.[3] Bacteria and fungi support and flourish in biodiverse conditions and suffer adverse effects from deforestation and pesticides.

Wild plants, and especially mammals, have been killed off in huge numbers to be replaced by a few abundant crops and domesticated animals such as cows and pigs. But all the earth's livestock still only comes to a tiny 0.1 gigaton of carbon, living at the expense of the much larger mass of plants, insects, reptiles, birds, wild land, and marine mammals that it has displaced.[4]

A brief history of ecocide

Ecocide is the destruction and reduction of energetic dissipation on the earth. It is also a *kinocide* because it produces a generalized slowing down of planetary dissipation. Each trophic level (mineral, atmospheric, vegetal, animal) is part of an energetic pyramid, with higher total energies below (mineral) and faster rates of expenditure on top (animal).

The higher you go on the pyramid, the more the organization of matter is in flux. This is why the earth has had five mass plant and animal extinctions already. When a lower-level event—such as solar flares, meteor assaults from the Milky Way, super volcanoes, changes in atmospheric carbon, or changes in phytomass—occurs, everything above that on the pyramid is also radically and systematically affected, through a series of feedback loops. Certain groups of humans have now directly and indirectly affected the kinetic patterns of the earth at every trophic level.

At the end of the last ice age and the beginning of the Holocene, the earth exploded with new flora and fauna. Just from 16,000 BCE to 3,000 BCE, tropical forests tripled and cool-temperature forests expanded thirtyfold.[5] Total phytomass during this time was, in all likelihood, more than 1,000

Kinocene Ethics   245

gigatons of carbon—well over twice today's levels. From the rise of ancient civilizations and states (c. 3500 BCE) to the beginning of the nineteenth century, 20% of global phytomass had been destroyed (taking it down to 750 gigatons). And then, in just the last two centuries alone, more than 30% of all remaining phytomass was destroyed due to the rise of industry, capitalism, colonization, deforestation, desertification, and species extinction (taking it down to 500–550 gigatons) (see Figure 15.1).[6]

Since phytomass (the biomass of plants) accounts for 80% of total biomass, it also accounts for 80% of the total energetic value of planetary biomass. One gigaton of carbon, for example, is 1×10^{21} joules of energy. We can, therefore, see a clear and dramatic net decline in planetary expenditure directly related to the decline just in stored phytomass alone—which, as we have seen, is only around 1% of plant energy expenditure. See, again, figure 15.1.

The history of ecocide is long and has already been the subject of numerous books.[7] This body of work, including two of my own books on migration and borders,[8] shows clearly, in my view, that the history of biomass destruction and increasing energy consumption in human history is unquestionably dependent on social inequality and forced migration. This is especially true with the advent of the capitalist-driven fossil fuel extraction, deforestation, colonization, slavery, and energy wars that began in earnest in the 19th and 20th centuries.

Every major social formation in Near Eastern and Western history expanded its scope and speed of circulation through the kinetic expulsion of human and nonhuman migrant populations. As far back as the Neolithic, some humans were expanding agriculture by expelling nomads, wild plants, and animals into forced migration—ultimately deforesting and depleting the Fertile Crescent. In the ancient world, the Sumerian, Egyptian, Greek, and Roman empires built themselves on the backs of "barbarian" slaves who were forced into migration from other areas in order to mine fuels, fight resource wars, and build the infrastructure of empire.

The early modern legal and penal system emerged on the backs of feudal serfs and the criminalization of vagabonds. Finally, the capitalist economic system expanded private property through the forced expulsion of peasants around the world (the English, the Irish, enslaved Africans, American Indians, and others) from their land and into waged or unpaid labor. Wild

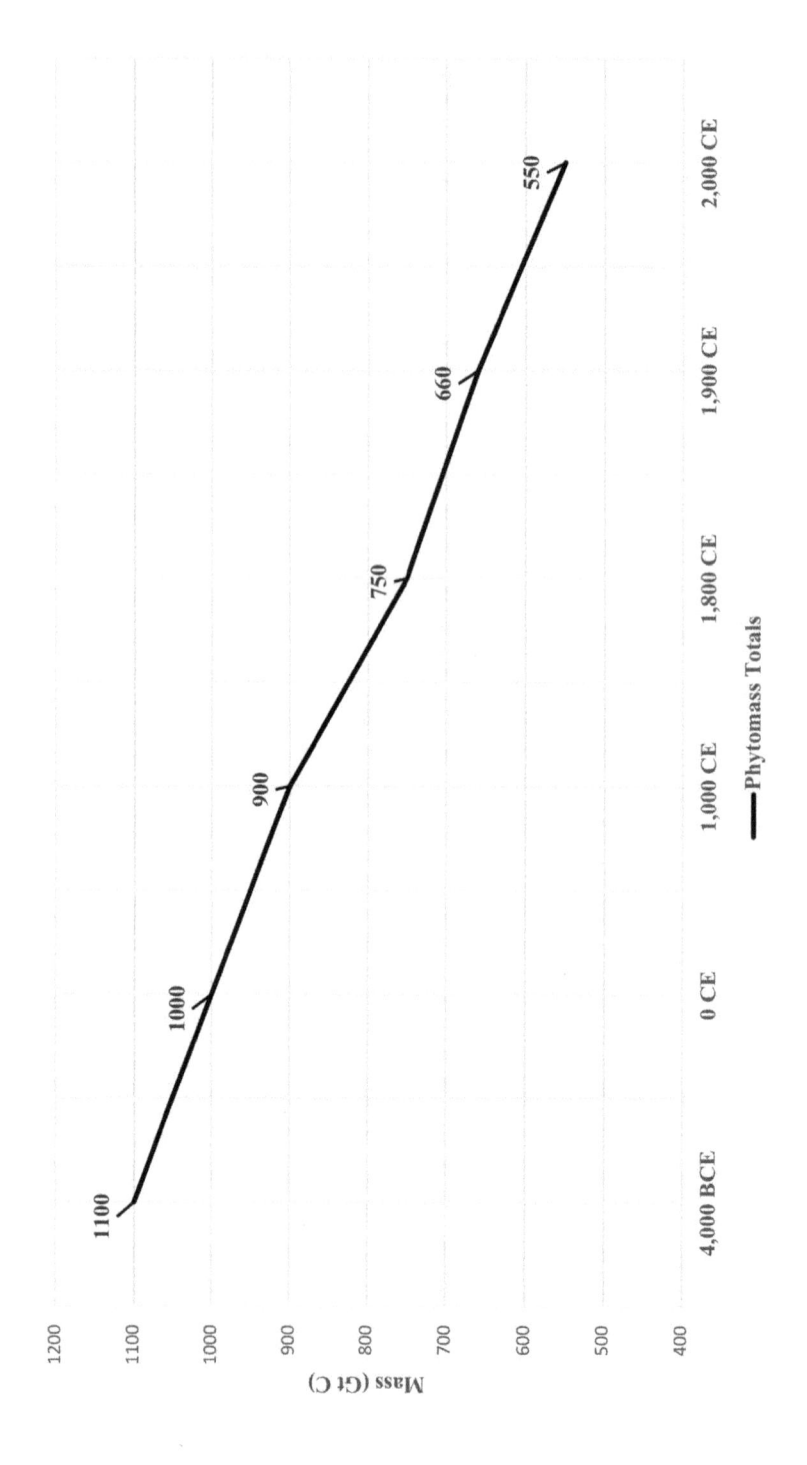

FIGURE 15.1 Declining phytomass totals through history in gigatons of carbon

Based on data from Vaclav Smil, "Harvesting the Biosphere: The Human Impact," *Population and Development Review* 37 (4): 613–36 (December 2011).

"wasted" lands were clear-cut, and biodiversity devastated in favor of grassy "sheep walks." Social expansion was always a social and ecological expulsion of migrants, both human and nonhuman.[9]

Today the fossil-fuel-using classes and their scientists talk about our epoch as one of "human" geological agency—as if all humans played an equal role in the disaster. In fact, in every major age of Western history, it has been a proportionally small group of powerful consumers that forced the rest of the planet to destroy the biosphere and one another through enslavement, migration, and genocide.

The data on this is clear: the history of phytomass destruction is directly correlated to the history of energy consumption by the fossil-fuel-using classes, the rise of capitalism, and colonialism. We can see in Figure 15.2 that the sharp rise in human energy expenditure is inversely related to the sharp decline in biomass in Figure 15.1.

Together, Figures 15.1 and 15.2 reveal a clear image of the decline in the capacity for planetary kinetic expenditure connected to the rise in human energy use, which required ecological destruction.

A geology of the present

One incredible consequence of the massive terraforming project of capitalist globalization is that it has resulted in one of the most mobile geological strata of all time. Humans have always been geological agents, but to varying degrees. All the patterns of motion from throughout the earth's history (centripetal, centrifugal, tensional, and elastic) are thus still present today, but in a new mixture.

A new minerality

Many human groups have thrown a wide range of different minerals into circulation in the form of vehicles (cars, trains, planes, boats), living bodies (humans, animals, plants), and space debris (rocket parts, satellites) across the surface, the sky, and space. Capitalists have dug minerals out of the earth, turned them into technologies, and moved them all over the world in a massive metastable geological stratum. Bones, especially those of farm animals, now compose an entire fossil mineral record. There is also a sedimentary layer of plastic and petroleum products buried below our feet and coursing through our bodies. There are more than 500,000 pieces of debris,

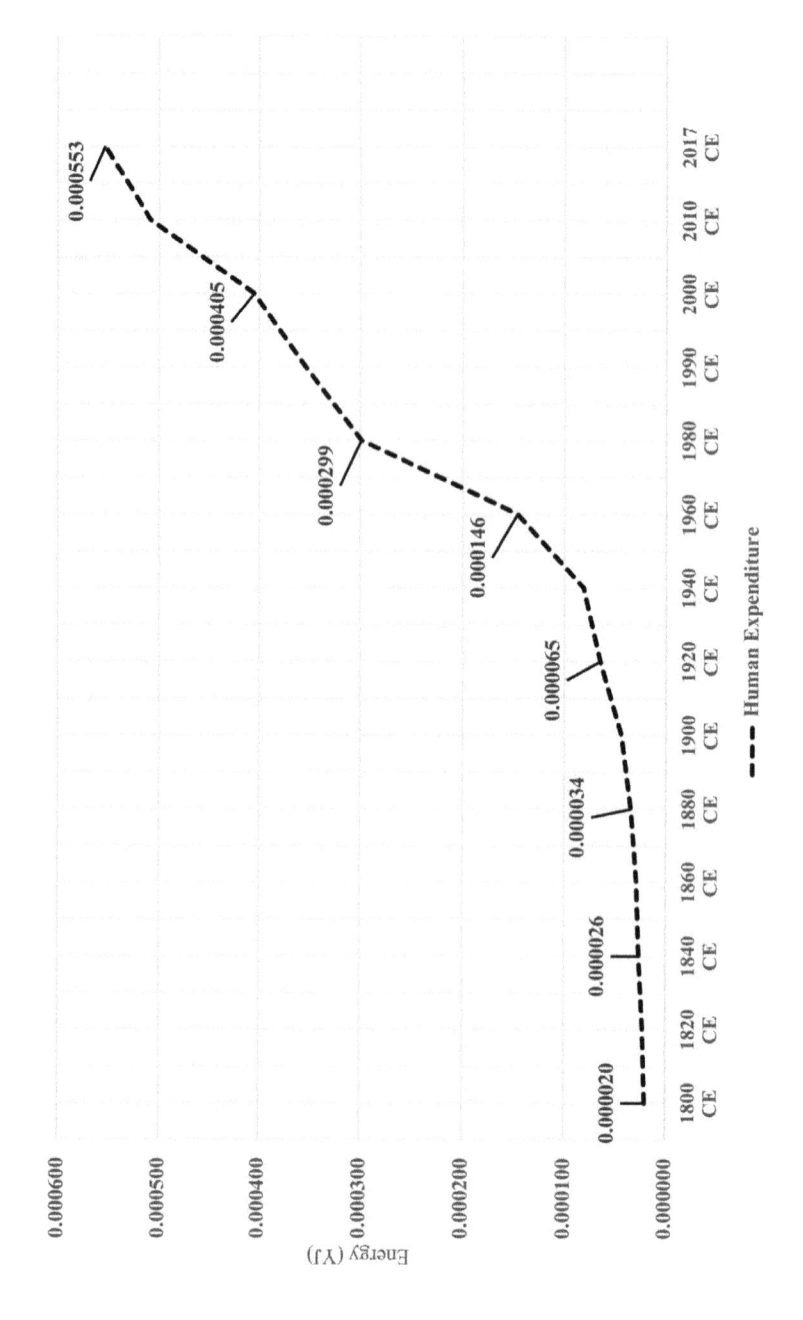

FIGURE 15.2 Human energy expenditure through history in yottajoules

Based on data from Vaclav Smil, "Harvesting the Biosphere: The Human Impact," *Population and Development Review* 37 (4): 613–36 (December 2011).

traveling at speeds of up to 17,500 mph, orbiting the earth. This is a new, highly mobile mineral stratum.

Additionally, as ice sheets melt from global warming, they take the weight off the land and trigger earthquakes. Seismic activity is increasing in Greenland and Alaska as a result. As seawater rises, the new weight bends the earth's crust and can increase volcanic activity, submarine landslides, and tsunamis.[10] Climate change is affecting the geological strata of the mineral earth.

A new atmosphere

The new atmosphere has layers of radio waves, airplanes (along with flying human bodies), various balloons, carbon dioxide concentrations, and ozone holes. Weather patterns are becoming more irregular, and some scientists now worry that climate change will lead to "cloud extinctions," fundamentally altering the atmospheric strata. We could soon live in a world without stratocumulus clouds.[11]

For every degree Fahrenheit warmer that the planet gets, the number of lightning strikes increases by 7 percent.[12] Increasing global temperatures also mean that the water cycle is moving faster, which means rising evaporation, heat waves, droughts, fires, and cyclones that try to dissipate that heat. As the atmosphere and oceans increasingly store our expended carbon, they also try to help dissipate that energy through massive storm systems and fires. But because they are trying to expend too much too quickly, they initiate feedback loops that decrease their long-term efficiency. Fires can be healthy for forests, but only when they are not too large or intense. Fires stimulate new growth in forests, which store more carbon. However, if the fires are too frequent or widespread, they will increase global temperatures faster than those temperatures can be reduced by new forests.

A new phytosphere

Climate change is not only destroying plant life; it is forcing it into migration. Plants are the first economic migrants. They, like the industrial proletariat, were seen by early English capitalists as unproductive, idle, squanderers of energy. The primitive accumulation or forced expulsion of peasants and indigenous people from their land went hand in hand with the forced expulsion of biodiverse plants and animals. The rise of mass industrial agriculture was both an ecological and a human genocide at the same time. The invention of the capitalist plantation was an enslavement of the ecosystem and of black and brown human beings at the same time.

The philosophical belief in the inferiority of "matter" is related to the treatment of the earth as passive, unproductive, and wasteful. This, in turn, is connected to the treatment of black people as coal and brown people as dirt.[13] For the English, colonization was not just a project of accumulation—it was a plan for the natural development of the productive forces of the earth that indigenous people, without good plantations to make use of them, were "squandering."

Plants, like numerous indigenous people, migrants, and refugees, are now in forced migration.[14] Capitalists ship plants around the world as food and fuel. They even burn their dead bodies as fossil fuel. A new phytosphere of "power-*plants*" and "*plant*-ations" is replacing the kinetic structure of vegetal life with smokestacks that release carbon but do not absorb it again.

Increasing carbon dioxide is also resulting in less nutritious food.[15] An entire stratum of vegetal life is now changing its nutrient composition, making it less and less able to support higher trophic levels.

A new animality

Contemporary animality has suffered a similar fate. Various human groups have destroyed more than 80% of wild land and sea mammal biomass, leaving only the most "productive" domesticated animals, locked into the factory farm: the *Plantationocene*. Wild animals, like plants, are being forced to migrate north to avoid rising temperatures and tides—or die. The bodily morphologies and organs of animals that are meant to increase kinetic expenditure are now being bred and sculpted into "productive" meat machines. As the elasticity of the human animal expands across the globe, the elasticity of wild animals is contracting into extinction.

Beyond anthropocentric energetics

Here is the problem: all the "increased" planetary mobility of this new geology is only an increase relative to human history and scale. Relative to human history and human energy use, there is certainly a dramatic increase in planetary and even energetic mobility. But humans are not the only mobile users of energy. They make up only .01% of the biosphere, .0000000000082% of the earth, and an unfathomably small percentage of the cosmos.

From the broader and more inclusive perspective of cosmic and terrestrial expenditure, human energy use/ecocide is causing a net reduction in motion and energy expenditure on the earth as a whole. The current global consumption of energy by some humans more than others, including the relatively minor increases in mineral, atmospheric, vegetal, and animal motion caused by climate change, is incredibly tiny compared to total planetary expenditure, which dwarfs human energy use by several orders of magnitude. In the figure below, human expenditure is not even visible on the graph of planetary expenditure (see Figure 15.3). This does not mean, however, that the consequences of ecocide have been negligible.

Ecological energetics, ecological systems theory,[16] energy accounting,[17] embodied energy theory,[18] and thermodynamics have done extremely important work to help put together the big picture of planetary energy use. Unfortunately, when it comes to thinking about human energy expenditure, energy scientists have entirely misinterpreted the ethical and philosophical consequences of their data. For example, physicists have interpreted the history of increasing energy consumption by certain groups of humans to mean that human animals are smart and energetically more "efficient" than nature. They treat human beings as evolutionarily, socially, cognitively, and technologically more advanced than other animals because of the sheer quantity of energy they expend.[19] This is a terrible mistake.

Fossil fuel capitalism is understood by some to be the highest form of human social organization because it allows for the largest possible consumption of energy.[20] Many physicists and economists even describe capitalist economics as following directly from the laws of thermodynamics.[21] Political liberalism is understood to be the highest form of political government because it maximizes the free energetic movement of all people toward increasing collective efficiency, productivity, consumption, and social-technological progress.[22] What energy theorists fail to mention is that, at the same time, liberal governments have committed some of the most egregious human and ecological atrocities—and in just the past thirty years![23] Air pollution alone, coming disproportionately from liberal capitalist countries, is the cause of seven million deaths

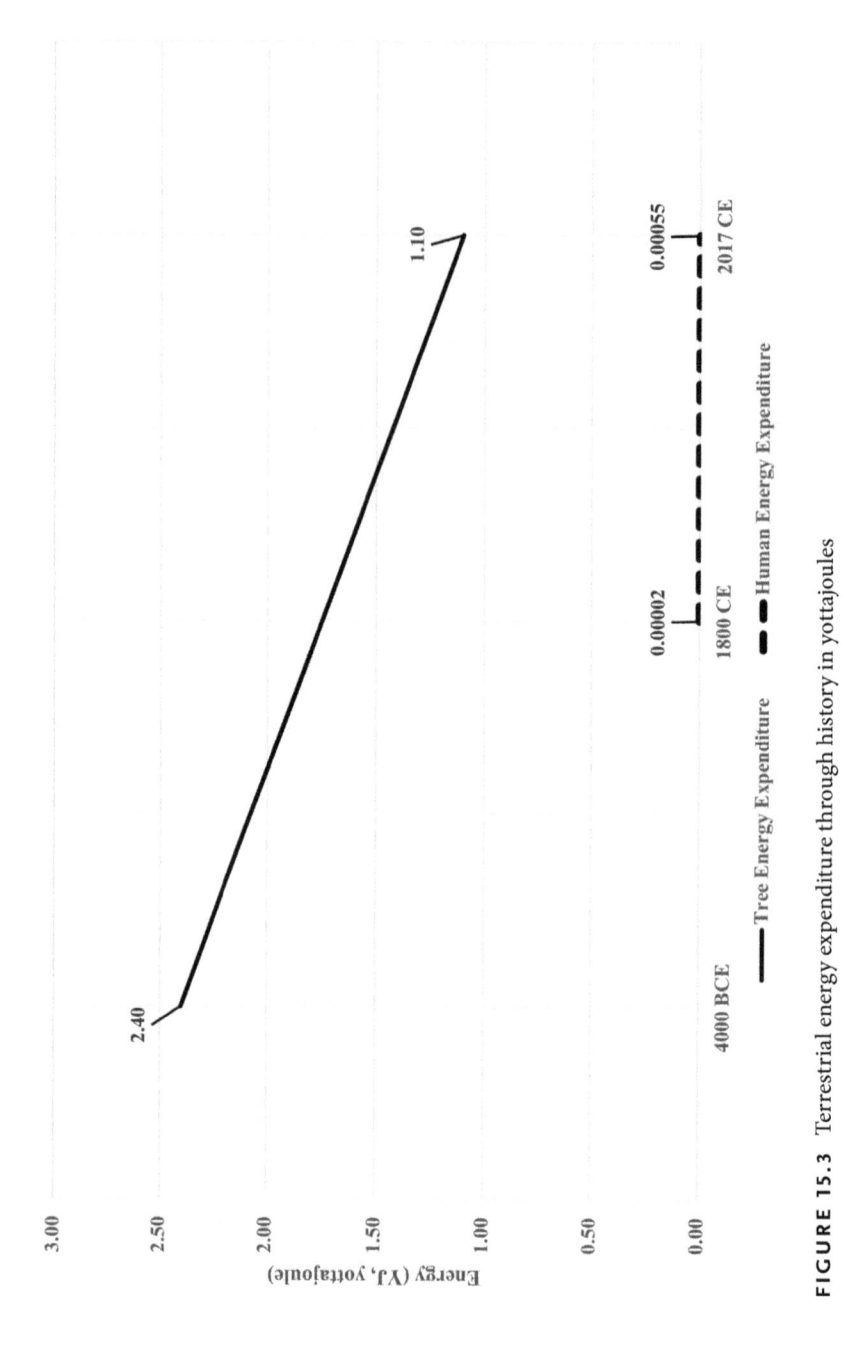

FIGURE 15.3 Terrestrial energy expenditure through history in yottajoules

Based on data from Vaclav Smil, "Harvesting the Biosphere: The Human Impact," *Population and Development Review* 37 (4): 613–36 (December 2011).

every year now and rising, largely killing people in developing countries.[24] Is this the "success" to be reaped from liberal capitalist energy expenditure?

Even Georges Bataille, the great philosopher of expenditure, was deeply anthropocentric in his analysis of *human* expenditure. This led him in the wrong direction, toward the valorization of warfare, human sacrifice, and, implicitly, capitalist destruction.[25] Humans are unique, for Bataille, because they expend the most energy and do so consciously. Religion is Bataille's name for the uniquely human awareness of our cosmic destiny to maximize entropy. But Bataille was triply wrong: 1) humans, as we showed in the previous chapter, do not (by far) expend the most energy (trees do); 2) other living and non-living material processes have made active mistakes that have caused mass extinctions, and humans are thus not unique in their destruction or salvation of the cosmos; and 3) the historical tendency of the cosmos so far is not to merely increase entropy, which will happen no matter what we do, but to increase the rate of kinetic expenditure—which is a unique challenge faced by all of nature.

Due to these three errors, Bataille and others have ignored the declining rate of the earth's total expenditure due to ecocide. This has allowed them to come to shortsighted, dangerous, and anthropocentric conclusions.

The historical tendency of nature to dissipate motion and energy is valid not for each aspect of nature in isolation, but only together, as entwined regions of an indeterminately changing and open whole. Humans are the cosmos continued by other experimental means to increase the expenditure of the planet and the cosmos, but not alone. There is thus absolutely no reason, other than pure anthropocentrism, for chaos and complexity theorists to treat humans as "successful" expenders of energy independent of their net effect on the planet and cosmos.

Energy theorists have understood the situation completely upside down. Biologists, on the other hand, understand well that each time an organism consumes energy, it expends an average 90% of that energy as waste. Each trophic level lives on the one below and expends nine-tenths of what it consumes each time, retaining only around 10% for what consumes it next. What organisms store as biomass is only a tiny fraction of a fraction of what life expends. Life tends to expend energy because the material cosmos does so more broadly. Nature loves to waste and to share its waste as much as possible. This is something our bodies "know" how to do. However, we tend not to consider it a real form of knowledge.

Unfortunately, most physicists and energy scholars are mostly interested in counting the tiny bit of energy or motion that is conserved and stored. Thus, ancient trees, at only 1% efficiency, are seen as grossly inefficient. Humans, at 16–20%, are understood to be better, and fossil fuels, at 30%, are considered ultra-efficient. Western culture and science have valorized efficiency, accumulation, and productivity under the assumption that nature and matter must be inefficient and stupid—thus requiring development and plantations to reach their true energetic potential. Kinetic expenditure is seen as an expression of the inferiority of the cosmos and evidence of the extraordinary superiority of human energy accumulation. We alone, with reason as our guide, can become efficient *users* of what nature so rapidly *wastes*. This is completely backward.

Notions of energy efficiency and environmental conservation are symptomatic of this same flawed framework, and we should alter them accordingly. Humans alone, in this model, are thought to be the unique consumers and conservers of nature. This is the prevailing definition of the Anthropocene, even though history and ecological energetics clearly show that numerous earth processes were destructive and restorative millions of years before humans.

When organisms and systems of all kinds suddenly begin to draw on a new dense energy source, there tends to be a period of extreme experimentation. What are the relative limits of how quickly and intensely organisms can consume energy without damaging the kinetic patterns that support them? Nature does not proceed along an a priori path to the fastest expenditure of energy, nor along a purely random one either. Nature experiments. Its first attempts are often destructive, as in the great die-off at the end of the Archean Eon caused by cyanobacteria expending too much energy in the form of oxygen. Certain human beings have also learned how to unlock relatively large amounts of energy from fossils, causing vast ecocide.

Planet-wide ecocide is the result not of too much planetary expenditure but of precisely the opposite: not enough! By maximizing total energy use for humans, the fossil-fuel-using classes on the earth have damaged the planet's capacity for longer-term optimal kinetic expenditure.[26] Their anthropocentrism has blinded them to the bigger picture of planetary expenditure. The effect is that they have ended up reducing the total net entropy of the planet, by destroying half the biomass on the earth, and, at the same time, have reduced the optimal rate of kinetic expenditure. As the earth's temperature heats up, it disrupts the kinetic patterns, such as water and carbon cycles, that are much more rapid systems of expenditure than anything humans could ever dream of.[27]

What is to be done? In the next section, I would like to argue for a dramatic change in perspective and ethical practice, away from this human-centered energetics. A different Kinocene is possible: a Kinocene not of extinction, but of *expenditure*.

KINOCENE EXPENDITURE

Expenditure is a defining feature of the cosmos as we know it. Kinetic expenditure is the tendency of matter in motion to spread out from higher concentrations to lower ones. It is a movement of death that occasionally makes life so that life may know death as well. The Kinocene extinction, as I described it above, is not the same as death. Extinction is the death *of death*. It is the loss of life's ability to die. In this final section, I would like to argue that what is needed more than ever today is an ethics of expenditure.

The ethics of death

The ethical valorization of life above death has wholly misunderstood the much larger cosmic situation in which our life occurs. Life is a kinetic by-product of non-living matter's movement toward increasing expenditure. The material meaning of life is to help improve the expenditure of the universe. The tendency of life is thus to die well together with others. An ethics of life, however, will not help us respond ethically to most of the earth and the universe—which are not alive. Furthermore, since we lack complete knowledge and control of even just our unconstructible planet, it is foolhardy to assume that we know or could know how to manage it, let alone the universe, properly, for the sake of life. And in fact, in the case of most of Western civilization, it has already been demonstrated that we do not know how to do that.

Death is different. Our bodies materially *know* how to expend energy and to die by wasting most of the energy they consume. Instead of trying to manage the world, they give themselves slowly and generously over to death. That is, they give themselves over to the rest of the world to consume and expend as best it can, without a pre-given plan for how to increase its rate of expenditure. In other words, the path toward increasing expenditure cannot be constructed in advance of our dying and living together as we go.

The ethics of kinetic expenditure is a complete inversion of the ethics of life. However, it is not a simple inversion. The inversion changes the relationship from one of increasing accumulation and instrumentalization to one of loss and generosity. Death is not an unfortunate part of life; instead, life is a part of death that reproduces itself—not in order to live but in order to die and help others die.

I do not mean to suggest with this ethical theory that the cosmos *should* die and that humans and other beings *should* help it. That would be a naturalistic fallacy. There is no such thing as an absolute should, anywhere in nature. The kinetic ethics of expenditure has a strictly hypothetical structure: if you want to survive and thrive in this particular universe as it has tended so far, then the best chance is to go with the flow of increasing your collective rate of expenditure.

Even this does not guarantee anything. Nothing controls nature, not even nature, which is continually experimenting, just like a bolt of lightning. A meteor can strike, or numerous organisms can cause mass extinctions. And perhaps some fossil fuel CEOs would rather live their individual lives in luxury at the expense of the long-term survival of the human species. But if the rest of us want to survive so our kind can expend itself with others into the future, then we have to quit our addiction to fossil fuels and abolish capitalism.

If we want to live and persist on the earth as one of its animals, then we have to give our lives generously back to the earth and help it hasten its dissipation. By undermining the efficiency of terrestrial expenditure (by biomass destruction, etc.), the fossil-fuel-using classes are not allowing the earth to die, nor to live so that it *can* die. The fossil-fuel-using classes are destroying the conditions of everyone's expenditure. They are accumulating energy to use in the name of life. The paradox is that the instrumentalization and preservation of life are now resulting in extinction.

The problem is not that too much of the biosphere is dying, but that not enough of it is dying in the form of energy dissipation. Life is capable of vastly more expenditure if it is allowed to live out its death and help the rest of the earth to do the same.

Again, this ethical inversion of life and death is not whimsy or polemic. Placing death first changes our orientation, toward both life and death, entirely. Life and death take on wholly new meanings. Life is no longer about accumulation and preservation, but expenditure, and death is no longer an end state, but the process of expenditure itself.

We can either become what we are (expending animals), or we can act as if our purpose is to conserve, accumulate, and instrumentalize life. The latter tends to produce extinction, the former to extend life. The paradox is striking and deeply counterintuitive from within the Western tradition. To live, we must give ourselves over to death, expenditure, and loss, trusting others to give back. The valorization of life (biopolitics) produces its opposite: extinction and death. Pascal could not have been more wrong in saying that "our nature consists in movement. Absolute stillness is death."[28] He should have said that "our movement lies in nature; we are its death."

If 6 trillion trees were still alive on this planet today, the earth would be expending twice as much energy as it is currently. The fossil-fuel-using classes are selfishly holding back the earth from its collective dissipation—its generous waste back to the cosmos. Whether we like it or not, we are directly participating in a larger cosmic drama. The cost of ignoring it and trying to hoard our own life will quite likely be our extinction. Perhaps the next round of post-extinction species will go with the flow instead of swimming against it, as certain human groups have—insisting that fighting against the current of nature is "knowledge" and "progress."

The gift that the Western fossil classes can give back is to surrender their arrogant belief that they know more about nature than it does. They can sacrifice homo economicus. They can generously give back their death, without purpose, without expectation of an instrumental return on their investment.

The question is not whether total human energy use is increasing or decreasing, but rather whether the rate of terrestrial and cosmic expenditure is doing so. Mere quantitative increases in human energy, above a certain level, are the cause of the current decline in the *rate* of expenditure. The destruction begins first for nonhuman life, then spreads to populations affected by extraction, and, eventually, affects the biosphere as a whole.[29]

Higher energy consumption by the fossil-fuel-using classes does not increase the global rate of expenditure but lowers it, because the energy used is not reused, composted, or further degraded.[30] Instead, the expended carbon is stored in the atmosphere and oceans, from where it is released by storms and fires that further destroy the earth's ability to efficiently increase the rate of expenditure.

Against conservation

This is why the idea of conservation in all its forms (ecological, economic, epistemological, etc.) is "upside-down reasoning," as Lucretius says.[31] The key is to increase the rates and patterns of kinetic expenditure in such a way that the patterns sustain themselves, so that they can keep on experimenting and expending, like an ancient forest. In other words, nuclear annihilation, fossil fuel capitalism, and any process that destroys efficient kinetic patterns are ultimately self-destructive. They are bursts of expenditure that cause a net reduction in planetary kinetic experimentation, diversity, and dissipation.

This is a counterintuitive and challenging notion: the survival of the biosphere and of human animals is not the primary meaning of their material, kinetic, and historical existence. The continued survival of life, this tiny portion of nature, is an aftereffect of cosmic and planetary expenditure. Life survives, strives, and reproduces, not directly for itself or its own utility, growth, expansion, and progressive evolution, but for the sake of something else entirely: the expenditure and death of the cosmos. The emergence and evolution of life with all its kinetic morphologies is not directly for the sake of development. Life is a gift given to the organism to intensify the expenditure of the cosmos and itself. By dying well, it gets to live well.

Conservation and accumulation are not the aim of life but are, rather, the consequences of death and expenditure. We preserve only to expend the more. This inversion is not a "new instrumentalism" that merely uses life "for" expenditure. Expenditure itself has no purpose or utility precisely because it is not a reinvestment with a necessary guarantee or with any equality of exchange. Death may give back the aftereffect of life, or it may not, depending on the greater relational experiment. One organism's gift of death does not decide what will receive the gift of life in return. This is a significant reversal that flips Western thinking on its head.

The gift of death thus has a dual meaning: life gives itself over to the cosmos in death, but death (non-living matter) gives itself, as a gift of life in cosmic reciprocity. Reciprocity is never an equal exchange, because the cosmos tends to move toward increasing rates and diverse forms of expenditure: *toward death*. There is no symmetrical balance. At the end of the universe, there is not necessarily homogeneous heat death, because matter is fundamentally in pedetic motion.

The cosmos, by giving itself the gift of death, also gives itself the gift of life by possibly contracting back on itself, reversing entropy, and producing a big crunch and another big bang. But since each life and death of the cosmos is related to its prior expansions and contractions through the initial and final states of quantum indeterminacy, the relation remains asymmetrical and indeterminate. The cosmos is relational but not determined, proportional, or symmetrical. The life and death of a cosmos thus mirror the life and death of all cosmoi in ongoing cycles of creation and destruction.

The logic of conservation, as a primary goal of ecological ethics, urges us to live a life of austerity, simplicity, and personal virtue. It rejects our urge to destroy, to waste, and to expend without utility or purpose. If humans want to survive, conservationists say, we must ration our resources, tighten our belts, renounce our desires for senseless expenditure, and further develop our higher rational and moral selves.[32] Deep ecologists and Earth First!ers even renounce human beings themselves, as a virus upon the earth. They say that we must depopulate so that the intrinsically valuable and naturally conservative biosphere may thrive for its own sake.

Thinking about ethics in terms of expenditure, however, is not a merely formal change in interpretation that leads us to the same practical activity of conservation. The logical conclusions of conservation ethics—such as population management, the preservation of wildlife at the expense of Third-World lives, the "voluntary human extinction project" of human sterilization, and the reduction of human activity to mere survival—do not follow from the ethics of expenditure. The Kinocene ethics of expenditure is quite different from the conservationist logic. The logic of conservation is a logic of renunciation, asceticism, and negation.

Moderate and radical environmentalists alike, even when they rail against anthropocentrism, remain fundamentally human-centered and rationalistic in a sense. They project onto all life the ideal of conservation, self-preservation, and biological survival because humans are alive and seem to culturally value conservation. However, conservation and preservation are simply not what nature does, 90% of the time.[33] What we call "conservation" is an indirect byproduct of expenditure.

Life is one of the most profligate expenders and consumers of energy per mass that we know of—at a rate per area that is many times greater than that of the sun itself! Trees, for example, show us precisely what they think of the idea of conservation, consuming yottajoules of energy just to excrete 99% of it out again. Trees "conserve" 1%—just enough to preserve themselves so that they can keep wasting. Naturalism is not consistent with conservation.

Capitalism and anthropocentrism are destroying the planet, and we ought to stop them if we want to survive. However, the logic of conservation follows the same instrumental logic as capitalist bio- and anthropocentrism, only directed toward other ends. The results of deep ecological, Neo-Malthusian anti-humanism have been and will be devastating for women and people of color around the world.

Instead of doing what we *think* trees do (conservation), we should (if we want to survive) be doing what they and our own bodies are *truly* doing. Trees are all wasting themselves into the world in acts of generosity, only taking what is given back to them, in reciprocity, to squander it again. It is hard to let go of the drive to manage and construct nature toward what we imagine is a better goal, or what we believe is consistent with the benevolence of Gaia. However, that is what the ethics of kinetic expenditure asks of us.

In Greek mythology, Python, the dying serpent, was the voice of Gaia. The critical idea here is that the truth of the earth is spoken and known through the act of dissipation (the rotting serpent Python). Like the priestesses of Delphi, we have to listen to the earth and follow its lead if we want to survive.

Reciprocity

Reciprocity in this context means giving as much as we can in expenditure and taking back what nature gives in return. Conservation is the tiny aftereffect of this more primary kinetic process, which saves just enough to keep on expending. This means that reciprocity is fundamentally collective, or held in common. It means that a kinetic ethics for the Kinocene is about dying or expending well together. It means that everything on the earth should waste or throw into circulation as much as it can, but not so much that it undermines the larger capacity of the earth and others to do the same. Everything on the earth, living and non-living, helps the rest of us to dissipate well together. That is the generosity of death, which gives the gift of life.

The problem is not that fossil capitalism wastes too much, but that it does not waste enough, and that it does not take enough from the earth. Pumping carbon into the atmosphere and destroying the earth's biomass destroys the conditions of increased intensity of expenditure, and it does not take the gifts of carbon offered back in return. Fossil fuel consumers are expending carbon into the atmosphere, and into the oceans, in enormous quantities: 2×10^{23} joules total over the past twenty-five years into the sea alone. This is the energetic equivalent of three atomic bomb explosions every second for twenty-five years.[34] This prevents the earth both from expending and from receiving. It disrupts the reciprocity of the carbon cycle, whose ultimate aim is to release energy back to the cosmos.

Reciprocity means giving to others so that they can again give to yet others—not destroying the conditions of giving. Reciprocity means taking what nature gives, not refusing it, or destroying the conditions of the giving. It is a cruel irony that the very stuff of life, carbon, has now become the great herald of extinction: the death of death.

Against capitalism

Capitalist economics is premised, in direct contrast to this kinetic reciprocity, on the fundamentally false notion of equality of exchange: equivalence. Note that in nature, there is no such thing. Nature is neither identical to itself at any point nor identical between points of itself, if for no other reason than because of spatiotemporal difference. Matter always flows asymmetrically, entropically, and in metastable patterns of increasing disequilibrium. Equivalence and equilibrium are, physically speaking, violations of the historical tendency of the universe, which is to dissipate and spread out kinetically.

By acting as if equivalence, equilibrium, identity, and exchange were real aspects of nature, economics and capitalist economics, in particular, have significantly damaged the earth's capacity to increase its rate of expenditure. When one acts as if nature moves in one way, when in fact it moves in another, massive disruptions in those motions occur. Capitalist constructivists have acted as if they could invent or construct a set of rules on top of nature and live in their own reality. They are like someone swimming upstream who insists that's the easiest, most natural way.

Classical, neoclassical, and orthodox economic theory also act as if economic exchange were a reversible process, when, physically speaking, it is not. The philosophical assumption of economics since David Hume has been that scarcity is the basis and starting point of economics, when in fact, as we have shown, nothing of the sort exists in nature. The ideas of equivalence, equilibrium, reversibility, and scarcity are false—meaning they have not been physically found in nature so far.[35]

Classical and neoclassical economics also believe that production and consumption are circular. They think that what humans produce is what they consume, exactly, without expenditure, externality, waste, or degradation. This is, again, thermodynamically impossible. Movement always tends toward increasing dissipation. As Harold Morowitz says, "Energy flows; matter cycles."[36] As matter cycles, energy flows through it. What is consumed is never identical to what is produced.[37] Therefore, there can be no equality of exchange, physically speaking.

By acting as if a commodity were strictly identical to its exchange value (how much money someone exchanges for it), capitalist economics has failed to consider the environmental impacts of deforestation, pollution, and climate change. It has also ignored the human implications of social devalorization (in such forms as racism, sexism, and classism) as integral and constitutive aspects of the economic process. These aspects have no value in capitalist economics.

As Marx rightly says, capitalists act as if the product were abstracted or independent from the process that produced it. The monetary value of an old-growth tree is reducible to 1% of its energy, which is usable for humans as biomass or "lumber." The rest of the 99% of its expenditure is "wasted," without any value. If we assigned even a modest monetary value to the tree's energy expenditure, or to women's domestic labor, or to migration and human displacement (for example), profit would be impossible. This does not mean that we should try to assign monetary values to natural processes! It means that the whole idea of capitalist economics and "value" is utterly absurd. Value necessarily requires the constitutive exclusion of the material-kinetic conditions that support its abstract exchange process.[38]

Because it privileges life, accumulation, conservation, and utility, capitalism excludes and destroys everything that it associates with death, expenditure, reciprocity, and non-useful waste. Hence the apparent and logical necessity of ecocide, indigenous genocide, slavery, patriarchy, forced migration, and biopolitics.[39]

Compost ethics

Composting works by a different ethos. It is the expenditure of expenditure. In this sense, it is the gift of death: the opposite of capitalist utility and accumulation. Composting shares used and wasted matter with mineral, atmospheric, vegetal, fungi, and animal others, to be used and expended again and again. Each compost cycle further degrades and dissipates energy as it increases the capacity of the earth for further degradation.

For example, the composting of garden and food scraps allows for the creation of new fungi and bacteria. Once they break the compost down, it is returned to the soil to increase the symbiotic relationships among fungi, bacteria, and plants in the ground. Humans then eat a portion of the compost that becomes food. They burn most of it into the air as heat when they move, and their solid waste can return to the earth to be broken down again by bacteria.

The problem is that when we bury our "trash" in landfills, it actually slows down or even arrests the composting process. The invention of non-biodegradable petroleum products is the modern expression of our fear of death and our quest for immortality. We will not even let our waste die. Landfills deny the gift of death and, thus, the gift of life. They prevent expenditure by preventing the expenditure of expenditure.

Compost is not about conservation; it is about waste, shared among a wide community of consumers who thoroughly and efficiently, and faster, break down and expend the energy of the waste back to the cosmos, as heat loss into space. Recycling and reusing is not conservation; it is a way to extract and expend and share matter with the earth more generously, so that we can all die and dissipate well together.

Carbon sequestration is also composting. The oceans, atmosphere, plants, and animals all sequester and compost carbon. Biological sequestration may even be responsible for regulating the earth's climate over billions of years, as the sun has continued to get larger and hotter.[40] The problem is that the fossil-fuel-using classes are destroying our compost piles by overfilling them. Every backyard composter knows that active compost produces heat, but dropping an atomic bomb into your compost would produce so much heat that it would destroy the entire metastable process of kinetic degradation. This is what we are currently doing.

As the ocean stores more heat, it begins to swell by thermal expansion, causing sea-level rise, increasing erosion, and increased nitrogen and thus further increasing temperature in a terrible feedback loop. Eventually, as in every historical greenhouse age, the cycle will run its course and reach a new metastable state. But if we want to live through the process, we need to start composting much more, much faster. This effectively means that we need to facilitate an increased rate of expenditure for existing waste, not conserve it.

Diversity
Social and ecological diversity both increase the rate of kinetic expenditure. Old climax-stage forests are the largest dissipative systems on the planet *because* they are the most diverse. In old, diverse forest and jungle ecosystems, each organism lives from the waste products of others. The more ways of living there are, the more ways there are to consume and expend different energy sources: thus the very rapid intensity of expenditure. No single organism can use all energy sources, so nature requires an army of specialized levels, relations, and singular techniques in order to capture and fully degrade and dissipate 99% of all incoming solar energy.

However, high rates of energy expenditure, as in climax ecosystems, should not be confused with "high" *total* energy consumption, as with fossil fuels. Fossil fuel use, for example, does not necessarily require or create social or cultural diversity. Choosing between different kinds of commodities is not the same as social or ecological diversity. Where fossil fuel energy expenditure is the largest in quantity, we see some of the highest gender, race, and class disparities in wealth and education.[41]

Humans are natural, and thus human diversity is part of ecological diversity. Technologies created by humans are also natural and, therefore, also part of environmental diversity. Human plasticity is at the same time social, technological, and ecological. The more kinds of people and technologies there are, living in the more diverse ways, the more thoroughly we will be able to degrade and expend planetary energy. However, this is true only as long as some of these ways do not reduce the capacity of others (living and non-living) for diverse dissipation.

Human diversity is also like a dendritic lightning strike that tries out living in different ways, in different places, with different techniques, cultures, and technologies. The more forms of thinking, doing, and being there are, the more varied the methods of energetic expenditure. Just as there is no one organism or way to extract energy in an ecosystem, there is no one human social ecology either.

If humans want to survive, they need to increase the *rate* of long-term planetary expenditure. To do that, they will need a plurality of techniques that respond to the singularities of their geography. In other words, humans need to multiply their values, habits, technologies, and cultures with and as part of ecological diversity more broadly.

Global urbanization, extractive capitalism, and climate change are all forcing people to migrate from diverse geographies and ways of life into cities with common economic, legal, and political forms. A new ethics for Kinocene expenditure must also be one that supports the free movement of people and their right to stay home and resist extractive capitalism. Decreasing biodiversity and rising global temperatures due to extraction and plantation agriculture also increase political conflict, social instability, and forced migration.[42]

Colonialism has eradicated and continues to eradicate diverse indigenous peoples, who also tend to live in some of the most historically diverse and dissipative ecosystems on the earth. Ecocide is thus fully entangled with the genocide of indigenous peoples who rely on and identify themselves with their land. The killing off of the buffalo in North America, for example, affected the entire ecosystem and the indigenous people on every level.[43]

Fossil-fuel-based plantation agriculture also reduces the net expenditure of the earth. The homogenization of seeds by Monsanto and industrial farming not only limits the diversity of farming conditions but also creates mono-crop plantations that are vulnerable to disease and climate change.[44]

Rural and indigenous women are particularly vulnerable to ecological destruction because they depend on forests and common lands for food, fodder, water, and building materials.[45] For many women, destroying their forests is equivalent to killing them and their families.[46] When crops fail, it is women and girls who are disproportionally affected by famine.[47]

In addition, the institutions and habits of patriarchy, by restricting what everyone, but especially women, is allowed to do with their bodies and lives, homogenize a range of gendered and sexual diversity and ways of expending energy.

Art, culture, and sexuality are also techniques that expend energy. When we restrict freedom and experimentation in those realms, we end up with, again, fewer ways for humans to waste their metabolic energy. Nature is queer and loves to multiply sexes, genders, and sexualities to see if new ways of life might increase the rate of expenditure.[48] If we want to survive, we need to find new and diverse paths to enjoy the expenditure of our energy—as long as these ways do not destroy the conditions for further experimentation and expenditure for others.

The subsistence farming, hunting, and gathering that takes place in many countries looks like poverty to capitalists because it has little or no monetary value on the world market. However, by using, expending, and composting on traditional lands, indigenous people support higher rates of ecological and planetary expenditure.

If we want to find new sources of compostable energy, we need to support as many different ways of living as we can that do not destroy the possibility of this "kinetic pluralism" itself. The reduction of ecological diversity also destroys human ways of living, speaking, and knowing. Fossil fuel migrants are forced out of their homes to make way for plantation agriculture and energy extraction and, at the same time, are forced into the fossil fuel economies of Western receiving countries. There is thus a double destruction of human and ecological diversity, and a double decrease in planetary expenditure, as sustainable human/natural ecologies and the people who live in them are forced to migrate and reproduce fossil fuel societies.

By allowing more of the world's population to eat, and not starve, supporting as many diverse ways of living as possible would mean an increase in the planetary rate of expenditure, since human metabolism is incredibly efficient relative to its mass. Poverty, capitalism, and colonialism are thus antithetical to human diversity and human expenditure, since they keep people from dissipating a surplus of energy. The real primary problem is not human population levels, as some environmentalists claim, but fossil fuel capitalism and the patriarchal control over women's reproduction.[49]

Social domination, from a kinetic and thermodynamic perspective, is a reduction of the human ability to experiment freely with new and different ways of expenditure. There is nothing in nature (no great chain of being) that justifies the social domination of others. In nature there is only the tendency to allow increasing experiments in motion and expenditure that do not undermine further experiments toward increasing expenditure.

Metabolic communism
The ethics of expenditure in the Kinocene also entails a new relationship with the "commons." Historically, humans have understood the commons to be the common pool of resources to be managed collectively. Under capitalism, these resources were directly appropriated through primitive accumulation (by theft, murder, colonialism, extraction, and privatization). Although quite different, both of these approaches to the commons have understood the commons to be a "pool of resources" for *human* use (whether managed, conserved, or devastated).

The kinetic theory of the earth requires a much deeper and broader approach to the commons. If we want to survive, we need a materialist commons, where we treat each main pattern of the earth's motion as a commons for the others. Together, all the earth's material processes are sympoietic with one another and thus "manage" one another and themselves as commons for others. We need a reciprocal commons, in which each body becomes the commons of others, to be directed together toward increasing expenditure. Thus, instead of "the commons," we should be thinking about the kinetic process of "commoning" as the making and remaking of a *material* (and not just multi*species*) commons that we are continually expending, composting, and transforming.[50]

This new commons, or what I am calling "kinetic communism," would be based not on conservation but on expenditure. What if we all managed and shared one another's food and waste not for the sake of hoarding, rationing, or saving it but for the sake of wasting it into space? Conservation is only a tiny and temporary stage required to sustain the most optimal expenditures our planet can support.

This is not only a terrestrial task but a cosmic one. The metabolic commons is also a cosmic commons in which the sun, solar system, and broader universe participate directly in self-degradation. The cosmos is a commons for itself in order to unravel itself. The earth is a commons used by the sun to degrade its energy, while the sun is a commons for the earth to expend back into space.

There is no single "best" way to manage our commons. The more ways we try together and the more we allow new material evolutions to experiment, the more likely we are to optimally expend energy. The more tributaries of a river, the more rapid its diffusion. Furthermore, since the earth is always changing, so too should our strategies for diffusion. Kinetic or metabolic communism is the process of preserving the commons just enough to optimize its expenditure and increase its processes of differentiation. Nature continually struggles to become other than it is, in order to expend itself faster.

Kinetic communism would entail the practical collective metabolic management of the whole planetary form of motion. Kinetic communism is nothing other than the practical and commonsense joint management of the threefold (natural, human, and social) metabolic and material process in such a way as to avoid destroying its own conditions for collective planetary expenditure.

If this kinetic ethics had to be put into three memorable and straightforward ethical imperatives, in the fashion of Epicurus, here is what they would be:

If we want to survive, *then* our best chance is to follow these guiding maxims:

1) Increase planetary expenditure.
2) Compost everything!
3) Increase diversity.

Across the material spectrum, these are the conditions under which increased planetary expenditure takes place.

* * *

I am sure people will misunderstand this ethics of expenditure, so let me try to clarify two points:

1) I am not deriving a universal "ought" from a naturalistic "is." The ethics of expenditure is strictly hypothetical and experimental: "*if* we want to survive, *then* our best chance is to try to increase collective planetary expenditure."

2) I am also not saying that all human ethical practices can be explained by or reduced to the ethics of expenditure. Human ethical practices are not unimportant or irrelevant in the face of cosmic and planetary processes; quite the opposite. Human ethics are part of nature and its processes—something we often forget. Increasing natural diversity also means increasing the diversity of human values, ethics, and ways of being. The ethics of expenditure is thus not meant to resolve every ethical dispute among humans. It aims to situate human ethical activity itself in nature, in order to curb our present problem of rampant human-centric behavior and constructivism.

The ethics of kinetic expenditure is not a universal ethical ground but a hypothetical ethical ground. It allows us to say not only that capitalism is descriptively wrong about nature, but also that it is unethical (assuming that we want to survive), because it leads to the reduction of planetary expenditure (including the reduction of human and ecological diversity and flourishing).

Furthermore, the ethics of expenditure relates to the material conditions of all human society as such. If we even want to have humanistic ethics in the first place, there must be humans alive to practice it. Thus, implicit in all humanist ethics is the assumption of planetary existence and survival. In short: If we want human ethics, then we need to be alive and survive, and if we're going to survive, then we need to try to increase planetary expenditure (with all that entails). Kinocene ethics, then, is perhaps a "metaethics."

I am not interested in valorizing death over life or nature over humans in a way that would reduce everything to merely meaningless natural processes—thus undermining social critique and human ethics. I am trying to bring naturalism and humanism back together in order to avoid the false dichotomy between human values and valueless nature. Human ethics does not occur in a constructivist vacuum but under specific natural and historical conditions that are directly relevant to our living, and living well. This is what I have tried to show.

Conclusion
The Future

THE WEST'S HISTORICALLY mistaken belief in a stable and predictable earth is one of the biggest mistakes ever made. This mistake is symptomatic of a similar preference for stasis in politics, ontology, science, and the arts.[1] Together, the preference for stasis and stability across the major domains of human knowledge and activity is the source of all contemporary global crises.

Only now, living in the face of such a dramatically unstable and mobile planet, are we forced to admit that the West was wrong. What we called "progress," the triumph of humans over nature, form over matter, and stasis over mobility, was just wanton destruction and domination. Movement and expenditure had always been primary. Human history was not the progressive realization of static forms. Narratives of progress and development in the Western tradition are dead now. We can now see human history for what it is: a series of kinetic patterns formed in the material diffusion of the cosmos.

For too long, the Western tradition has treated philosophy as occurring only in the minds of humans (and mostly white male humans at that). It is time to return thinking not only to all its dispossessed but to its cosmic vocation as well. Philosophy and thinking are a technique of expenditure. Human brains and bodies materially evolved to increase expenditure in profligate activities like philosophy. This brings philosophy closer to its shared condition of material expenditure with the rest of the cosmos. It makes philosophy a profoundly cosmic activity at the same time as it makes the cosmos itself philosophical—that is, capable of philosophy.

The critical history put forward in *Theory of the Earth* completes my six-volume series on the philosophy of movement. The first five volumes dealt with the theme of matter in motion in Western and Near Eastern human history, but the sixth situates them all in the much deeper and broader history of the cosmos. It brings them all together in material and cosmic history, situating them as kinetic patterns of diffusion evolving with everything else in the universe.

A MATERIALIST THEORY OF DEEP TIME

"I love the ignorance concerning the future," Fredrich Nietzsche wrote.[2] The ignorance of the future is not ours alone as humans. We are ignorant of the future because the cosmos itself is ignorant of the future. We do not and cannot know in advance the fastest way to reduce a kinetic energy gradient. The patterns of motion we witness in the cosmos (spiral, dendritic, centripetal, centrifugal, tensional, and elastic) are not forms in the mind of God but a kind of material knowledge in the body of the cosmos.

Nor are these patterns absolute. There is no a priori reason why nature could not invent new fields or patterns in the future. The continual iteration and inventiveness of the cosmos is not mental but material and kinetic. The cosmos knows and remembers how to move toward dissipation. It is a truly material memory, stored everywhere and in everything.

The past

The cosmos knows and remembers because it is nothing other than its past transformed and arranged differently. If matter is neither created nor destroyed, but only rearranged, then the matter of the deep and cosmic past is still around. If the universe is expanding. but not into some as-yet-unpopulated empty region or time, then its expansion is nothing more than a fully immanent transformation of the whole universe.

The cosmos is the material kinetic process of "spacetiming"; it is the kinetic producer of space and time themselves. If spacetime is a feature of the quantum world and not a static a priori fabric, then time is neither linear nor cyclical, but *kinotopological*. If time is entirely material, then the cosmos folds and unfolds itself like a crumpled piece of paper, filled with differentiated regions that we call past, present, and future.

The present is not a point on a line separate from a previous point. Where does the past go when the present happens? The materialist answer is: it does not go away outside the cosmos; it does not lose its spatiotemporality, but nature retains it in a different kinotopological pattern. History and historical ontology are possible precisely because of this. The act of human inquiry is not separate from nature. We *are* the past. We are what the past is doing, not "now" (in time) or "here" (in space) but "in this way" (in motion).

The kinetic theory of the earth and the cosmos leads directly to this strange and radical conclusion: that time is not a linear flow from low entropy to high entropy, but rather an ongoing evolution of kinetic patterns tending toward dissipation and dispersal. The past is not a previously less entropic state of randomly moving discrete particles. The present is a region of the past itself.

Of course, no dinosaurs are alive today. However, we can say that the material that made up the dinosaurs and their elastic patterns of motion are still here with us. They are material memories in us. They compose our bodies, fuel our cars, are destroying the biosphere as fossil fuels, structure the bodies of birds, and so on. Dinosaur bodies did not go away but, as the material past, still compose the present. Our ancestors, both human and more-than-human, are really and materially with us, as us, in us.

The future

The future is also not an abstract point on a thermodynamic line. Kinetically speaking, we are currently performing our future now. It is a metaphysical abstraction to think of the future as a point like the present, but beyond the present. It does not exist. However, the future, like the past, is merely a kinetic reconfiguration of us. The future is relational without being entirely determined by the past or present. The future is not outside the cosmos. There is nowhere for it to be right now because it is a continual deformation, dissipation, and transformation of the cosmos itself.

Our ignorance concerning the future is, therefore, not a metaphysical ignorance about a future moment currently not in space and time. Our ignorance of the future entails an immanent ignorance concerning our present, which is also the future, or is becoming the future, I should say. The so-called "arrow" of time is thus more like an "eddy" of time, where matter flows in and out transforming, deforming, and dissipating itself into asymmetrical regions.

Time is neither created nor destroyed, but rearranged. Even black holes, those cosmic compost piles, radiate back out again. Time is not an ideal image of a line in our heads; it cannot be anything other than immanent and continually rearranged, like a vortex spinning itself out.

The care of the future

The Western idea of the future as an abstract point further along an arrow or line is a metaphysical one—and is perhaps one reason why ethical reciprocity and relation seem so challenging. Only in a material-kinetic and relational theory of time is it possible to care for the future. The future is not beyond us, but immanent and within us. Therefore, caring for the future means caring for ourselves and all our relations now. Caring for the earth and the cosmos by composting and reciprocating our gifts of death is inextricable from the care of the future and the past. Material evolutions and transformations now are changing the immanent future of those bodily components.

An ethics for the Kinocene is relational because time itself is relational. Caring for unborn grandchildren is a metaphysical and moral abstraction unless we have a thoroughly materialist theory of time that is fully relational and kinetic. We are both our ancestors and our unborn children. What increases planetary rates of expenditure for us now (in the broadest possible planetary sense) is also immanently part of the future. The care of the future is thus ongoing, relational, processual, and kinetic.

Limits and limitations

It is essential to observe limits, if only to then cross them even more gloriously and extravagantly. That is the case with this book, which has several notable limitations that I wish to acknowledge as limits I have observed here, which are yet to be crossed.

The first is that I have drawn on numerous sciences and scientific studies without thoroughly interrogating each science or study in depth. If I had, this would have been a completely different book. This is bound to raise the eyebrows of science-leery humanists, who may have preferred to read a critical history of science before accepting any scientific claim at face value, no matter how universally accepted it might be in scientific communities.

For these readers, I have already written a book on this topic, entitled *Theory of the Object,* which provides a kinetic theory of the history of science, mathematics, logic, and technology. My usage of scientific literature here is not naive or uncritical. It is based on a kinetic interpretation of science and the techniques of objectification developed in *Theory of the Object.* Nor is my usage of scientific literature ontological. I do not believe that the sciences or physics, in particular, have all the final answers about nature. My usage and interpretation of science are strictly *historically* ontological, meaning that I treat science and *Theory of the Earth* as real immanent knowledge practices *of their time.* Thus, I claim for *Theory of the Earth* nothing more than that it is a natural history for its time, which describes one vantage point on the history of the cosmos, nothing more. This is not a metaphysical project.

Second, *Theory of the Earth* is limited to the study of only the most dominant historical patterns of motion on the earth, considered separately. In real history, by contrast, all the regimes and fields coexist and mix to one degree or another. To show all such mixtures and degrees for each historical period would have been too large a task here, however, and must be reserved for future studies. This book, therefore, considers only the dominant distributions of the earth, and only during the period of their historical rise to prominence. The purpose of this book was not to tell every cultural story about the origins of the cosmos and the earth, but explicitly to unsettle and disrupt my own: the dominant Western one.

Third, this book is not attempting a systematic chronology of the earth's history. Numerous textbooks have already done this well enough. My aim, instead, was to look at broad tendencies and kinetic patterns that have emerged, without being accountable for every single event in the earth's history. What is unique about *Theory of the Earth*, and where I hope readers will evaluate its contribution, is its focus on the hidden kinetic structures operating within the history of the earth, which reveal a subterranean geokinetics.

This limitation is not reductionistic. My argument is not that the mobility of the earth is the only or best way to understand it. Instead, the argument here is historical and aims to add another interpretive dimension to others already out there, from the situated perspective of the early twenty-first century and specifically in the context of the advent of climate change.

Future directions

Future work is needed on all these limits and limitations and perhaps more. For example, one could take a closer historical approach to the scientific and geological practices themselves that lead to the knowledges we currently have about the earth's history. Here we could look at the ontological, political, scientific, and aesthetic dimensions of the human study of the earth and the cosmos itself.

Another direction would be to show in much more detail how all these prior kinetic patterns in the earth's history now mix in the present. I have only intimated a bit of this in my description of the Kinocene and of a few of the significant ways it has changed each mineral, atmospheric, vegetal, and animal pattern of motion. But much more is required for a full contemporary geokinetic analysis.

Finally, as history unfolds, our knowledge of it changes along with our situated perspective. No kinetic history can be a final one. It is only one of many perspectives glimpsed in a multidimensional gestalt switch, where instead of just two (rabbit and duck), there is a vast multiplicity of aspects to be found—all of them real aspects of the cosmos.

My hope with this book is not to have finished something but to have started something—to have seen some patterns previously unseen, which may be important to our contemporary moment and future. Just as someone might show a new constellation by connecting star to star up in the sky with their finger, I hope *Theory of the Earth* has shown the reader some old stars in a new way. My aim was to start with our contemporary ecological crisis but then to resituate it in a much larger, deeper, and stranger philosophical framework, such that the set of relevant problems and solutions might change entirely, and humanity might see a way to live up to its cosmic potential.

Notes

Introduction

1. See chapter 15 of this book for the contemporary political consequences of the earth's instability.

2. Craig Welch, "Half of All Species Are on the Move—And We're Feeling It," *National Geographic*, 27 April 2017. https://news.nationalgeographic.com/2017/04/climate-change-species-migration-disease/

3. For a review of the various "-cene" designations and their shortcomings and strengths, see Jairus Grove, *Savage Ecology: War and Geopolitics at the End of the World* (Durham: Duke University Press, 2019).

4. Paul Crutzen and E.F. Stoermer, "The 'Anthropocene,'" *Global Change Newsletter* (2000) 41: 17–18. The rhetoric of the Anthropocene often makes it sound like all humans are equally responsible and equally vulnerable when they are not. See Dipesh Chakrabarty, "Postcolonial Studies and the Challenge of Climate Change," *New Literary History* 43, no. 1 (Winter 2012): 1–18.

5. See Kathryn Yusoff, "Anthropogenesis: Origins and Endings in the Anthropocene," *Theory, Culture & Society* 33, no. 2 (2016): 3–28; Timothy LeCain, "Against the Anthropocene: A Neo-Materialism Perspective," *International Journal for History, Culture and Modernity* 3, no. 1 (2015): 1–28; Jason W. Moore, *Anthropocene or Capitalocene? Nature, History, and the Crisis of Capitalism* (Oakland: Pm Press, 2016); Astrida Neimanis, Cecilia Åsberg, and Johan Hedrén, "Four Problems, Four Directions for Environmental Humanities: Toward Critical Posthumanities for the Anthropocene," *Ethics & the Environment* 20, no. 1 (2015): 67–97; Eyal Weizman and Fazal Sheikh, *The Conflict Shoreline: Colonization as Climate Change in the Negev Desert* (Göttingen: Steidl, 2015); McKenzie Wark, *Molecular Red: Theory for the Anthropocene* (New York: Verso, 2016); Jan Zalasiewicz, Mark Williams, Will Steffen,

and Paul Crutzen. "The New World of the Anthropocene," *Environmental Science and Technology* 44, no. 7 (Feb 2010): 2228–31; Arianne Conty, "The Politics of Nature: New Materialist Responses to the Anthropocene," *Theory, Culture & Society* 35, no. 7–8 (Oct 2018): 73–96; Heather Davis and Etienne Turpin, *Art in the Anthropocene: Encounters Among Aesthetics, Politics, Environments and Epistemologies* (London: Open Humanities Press, 2015); Richard Grusin, *Anthropocene Feminism* (Minneapolis: University of Minnesota Press, 2017); Adam Bobbette and Amy Donovan, eds., *Political Geology: Active Stratigraphies and the Making of Life* (London: Palgrave Macmillan, Cham, 2019).

6. See chapter 15 of this book for more on the ethics of the Kinocene.

7. Kathryn Yusoff, *A Billion Black Anthropocenes or None* (Minneapolis: University of Minnesota Press, 2018); Dipesh Chakrabarty, "The Climate of History: Four Theses." *Critical Inquiry* 35, no. 2 (Winter 2009): 197–222; Andrew Baldwin, *Life Adrift: Climate Change, Migration, Critique* (London: Rowman & Littlefield International, 2017).

8. "For most of the last two centuries, with some exceptions, social thought has not given serious attention to the earth sciences. While the social sciences and humanities have conversed productively with biology, linguistics, psychoanalysis, complexity studies and even mathematics, the geosciences seem to have offered less fertile ground for engagement." Nigel Clark and Yasmin Gunaratnam, "Earthing the Anthropos? From 'socializing the Anthropocene' to geologizing the social," *European Journal of Social Theory* 20, no. 1 (Aug 2016): 146–63; 147. See also Nigel Clark, *Inhuman Nature: Sociable Life on a Dynamic Planet* (London: Sage, 2011).

9. For a literature review of the limited work done on the philosophy of geology, see Claude C. Albritton, "Philosophy of Geology," in *General Geology. Encyclopedia of Earth Science* (Springer, Boston, MA: 1988). https://link.springer.com/referenceworkentry/10.1007%2F0-387-30844-X_45

10. Ecofeminists have been writing about this for a long time. See Carolyn Merchant, *The Death of Nature: Women, Ecology, and the Scientific Revolution* (San Francisco: Harper One, 2008).

11. "With some half a century of developments in the geosciences converging on the idea of earth systems with multiple possible operating states, the very nature of 'the ground' needs major overhauling." Clark and Gunaratnam, "Earthing the Anthropos?," 159."Here Arendt rediscovers Heidegger's analyses that we cited in the first part of this present work: The Earth viewed from outer space is no longer the 'earth on which man lives,' it is no longer the homeland (*Heimat*). In the same way that for Husserl, the Earth is not a heavenly body in motion among many others but first and foremost a 'ground,' it 'does not move and does not rest,' since 'only in relation to it are movement and rest given as having their sense of movement and rest.' It is in this sense that the Earth is the 'arche-dwelling,' the 'ark of the world' that nothing can replace and that we reference and intellectually give as an answer each time we imagine that the Moon or an airplane could constitute another foundational basis or 'ground.' 'Arche-dwelling,' 'homeland,' 'habitat,' for Arendt,

Heidegger, and Husserl, the thesis is clear: humanity is under condition of the Earth, understood as that which can't be reduced to an object, or a subject—in other words, a transcendental nonobjective form." Frédéric Neyrat, *The Unconstructable Earth: An Ecology of Separation*, trans. Drew S. Burk (New York: Fordham University Press, 2018), 170."In a certain way, Arendt's concrete transcendental is neither concrete nor historical enough, just as Husserl's arche-dwelling does not take into account the fact that the Earth has not always been for humanity. And Heidegger failed to make a cosmic event out of the 'homeland.' In order to concretize the transcendental of the Earth, we must not consider it as an object (that we can capture from outer space thanks to a camera) or as a quasi-subject (such as Gaia, a rather local expression of naturing nature) but rather as a *trans-ject* or perhaps more specifically and simply a *traject*, as an interval spanning space-time. In fact, the Earth is not merely a 'ground' upon which we stand, not simply a planet surrounded by a moon and artificial satellites; it's also a long-term event that began 4.54 billion years ago, the historical trajectory of an entity that will disappear in several billion years." Neyrat, *The Unconstructable Earth*, 171.

12. See Thomas Nail, *Being and Motion* (Oxford: Oxford University Press, 2018), chapter 14.

13. See Nail, *Being and Motion*, chapters 17-20.

14. See Stephen J. Gould, *Time's Arrow, Time's Cycle: Myth and Metaphor in the Discovery of Geological Time* (Cambridge: International Society for Science and Religion, 2007).

15. Vladimir Ivanovich Vernadsky, *The Biosphere*, eds. David B. Langmuir and Mark A. McMenamin (New York: Springer Science, [1998] 2013)."As historian John Brooke recounts, the years 1966-73 alone saw the emergence of four major new perspectives on the dynamics of the earth: (1) the confirmation of the theory of plate tectonics; (2) a new appreciation of the role of extra-terrestrial impacts in shaping earth history; (3) the thesis that evolution is punctuated by catastrophic bursts linked to major geophysical events; and (4) the beginnings of the idea that the different components of the earth function as an integrated system—as expressed in the Gaia hypothesis and earth systems theory. Look beyond the immediacy of Anthropocene debates into the encompassing field of contemporary geosciences and we are soon reminded that such processes as cyclical changes in the planet's orbit and axis, the openness of the earth to solar radiation and astronomical events, the magma-driven movements of tectonic plates, the stratal composition of the earth's crust, the deep structures of biological life and functioning of the biosphere continue to set the broad parameters for the functioning of the earth system." Clark and Gunaratnam, "Earthing the Anthropos?," 156."If we take seriously evidence from the earth sciences that the main driver of the Mid-Holocene Climatic Transition was variability in the earth's axis and orbit, then there may indeed be a trace of 'universality' woven into the fabric of human cultural-historical difference." Clark and Gunaratnam, "Earthing the Anthropos?," 157."Neocatastrophism has enlivened modern geo-science by dispatching the belief that the planet took on its current shape only through the

gradual and continuous operation of familiar processes like erosion and sediment buildup. The new geology lets into the picture abrupt die-offs and bursts of species formation, climatic and geomorphological upheavals, and high-speed collisions with extraterrestrial bodies." Jeremy Davies, *Birth of the Anthropocene* (Oakland, CA: University of California Press, 2018), 9."Neocatastrophism has introduced us to a whole list of geophysical forces—asteroids, ocean currents, volcanoes, and the like—that, under the right circumstances, can suddenly come to exert a much greater and more destabilizing influence than usual on the workings of the earth system." Davies, *Birth of the Anthropocene,* 10.

16. Aristotle, *Physics*, trans. Joe Sachs (New Jersey: Rutgers University Press, 2001), Book VIII, page 188.

17. Several contemporary thinkers have conceived of a world "without us" (Alan Weisman, *The World Without Us* [New York: HarperCollins, 2014]; Eugene Thacker, *Cosmic Pessimism* [Minneapolis: Univocal, 2015]), or an "earth after us" (Jan Zalasiewicz, *The Earth After Us: What Legacy Will Humans Leave in the Rocks?* [Oxford: Oxford University Press, 2014]), or an "ancestral" earth revealing a "world where humanity is absent" (Quentin Meillassoux, *After Finitude: An Essay on the Necessity of Contingency* [London: Bloomsbury, 2017])."This fiercely unilateral conception maintains the divide between 'us' and the 'world,' between the human 'subject' and the 'ancient' realities that can only be accessed by pure science." Neyrat, *The Unconstructable Earth*, 172.

Meillassoux is right that the past-being, before humans, is not related to humans (who did not exist yet). Nevertheless, this does not mean that our thoughts about this being are unrelated to that past-being. The past makes human thought and existence possible. This is not a non-relation but rather an asymmetrical relation.Furthermore, by making thought and being non-relational, Meillassoux is anthropocentric and rationalist, because he says that nature cannot think. Nature, however, is already mathematical and thoughtful to some degree. Humans are not a break with nature; they are a regional expression of processes already present in nature.

Meillassoux says that humans are mere matter, but that we are unique matter that is radically different than what nature has. However, how does such a break occur in matter? It cannot.

18. Uniformitarianism still holds sway over many geologists and remains a matter of lively debate. See Marcia Bjornerud, *Timefulness: How Thinking Like a Geologist Can Help Save the World* (New Jersey: Princeton University Press, 2018), 61.

19. Bjornerud, *Timefulness,* 61.

20. Bjornerud, *Timefulness,* 63. See also John L. Brooke, *Climate Change and the Course of Global History: A Rough Journey* (New York: Cambridge University Press, 2014); Mike Davis, "Cosmic Dancers on History's Stage? The Permanent Revolution in the Earth Sciences," *New Left Review* 217 (1996): 48–84. "One of the crucial breakthroughs, hinging on a wealth of empirical evidence and a deepening appreciation of the way feedback operates in complex systems—was the discovery that climate change

in the past has often been abrupt rather than incremental." Clark and Gunaratnam, "Earthing the Anthropos?," 153.

21. "Just think, as I write this in 2008, more than one thousand of the world's best climate scientists have worked for seventeen years to forecast future climates and have failed to predict the climate of today. I have little confidence in the smooth, rising curve of temperature that modelers predict for the next ninety years. The Earth's history and simple climate models based on the notion of a live and responsive Earth suggest that sudden change and surprise are more likely." James Lovelock, *The Vanishing Face of Gaia: A Final Warning* (Vancouver: Langara College, 2013), 7.

22. Peter D. Ditlevsen and Sigfus J. Johnsen, "Tipping Points: Early Warning and Wishful Thinking," *Geophysical Research Letters* 37, no. 19 (October 2010). DOI: 10.1029/2010GL044486

23. Neyrat, *The Unconstructable Earth,* 41.

24. See Bjornerud, *Timefulness,* chapter 6, for a full description and critique of each technological fix.

25. Georges Bataille, "Corps célestes," *Oeuvres Complètes, Volume* 1 (Paris: Gallimard, 1970), 516.

26. "Where did we ever get the strange idea that nature—as opposed to culture—is ahistorical and timeless? We are far too impressed by our own cleverness and self-consciousness. . . . We need to stop telling ourselves the same old anthropocentric bedtime stories." Steve Shaviro, *Doom Patrols: A Theoretical Fiction about Postmodernism* (New York: Serpent's Tail, 1997).

27. "One can have interesting thoughts about the long intervals between such revolutions, about the more profound revolutions caused by alterations of the earth's axis, and also those caused by the sea. They are, however, hypotheses in the historical field, and this point of view of a mere succession in time has no philosophical significance whatever." Georg Wilhelm Friedrich Hegel, *Hegel's Philosophy of Nature: Being Part Two of the Encyclopedia of the Philosophical Sciences* (1830), translated from Nicolin and Pöggeler's edition (1950) and from the *Zusätze* in Michelet's text (1847) by A.V. Miller (Oxford: Oxford University Press, 2004), 283.

28. See Immanuel Kant, "Idea for a Universal History from a Cosmopolitan Perspective," in *Toward Perpetual Peace and Other Writings on Politics, Peace, and History* (Yale University Press, 2008).

29. "One reason for this may be that our planet—as presented by the scientific disciplines specializing its study—has appeared to change so gradually that it can largely be taken for granted as the static backdrop of social existence. Perhaps more importantly, in its very obduracy, the earth has generally signified inertia and stability, so that any association with social life has usually been taken to imply a limitation or closure of the possibilities open to collective social action." Clark and Gunaratnam, "Earthing the Anthropos?," 147.

30. See Benjamin Lieberman and Elizabeth Gordon, *Climate Change in Human History: Prehistory to the Present* (London: Bloomsbury Academic, 2018).

31. See Clark, *Inhuman Nature*; Karen Bosworth, "Thinking Permeable Matter through Feminist Geophilosophy: Environmental Knowledge Controversy and the Materiality of Hydrogeologic Processes," *Environment and Planning D: Society and Space* 35, no. 1 (2017): 21–37.

32. See Isabelle Stengers and Andrew Goffey, *In Catastrophic Times: Resisting the Coming Barbarism* (London: Open Humanities Press, 2015), 44. "The intrusion of Gaia. It is a matter here of thinking intrusion, not belonging." Stengers and Goffey think of Gaia as a subject distinct from humans that intrudes on them. This assumes an anthropocentric division. For similar vitalist and subjectivist theories of the earth, see Bruno Latour, *Facing Gaia: Eight Lectures on the New Climatic Regime* (Polity: United Kingdom, 2017); Jane Bennett, *Vibrant Matter: A Political Ecology of Things* (Durham: Duke University Press, 2010); Isabelle Stengers, "Autonomy and the Intrusion of Gaia," *South Atlantic Quarterly* 116, no. 2 (2017): 381–400.

33. I agree with much of Gaian systems theory and think it pushes in the right direction on many fronts. In particular, Gaia theory's emphasis on the earth as a non-static and non-stable planet is key. I support "a perspective in which one inhabits not a static environment," as Dorian Sagan puts it. However, Gaian theory, as Sagan admits, "has occasionally served as a platform for a New Age joy slide into the muck of planetary personification. If biocentrism is currently a prime grove for the culling of noble fictions, then certainly the tree of Gaia, at the very best, bears some of the most tempting fruit" (Dorian Sagan, *Cosmic Apprentice: Dispatches from the Edges of Science* [University of Minnesota Press, 2013], 176). My main concern is with those who personify the earth and think of Gaia from a narrowly biocentric perspective.

34. See Thomas Nail, *Lucretius I: An Ontology of Motion* (Edinburgh: Edinburgh University Press, 2018); Thomas Nail, *Lucretius II: An Ethics of Motion* (Edinburgh: Edinburgh University Press, 2020).

35. "In this actual world there is . . . not much point in counterposing or restating the great abstractions of Man and Nature," wrote the cultural theorist Raymond Williams. "We have mixed our labour with the earth, our forces with its forces too deeply to be able to draw back and separate either out." Or, as the geographer Neil Smith asserted, surveying the cumulative effects of the ever-expanding forces of production: "No God-given stone is left unturned, no original relation with nature unaltered, no living thing unaffected." Quoted in Clark, *Inhuman Nature*, 8. See also Bill McKibben, *The End of Nature* (New York: Random House, 2006).

36. Clark, *Inhuman Nature*, 7. See Andrew Baldwin and Giovanni Bettini, eds., *Life Adrift: Climate Change, Migration, Critique* (London: Rowman & Littlefield International, 2017), for an excellent study of the "climate refugee."

37. See Rosi Braidotti, *The Posthuman* (Oxford: Polity Press, 2012).

38. Excellent work has been done on human-nature hybrids. See Bruno Latour, *We Have Never Been Modern* (Cambridge: Harvard University Press, 1993); Donna Haraway, *Staying with the Trouble: Making Kin in the Chthulucene* (Durham: Duke University Press, 2016); Eduardo Kohn, *How Forests Think: Toward an Anthropology Beyond the Human* (Berkeley: University of California Press, 2013). However, not

much has been said about nature-nature hybrids before there were humans. For all their talk of post-humanism, the humanities have been extremely hesitant to examine a time before the human."In a weird resonance with critiques by Donna Haraway, Bruno Latour, Philippe Descola, Jane Bennett, William Connolly, and others who have effectively demonstrated that there is no 'nature' external to humans, ecomodernists have conceded the point, declaring the need to invent precisely the human/non-human or culture/nature binary of modernist social theory via an unprecedented scale of global governance." Jarius Grove, "The Geopolitics of Extinction," in *The Anthropocene to the Eurocene in Technology and World Politics*, ed. Daniel R. McCarthy (New York: Routledge, 2018), 204–33; 213.

39. See Stacy Alaimo and Susan J. Hekman, eds., *Material Feminisms* (Indiana: Indiana University Press, 2008), Introduction. "If we acknowledged this energetic subsidy as lending of not just materials but capacities for the geologic within hominid corporeality, then the location of agentic power shifts. When we understand our being is mineralogical as well as biological, and that we already possess a capacity for the geologic, then the specific constellations of where and how we locate responsibility changes." "There can be no human that is other to these forces, because the human is an expression of the various constellations of this minerality. There is no telos or origins to this experimentation and mutation—it is just that." Yusoff, "Anthropogenesis," 12; 21–22."Geometry, in effect, is the science of what is absolutely objective—i.e., spatiality—in the objects that the Earth, *our* common place, can indefinitely furnish as our common ground with other men. But if an objective science of earthly things is possible, an objective science of the Earth itself, the ground and foundation of these objects, is as radically impossible as that of transcendental subjectivity. The transcendental Earth is not an object and can never become one. And the possibility of a geometry strictly complements the impossibility of what could be called a '*geology*,' the objective science of the Earth itself." Jacques Derrida, *Edmund Husserl's Origin of Geometry: An Introduction* [1962], trans. J. P. Leavey Jr. (Lincoln: University of Nebraska Press, 1989), 38 (emphasis in original).

40. "In summary, we suggest that one of the main provocations of contemporary earth science—within and beyond the Anthropocene thesis—is to push critical social thought's own insistence on locatedness, positionality and contextualization to its logical conclusion. From this perspective there are no societies that do not bear the trace of the geoclimatic conditions in which they emerged, no social formations that are not in some significant way shaped by the geological formations in which they are embedded, no cultures that are impervious to the flows or strata they tap into." Yasmin Gunaratnam and Nigel Clark, "Pre-Race Post-Race: Climate Change and Planetary Humanism," *Darkmatter* 9, no. 1 (2012). Available at: http://www.darkmatter101.org/site/2012/07/02/pre-race-post-race-climate-change-and-planetary-humanism/

41. Michel Foucault, Giorgio Agamben and others remain focused on human history. See Neyrat, *Unconstructable Earth*.

42. Clark and Gunaratnam, "Earthing the Anthropos?," 159.

43. See Nail, *Being and Motion*, chapter 4, for the theory of the kinetic transcendental.

44. Clark, *Inhuman Nature*, 11.

45. "The earth is characterized as living or quasi-living not to emphasize the organic interconnection between everything on earth, but to take into consideration the surplus that results from any project of technological dominance. It's precisely this surplus that makes the earth into a wholly full body—not a body filled with matter or organs, but with potentialities that no system—whether technical or living, artificial or organic—is able to contain." Frédéric Neyrat, "Eccentric Earth," *Diacritics* 45, no. 3 (2017): 4–23.

"No longer a static, rigid taxonomy; it becomes protean, upwelling, a vital force erupting forth, proliferating, unpredictable, and metastasizing. We may actually be facing the most extraordinary frontier—the frontier of nature as an ultimately creative, responsive, and transformative power, which regards human beings simply as a trace that is overcome and left behind." Michael Bess, "Deconstructing Nature," *Letters* 8, no. 1 (Fall 1999): 2.

"The long-held barriers between nature and culture are breaking down. It's no longer us against "Nature". Instead, it's we who decide what nature is and what it will be." Christian Schwägerl, *The Anthropocene: The Human Era and How It Shapes Our Planet* (Santa Fe: Synergetic Press, 2014).

46. The unpredictable earth is what Frederic Neyrat calls the *traject*. "As unconstructable traject, the earth can neither be controlled nor dominated. From its dark origins to its glacial ends, the earth will always love to hide." See Neyrat, "Eccentric Earth," 11.

47. Neyrat calls this the "concrete transcendental" ("Eccentric earth," 9). "This proto-anthropogenic subject whose death is signaled through this epochal shift heralds a new philosophy in which the earth returns not to ground the figure of thought, but as a condition of its labour; thought must continually move through and with the inhuman, before, during and after subjectivity. So there is a shift in register from humanist thought, which characterized the inhuman as a dehumanizing force, to a concept of the inhuman as materially constitutive of the possibilities of life." Yusoff, "Anthropogenesis," 7.

48. The increasing mobility of the earth is not an "epochal concept" of "our era" in a univocal or exclusive sense. It is only one of the most popular and powerful features of the present, among others. See Gabriel Rockhill, *Interventions in Contemporary Thought: History, Politics, and Aesthetics* (Edinburgh: Edinburgh University Press, 2016), 51–52. "Such an ontological framing draws upon the new geoscience notion that the earth system has, at any stage, the potential to shift into other, not yet actualized states—though we should be mindful that, as a philosophical or cultural thematic, this is an extrapolation from scientific findings that may exceed the concerns or priorities of these sciences themselves." Clark and Gunaratnam, "Earthing the Anthropos?," 159.

49. We cannot know everything about the earth, but we can learn something new about it.

50. See Lovelock, *The Vanishing Face of Gaia*, 7, on climate change feedback and unpredictability.

51. Geophilosophy is not nearly deeply historical enough. Deleuze's geophilosophy only goes back to the Greeks. See Gilles Deleuze, *What is Philosophy?* (New York: Columbia University Press, 1994), 87–89. There is thus a kind of implicit anthropocentrism in geophilosophy. A true geophilosophy would start with the earth before humans and with the cosmos before the earth. See Gasché Rodolphe, *Geophilosophy: On Gilles Deleuze and Félix Guattari's "What Is Philosophy?"*; Anna Hickey-Moody and Timothy Laurie, "Geophilosophies of Masculinity: Remapping Gender, Aesthetics, and Knowledge," *Angelaki* 20, no. 1 (Mar 2015): 1–10; Gary Shapiro, "Nietzsche on Geophilosophy and Geoaesthetics," in *A Companion to Nietzsche*, ed. Keith A. Pearson (Chichester: Wiley-Blackwell, 2009), 707–31.

For an excellent survey and critique of the geophilosophy literature (Lefebvre, Kant, Husserl, Hegel, Bataille, Derrida, Deleuze), I can do no better than Nigel Clark has already done in chapter 1 of *Inhuman Nature*.

52. "If we are unable to enclose this involvement against an outside that purportedly has no language, and if the subject of interpretation is consequently also its object, then we are within the perverse desires of a geomancy, a geo-logy, whose figurations are strangely 'fitting.' Could the generalized origin of re-presentation, the hiccough of this subject/object shimmering as the 'always already not yet,' be thought as the Earth's own scientific investigations of itself?" Vicki Kirby, *Quantum Anthropologies: Life at Large* (Durham and London: Duke University Press, 2011), 34.

53. The study of human knowledge is important, but that is not the subject of this book. See Bruno Latour, *Politics of Nature: How to Bring the Sciences into Democracy* (Cambridge: Harvard University Press, 2009). Latour has not given nearly enough attention to nonhuman or prehuman networks. Graham Harman acknowledges this as well. Despite the hypothetical embrace of other-than-human autarchies, "only the most flickering hints of networks devoid of human involvement" can be found anywhere in the Latourian corpus. Graham Harman, *Prince of Networks: Bruno Latour and Metaphysics*. (Melbourne: re.press, 2009), 124. "This may be more than an accidental oversight. If it is not permitted for human interlocutors to speak of non-human worlds without documenting their own role in the description, translation and inevitable re-composition of these realities, then it is hard to imagine how a domain fully independent of the human can legitimately receive attention as anything more than an abstract possibility. To engage substantively with an inhuman region in and for-itself would by definition repudiate the entanglement that attends all such intervention, according to Latour's logic, thereby constituting an act of purification of the human presence. And yet, as I suggested in the introduction, if we pursue the injunction of actor-network theorists to follow the things themselves, it is inevitable that sooner or later we are going to be drawn into realms which precede, antecede or otherwise exceed human influence—as the

current understanding of issues like global climate change makes all too apparent." Clark, *Inhuman Nature*, 37.

On the possibility and status of doing a nonhuman philosophical geology and the question of asymmetrical relations between past and present See Ian Hodder and Gavin Lucas, "The Symmetries and Asymmetries of Human-Thing Relations. a Dialogue," *Archaeological Dialogues* 24, no. 2 (2017): 119–37.

54. Thomas Nail, *Theory of the Object* (Oxford: Oxford University Press, under review). See also Daniel Lee Kleinman and Kelly Moore, eds., *Routledge Handbook of Science, Technology, and Society* (New York: Routlege, 2019).

55. See Nail, *Lucretius II*.

56. The earth also continues to move today in ways that are not affected by what humans think or do.

57. There is no symmetry between the present and the past. However, that does not mean that things are radically unrelated. Ontology is not flat and symmetrical but curved and topologically heterogeneous. There can be no flat ontology in an irreversibly entropic cosmos. See Ilya Prigogine and Isabelle Stengers, *Order Out of Chaos* (New York: Verso, 2018); Ilya Prigogine, *From Being to Becoming* (San Francisco: Freeman, 1980).Graham Harman is right that matter is related to itself without us. However, it does not follow that a) we are unrelated to the past (which supports and conditions and persists through us) or b) that there is an irreducible, non-relational, vacuum-sealed, withdrawn essence hiding in every object. It just means that there are regional relations that occur within larger pedetic, entropic, or asymmetrical relations between the past and present of all objects. Harman calls this a "non-relation," but it is just a pedetic or asymmetrical relation.The problem with vitalism is that it treats all relations as having an equally flat ontology, as a democracy of things. However, this provides no tools for helping us to think about new things, or for politically clarifying the asymmetrical relations between and among things and humans. Vitalism poses a political problem because it assumes the equality of agency. Vitalism is also an ontological problem because it does not account for changing relationships of dependence.

58. "If this virtual 'geometry' requires no outside to access the human—for the genesis of humanness would be an internal articulation of and within itself—then, by implication, 'the human' would not be bound and restricted by some special lack of access to that same generative unfolding. 'The human' would certainly be a unique determination, yet 'one' whose cacophonous reverberations would speak of earthly concerns." Kirby, *Quantum Anthropologies*, 39.

59. "We are at a moment, I have been arguing, when ongoing developments in the earth sciences, some recent turns in philosophical inquiry, and a range of ethico-political issues arising out of our ecological predicament are gathering over the theme of an autonomous, dynamic, self-generating cosmos." Clark, *Inhuman Nature*, 211.

60. This is what Merleau-Ponty called the "flesh." See Maurice Merleau-Ponty, "Eye and Mind," in *The Primacy of Perception: And Other Essays on Phenomenological Psychology, the Philosophy of Art, History and Politics* (Chicago: Northwestern

University Press, 2015). See also Maurice Merleau-Ponty, *The Visible and the Invisible: Followed by Working Notes* (Evanston: Northwestern University Press, 1997).

61. Kathryn Yusoff challenges fellow critical social thinkers to "use the Anthropocene as a provocation to begin to understand ourselves as geologic subjects, not only capable of geomorphic acts, but as beings who have something in common with the geologic forces that are mobilised and incorporated." Kathryn Yusoff, "Geologic Life: Prehistory, Climate, Futures in the Anthropocene," *Environment and Planning: Society and Space*. 31 (2013): 779–95; 781.

62. There are several major historical conditions of the present. For a closer look at each of them see my following books: Thomas Nail, *The Figure of the Migrant* (Stanford University Press, 2015); Thomas Nail, *Theory of the Border* (Oxford: Oxford University Press, 2016); Thomas Nail, *Theory of the Image* (Oxford: Oxford University Press, 2019); Nail, *Theory of the Object*; and Thomas Nail, *Being and Motion* (Oxford: Oxford University Press, 2018).

63. For more on the methodology of kinetic philosophy see Nail, *Being and Motion*.

64. See Nail, *Being and Motion*, for a full description of the historical method. Or, as the historian Christophe Bonneuil puts it: "Anthropocene science offers 'a single grand narrative from nowhere, from space or from the species.'" Christophe Bonneuil, "The Geological Turn: Narratives of the Anthropocene," in *The Anthropocene and the Global Environmental Crisis: Rethinking Modernity in a New Epoch*, eds. Clive Hamilton, Christophe Bonneuil, and François Gemenne (London: Routledge, 2015), 29. "Anthropocene discourses will need to embrace 'a plurality of narratives from many voices and many places'" to avoid a new master narrative. Bonneuil, "The Geological Turn," 29.

65. See Karen Barad, *Meeting the Universe Halfway* (Durham: Duke University Press, 2007).

66. See Barad, *Meeting the Universe Halfway*, and Kirby, *Quantum Anthropologies*, for a critique of the notion of a single objective nature.

67. See Nail, *Lucretius II*. If the cannot be "false" or "unreal" to itself, then "all perceptions are true," just as Lucretius says.

68. Theory, however, also has its own material kinetic process of inscription. The study of this process of inscription warrants its own independent investigation. See Nail, *Being and Motion*.

69. See Nail, *Being and Motion*; Nail, *Theory of the Object*.

70. For a kinetic theory of quantum gravity see Nail, *Theory of the Object*.

71. See Nail, *Being and Motion*.

72. Speaking of climate, Michel Serres reminds us that "our lives depend on this mobile atmospheric system, which is constant but fairly stable, deterministic and stochastic, moving quasi-periodically with rhythms and response times that vary colossally." Michel Serres, *The Natural Contract* (Ann Arbor: University of Michigan Press, 1995), 27.

"A symphony of rhythms and temporalities thus underpins our development as humans and as living organisms. It marks us as creatures of this earth, as beings that are constituted by a double temporality: rhythmically structured within and embedded in the rhythmic organisation of the cosmos." Barbara Adam, *Timescapes of Modernity: The Environment and Invisible Hazards* (London: Routledge, 1998), 13."Beyond this short-term frequency, now relatively well-understood, climate scientists speculate about larger-scale periodicities that could range between decades, centuries or even millennia." Mike Davis, *Late Victorian Holocausts: El Nino Famines and the Making of the Third World* (London: Verso, 2001), 234.

73. Western culture has also taken itself to be universal, and critical theorists have shown the geographical, historical, gendered, raced, and classed particularity of that claim. But we have yet to appreciate the deeply geological and atmospheric particularity of all our planetary pretensions to universality and particularity.

74. In the example of the line AB, it is "already motion that has drawn the line" to which A and B have been added afterward as its endpoints. Henri Bergson, *Matter and Memory* (New York: Zone Books, 1991), 189.

75. I have made it clear in the conclusion to each of my books that limiting historical ontology to human history is not the same as an ontological commitment to anthropocentrism. I have always believed that these patterns are not the sole invention of humans, but exceed them. They are emergent patterns of the cosmos itself, as we will see in depth in this book.

76. See Vicki Kirby, ed., *What If Culture Was Nature All Along?* (Edinburgh: Edinburgh University Press, 2018).

77. "What would a human be without elephants, plants, lions, cereals, oceans, ozone or plankton?" Bruno Latour, "To Modernise or Ecologise? That is the Question," in *Remaking Reality: Nature at the Millennium*, eds. Bruce Braun and Noel Castree (London and New York: Routledge, 1998), 220–41; 231.

78. T.J. Demos, "Anthropocene, Capitalocene, Gynocene: The Many Names of Resistance," FotoMuseum.com, 6 December 2015. https://www.fotomuseum.ch/en/explore/still-searching/articles/27015_anthropocene_capitalocene_gynocene_the_many_names_of_resistance"Carbocene: an age of powerful carbon-based fuels that have helped to create ways of thinking and acting that humans now find exceedingly difficult to escape." LeCain, "Against the Anthropocene," 1.

79. LeCain, "Against the Anthropocene."

80. Donna Haraway's concept of the Chthulu scene is perhaps closest to recognizing the constitutive role of the earth in climate change. See Haraway, *Staying with the Trouble*. "Consider that none of the other officially recognized geological periods are named for a specific class or order of creatures, much less one species." LeCain, "Against the Anthropocene," 19.

81. Katherine Yusoff, "Anthropogenesis." "Yet in suggesting that humans were indeed powerful enough to cause such global ecological shifts, the Anthropocene concept also tends to encourage the hubristic modernist faith in the human ability to fix the resulting problems." LeCain, "Against the Anthropocene," 4."But rather than crediting humans alone, neo-materialism suggests that they accomplished these

things only at the price of throwing their lot in with a lot of other things, like coal and oil, whose powers they only vaguely understood and certainly did not really control. Likewise, once the partnerships were made, these powerful things began to shape humans and their cultures in all sorts of unexpected ways, many of them not necessarily for the better. In sum, neo-materialist theory pushes us to consider how the planet has made humans rather than the other way around." LeCain, "Against the Anthropocene," 5.

82. "Geologic corporeality is something that is inherited; it is before us and immanent within the condition of our being. If there is a response to be made to our fossil fuelled-being, it must acknowledge this condition, and seek to question its geosocial reproduction." "Only by learning to know and sense ourselves as geological (and accepting that this knowledge will never be complete), and as a being that is toward the geological, can we hope to move against coal-fired inheritances." Yusoff, "Anthropogenesis," 23.

Chapter 1: The Flow of Matter

1. "Through loss man can regain the free movement of the universe, he can dance and swirl in the full rapture of those great swarms of stars. But he must, in the violent expenditure of self, perceive that he breathes in the power of death." Georges Bataille, "Celestial Bodies," translated by Annette Michelson, *October*, Vol. 36, Georges Bataille: Writings on Laughter, Sacrifice, Nietzsche, Un-Knowing (Spring, 1986), pp. 75–78; 78.

2. "The crowning achievement of this tendency is anthropocentrism. The weakening of the terrestrial globe's material energy has enabled the constitution of the autonomous human existences which are so many misconceptions of the universe's movement." Bataille, "Celestial Bodies," 77.

3. See Nail, *Lucretius II*.

4. Helge S. Kragh and Dominique Lambert, "The Context of Discovery: Lemaître and the Origin of the Primeval-Atom Universe," *Annals of Science* 64, no. 4 (2007): 445–70.

5. Carlo Rovelli, *Reality Is Not What It Seems: The Journey to Quantum Gravity* (New York: Riverhead Books, 2018).

6. Rovelli, *Reality Is Not What It Seems*; Nail, *Theory of the Object*; Lee Smolin, *Trouble with Physics: The Rise of String Theory, the Fall of a Science, and What Comes Next* (New York: Penguin Books, 2008).

7. Martin Heidegger, *Being and Time*, trans. Joan Stambaugh (New York: SUNY Press, 1996).

8. For a full development of this triple parallel, see Nail, *Theory of the Object* on elasticity in modern objects and Nail, *Being and Motion* on elasticity in the modern ontology of time.

9. See Lisa Randall, *Dark Matter and the Dinosaurs: The Astounding Interconnectedness of the Universe* (New York: Vintage, 2017).

10. See Rovelli, *Reality Is Not What It Seems*; ChunJun Cao, Sean M. Carroll, and

Spyridon Michalakis, "Space from Hilbert Space: Recovering Geometry from Bulk Entanglement," *Physical Review D* 95, no. 2 (2017). DOI:10.1103/PhysRevD.95.024031; Lee Smolin, *Three Roads to Quantum Gravity* (New York: Basic Books, 2017); Nail, *Theory of the Object*.

11. What the ice cores showed were the signatures of sudden transformation. See Wallace S. Broecker, "Unpleasant Surprises in the Greenhouse?," *Nature* 328 (1987): 123–26."Each long wave movement in and out of an ice age turned out to be rent by multitudes of rapid warmings and coolings, vicious see-sawings that saw the temperature of Greenland transformed by around 15 degrees Fahrenheit in a decade, and global weather tipped into a completely different state in as little as a few years." Clark, *Inhuman Nature*, xi.

As the glaciologist Richard Alley reflects: "for most of the last 100,000 years, a crazily jumping climate has been the rule, not the exception." Richard Alley, *The Two-Mile Time Machine: Ice Cores, Abrupt Climate Change, and Our Future* (Princeton, NJ: Princeton University Press, 2000), 120. "The real point is that not only is climate change natural, but it's also easy to set in motion." Matthew Huber, "North Pole's Ancient Past Holds Clues about Future Global Warming," *Phys.org*, 31 May 2006, www.physorg.com/news68305951.html.

12. See Nail, *Being and Motion* and Nail, *Lucretius I*.

13. Bruno Latour warns us against the "view from Sirius." "From now on, it is from this Great Outside that the old primordial Earth is going to be known, weighed, and judged. What was only a virtuality is becoming, for the greatest minds as well as the smallest, an exciting project: *to know is to know from the outside*. Everything has to be viewed as if from Sirius—a Sirius of the imagination, to which no one has ever had access." Bruno Latour, *Down to Earth: Politics in the New Climatic Regime* (Cambridge: Polity, 2018), 68.

Latour is still committed to a human-hybrid *construction* of the earth. He fears the exo-terrestrial because it cannot be constructed by humans.

14. Even Carlo Rovelli has said that below the Planck scale "there is nothing"—and chosen to affirm a version of the ancient atomism of so-called "granules" of spacetime. "Below [the Planck] scale nothing more is accessible. More precisely, nothing exists there" (Rovelli, *Reality Is Not What It Seems*, 152). For a full critique, see Nail, *Theory of the Object*.

15. Robert M. Hazen, *The Story of Earth: The First 4.5 Billion Years, from Stardust to Living Planet* (New York: Penguin Books, 2013). See also Natalie Wolchover, "Physicists Debate Hawking's Idea That the Universe Had No Beginning," *Quanta Magazine*, 6 June 2019. https://www.quantamagazine.org/physicists-debate-hawkings-idea-that-the-universe-had-no-beginning-20190606/

16. Real indeterminacy is different from what is commonly understood as the influence of human observers on their observations. Quantum indeterminacy is real and ontological because observation is not something that only humans do. Everything observes. See Barad, *Meeting the Universe Halfway*.

17. Vladimir Ivanovich Vernadsky, *The Biosphere*, eds. David B. Langmuir and

Mark A. McMenamin (New York: Springer Science [1998] 2013); Randall, *Dark Matter and the Dinosaurs*; John Briggs and F. David Peat, *Turbulent Mirror: An Illustrated Guide to Chaos Theory and the Science of Wholeness* (New York: Harper & Row, 2000), 42–43.

18. In quantum gravity, the Big Bang is only one of several "big bounces" in which the universe expands and contracts, endlessly and uniquely each time.

19. "It is certainly not so that it moves in space, although it could move, but rather as we tried to show above: the earth is the ark which makes possible in the first place the sense of all motion and all rest as mode of motion. But its rest is not a mode of motion." Edmund Husserl, "Foundational Investigations of the Phenomenological Origin of the Spatiality of Nature," trans. Fred Kersten, in *Husserl, Shorter Works*, ed. Peter McCormick and Frederick A. Elliston (Indiana: University of Notre Dame Press, 1981), 222–33; 230.

20. Doreen Massey, "Landscape as a Provocation: Reflections on Moving Mountains." *Journal of Material Culture* 11 (2006): 33–48.

21. See Nail, *Lucretius I*; Nail, *Lucretius II*; Thomas Nail, "Matter and Motion," unpublished manuscript.

22. See Nail, *Theory of the Object*, chapter 16. See also Barad, *Meeting the Universe Halfway*.

23. "This then is one account of nature, namely that it is the primary underlying matter of things which have in themselves a principle of motion or change." Aristotle, Physics II, 193a28–193a29. "This is Motion. This becoming, however, is itself just as much the collapse within itself of its contradiction, the *immediately identical* and *existent* unity of both, namely, Matter." Georg W. F Hegel, *Hegel's Philosophy of Nature: Being Part Two of the Encyclopaedia of the Philosophical Sciences*, trans. and eds. Arnold V. Miller and Karl L. Michelet (Oxford: Clarendon Press, [1830] 1970), 41.

24. See Nail, *Being and Motion* for a full account of this history.

25. Karl Marx, *The Poverty of Philosophy* (New York: International Publishers, 1963), 78.

26. Georges Bataille, "Base Materialism," in *Visions of Excess: Selected Writings, 1927–1939*, trans. Allan Stoekl (Minneapolis: University of Minnesota Press, 2008), 45–48.

27. The shape of the earth's orbit varies between nearly circular and mildly elliptical. These variations occur with a period of 413,000 years. Other components have 95,000-year and 125,000-year cycles. They loosely combine into a 100,000-year cycle.

28. For a study of Saturn's chaotic orbit, see Briggs and Peat, *Turbulent Mirror*, 42–43.

29. Randall, *Dark Matter and the Dinosaurs*.

30. The Clay Mathematics Institute (CMI), http://www.claymath.org/millennium-problems/navier–stokes-equation

31. The quote is probably apocryphal, but the sentiment is striking.

32. For a non-epistemological interpretation of uncertainty and indeterminacy, see Rovelli, *Reality Is Not What It Seems*, 132–34. See Barad, *Meeting the Universe Halfway*, 301 for an interesting interpretation of uncertainty beyond Bohr's idea of limiting observation to the walls of the laboratory.

33. See Nail, *Lucretius I.*

34. Consider also the likelihood of certain recurring patterns in the case of a cosmic big bounce scenario put forward by Carlo Rovelli and others, or in the case of the multiverse put forward by Sean Carroll and others.

35. (299,792,458 m/s)

36. (9.81 m/s/s)

37. "The first system (left) converges on a steady state—a point in phase space. The second repeats itself periodically, forming a cyclical orbit. The third repeats itself in a more complex waltz rhythm, a cycle with "period three." The fourth is chaotic." James Gleick, *Chaos: Making a New Science* (New York: Viking, 1987), 50–51.

Chapter 2: The Fold of Elements

1. Randall, *Dark Matter and the Dinosaurs.*

2. Michel Serres develops a similar theory of vortices: "The vortex conjoins the atoms, in the same way as the spiral links the points; the turning movement brings together atoms and points alike." Michel Serres, *The Birth of Physics* (Manchester: Clinamen Press, 2000), 16. Deleuze and Guattari then further develop this under the name of "minor science" in Gilles Deleuze and Félix Guattari, *A Thousand Plateaus: Capitalism and Schizophrenia*, trans. Brian Massumi (London: Continuum, 2008), 361–62.

Schelling illustrates this in a footnote with the image of the whirlpool: "A stream flows in a straight line forward as long as it encounters no resistance. Where there is resistance—a whirlpool forms. Every original product of nature is such a whirlpool, every organism. The whirlpool is not something immobilized, it is rather something constantly transforming—but reproduced anew at each moment. Thus no product in nature is fixed, but is introduced at each instant through the force of nature entire." F. W. J. Schelling, *First Outline of a System of the Philosophy of Nature*, trans. Keith R. Peterson (Albany: State University of New York Press, 2004), 18.

3. David Grandy, *Everyday Quantum Reality* (Chicago: Indiana University Press, 2010); Johnjoe McFadden and Jim Al-Khalili, *Life on the Edge: The Coming of Age of Quantum Biology,* (New York: Broadway Books, 2016).

4. Steven Strogatz, *Sync: The Emerging Science of Spontaneous Order* (New York: Hyperion, 2003), 70–100.

5. Daniel Graham, *The Texts of Early Greek Philosophy: The Complete Fragments and Selected Testimonies of the Major Presocratics* (Cambridge: Cambridge University Press, 2010), 159. 62 [F39].

6. See Strogatz, *Sync;* Briggs and Peat, *Turbulent Mirror,* 42–43; Ilya Prigogine and Isabelle Stengers, *The End of Certainty: Time, Chaos, and the New Laws of Nature* (New York: Free Press, 1997).

7. K. Zioutas, M. Tsagri, Y.K. Semertzidis, T. Papaevangelou, D. H. H. Hoffmann, and V. Anastassopoulos, "The 11-Years Solar Cycle As The Manifestation Of The Dark Universe," *Modern Physics Letters A* 29, no. 37 (2014). arxiv.org/abs/1309.4021

8. See Thomas Nail, *Marx in Motion: A New Materialist Marxism* (Oxford: Oxford University Press, 2020).

9. For a philosophical theory of diffraction, see Barad, *Meeting the Universe Halfway*.

10. See Anatolii Burshteĭn, *Introduction to Thermodynamics and Kinetic Theory of Matter* (New York: Wiley, 1996).

11. As Hume argues. See Nail, *Being and Motion*.

12. See Nail, *Being and Motion*.

13. See Thomas Nail, "Matter and Motion," unpublished manuscript, chapter 1.

14. See Nail, *Lucretius I*.

Chapter 3: The Planetary Field

1. Randall, *Dark Matter and the Dinosaurs*.

2. Latour, Lovelock, and Stengers argue for the autonomous functioning of the earth.

3. See Hodder and Lucas, "The Symmetries and Asymmetries of Human-Thing Relations."

4. The debate between relation and non-relation in new materialism misses a third option: asymmetrical relations, or what Neyrat calls "separation," which is a relation of distance. See Graham Harman, "Agential and Speculative Realism: Remarks on Barad's Ontology," *Rhizomes: Cultural Studies in Emerging Knowledge* 30 (2016), https://doi.org/10.20415/rhiz/030.e10; Neyrat, *The Unconstructable Earth*.

5. The theory of knots is further developed in Nail, *Being and Motion*, chapter 3.

6. If "the forming of the five senses is a labour of the entire history of the world down to the present," as Marx says (Karl Marx, *Early Writings* [New York: Penguin, 1992], 353; translation modified), then Marxism requires a much deeper historical materialism of the earth itself. We need a Marxist geology and cosmology, a Marxism for the Anthropocene. Marx wrote extensively about geology, and the theory of the earth plays a role throughout his writings. See Nail, *Marx in Motion*.

Chapter 4: Centripetal Minerality

1. "In sum, my aim here is to interrupt the way that Derrida's 'nonconcept textuality' or 'language in the general sense' has been taken up, knocked into disciplinary shape, properly and predictably contextualized, and, inevitably, conceptualized." Kirby, *Quantum Anthropologies*.

2. See Kohn, *How Forests Think*. I disagree with Kohn that human thought is continuous with but *different* from the distinct network of habitual relays or relations between beings. Kohn, following Peirce, claims that only humans can use abstract symbols. Pierce and Kohn do not want to say that all of nature has "language" in the same way. They are committed to the idea of dividing nature into three kinds of sign: icon, index, and symbol. But this division cuts nature off from itself in a way that is materially and historically untenable. There are differences in distribution or arrangement of relation, but they are differences of degree, not of kind.

Furthermore, by leaving non-living matter out of language, Kohn and Peirce both end up relegating matter to mechanistic meaningless objects. Kohn directly rejects Jane Bennet's position and the new materialists on this point. For Kohn and

Peirce, icons = matter, indexes = life, and symbols = human. Kohn and Peirce thus end up reproducing the age-old great chain of being, couched in linguistic terms. Even though both say that they want a kind of ontological continuum, they end up undermining it with the tripartite division of semiotic levels.

3. Kirby, *Quantum Anthropologies*.

4. Jeffrey Jerome Cohen, ed., *Animal, Vegetable, Mineral: Ethics and Objects* (Washington, DC: Oliphaunt Books, 2012); Jeffrey Jerome Cohen, *Stone: An Ecology of the Inhuman* (Minneapolis: University of Minnesota Press: 2015); Jeffrey Jerome Cohen and Lowell Duckert, *Elemental Ecocriticism: Thinking with Earth, Air, Water, and Fire* (Minneapolis: University of Minnesota Press, 2015); Ömür Harmansah, ed., *Of Rocks and Water: Towards an Archaeology of Place* (Oxford and Philadelphia: Oxbow Books, 2014); Kohn, *How Forests Think*; John Sallis, *Stone* (Bloomington: Indiana University Press, 1994); Christopher Y. Tilley and Wayne Bennett, *The Materiality of Stone: Explorations in Landscape Phenomenology* (Oxford: Berg, 2004); David Macauley, *Elemental Philosophy: Earth, Air, Fire, and Water As Environmental Ideas* (Albany, NY: SUNY Press, 2011).

5. This is what vitalism does. See Bennett, *Vibrant Matter*; Cohen, *Animal, Vegetable, Mineral*.

6. Georges Bataille, *The Accursed Share: An Essay on General Economy* (New York: Zone Books, 1988).

7. See Nail, *Lucretius I*.

8. Hazen, *The Story of Earth*.

9. Hazen, *The Story of Earth*.

10. Cohen, *Stone*; Tilley and Bennett, *The Materiality of Stone*.

11. "Relations do not create things like rocks and mountains; things like rocks and mountains are what enable relations to flourish." Cohen, *Stone*, 3.

12. Ben Woodard, *On an Ungrounded Earth: Towards a New Geophilosophy* (Brooklyn, NY: punctum books, 2013), 18–25.

13. Nancy Tuana, "Viscous Porosity: Witnessing Katrina," in *Material Feminisms*, eds. Stacy Alaimo and Susan J. Hekman (Indianapolis: Indiana University Press 2008), 188–213; 194. Tuana contends that fluidity and flows overlook "sites of resistance and opposition" while viscosity "retains an emphasis on resistance to changing form." Thickness is not impenetrable nor immutable, but something that undergoes change in and through friction (188).

14. "Infinite productivity must counter itself, it must put the brakes on itself, it must slow itself down. Nature can't only be forward advancement, nature must also lag behind, it must also constitute its own 'constraint.' This lag can be defined as an 'antiproductive' tendency without which all productivity would be vain: There would only be the absolute, the infinite, endless movement, and as a result, nothing that would be finite, no object whatsoever." Neyrat, *The Unconstructable Earth*, 156.

15. Ted Toadvine, "Thinking after the World: Deconstruction and Last Things," in *Eco-Deconstruction: Derrida and Environmental Philosophy*, eds. Matthias Fritsch, Philippe Lynes, and David Wood (New York: Fordham University Press, 2018), 50–80.

16. "The mutation in technology changes not simply the archiving process, but

what is archivable – that is, the content of what has to be archived is changed by the technology." Jacques Derrida, "Archive Fever. A Seminar..." in *Refiguring the Archive*, ed. Carolyn Hamilton (Dordrecht, Boston, London: Kluwer Academic Publishers, 2002), 38–80; 46.

17. "Archivization produces as much as it records the event." Jacques Derrida, "Archive Fever: A Freudian Impression," *Diacritics* 25, no. 2 (1995): 17.

18. "Thus, rocks, we argue, queer archives." Stephanie Springgay and Sarah E. Truman, "Stone Walks: Inhuman Animacies and Queer Archives of Feeling," *Discourse: Studies in the Cultural Politics of Education* 38, no. 6 (2017): 851–63; 861.

19. "Poetry of molecular parsings whose alphabet is the periodic table—chemical vocatives for assemblage and dissociation. A metaphor of language? Or more instantiations of a language whose 'langue' involves the Earth itself?" Kirby, *Quantum Anthropologies*, 41.

20. See Nail, *Being and Motion*.

21. Robert Epstein, "The Empty Brain," *Aeon*, 18 May, 2016. https://aeon.co/essays/your-brain-does-not-process-information-and-it-is-not-a-computer. See also Nail, *Theory of the Object*.

22. "Which bring[s] the flow of events to a standstill, that history forms, at the interior of this flow, as crystalline constellation." Walter Benjamin, *The Arcades Project*, trans. Howard Eiland (Cambridge, Massachusetts, 1999), 854.

23. Meillassoux (*After Finitude*) thinks that organic life, reason, and God are ex nihilo emergences radically distinct from material nature.

Chapter 5: Hadean Earth

1. Arthur C. Clarke, *Greetings, Carbon-Based Bipeds!: Collected Essays, 1934–1998*, ed. Ian T. Macauley (New York: St. Martin's Griffin, 2000), 333.

2. The title of this section comes from Frédéric Neyrat's wonderful paper of the same name. Frédéric Neyrat, "Our Troubled Planet: Dialectics of Meteors," filmed 19 December 2018 at NYU Center for French Language and Cultures, video, 25:22, https://atoposophie.wordpress.com/2018/12/19/dialectics-of-meteors/

3. Centre National De La Recherche Scientifique (CNRS) Press Release, "Moon Thought to Play Major Role in Maintaining Earth's Magnetic Field," *Astronomy Now*, 1 April 2016. https://astronomynow.com/2016/04/01/moon-thought-to-play-major-role-in-maintaining-earths-magnetic-field/

4. "The experimental approach to deep water has focused on the possibility that the most common of minerals—olivine, pyroxene, garnet, and their denser deep-Earth variants—may be able to incorporate a small amount of water at mantle conditions. The study of water in 'nominally anhydrous' minerals, which became a major focus of high-pressure mineralogy in the 1990s, yielded astonishing results. It turns out that at high pressure and temperature, it's relatively easy for some minerals to incorporate lots of hydrogen atoms, which are the mineralogical equivalent of water (because hydrogen atoms combine with oxygen in these minerals). Minerals that are invariably dry in the cooler, low-pressure environments of the shallow crust—where

explosive volcanism releases any water—can become rather wet in the deep mantle." Hazen, *The Story of Earth*, 91.

5. Andy Coghlan, "There's as Much Water in Earth's Mantle as in All the Oceans," *New Scientist*, 7 June 2017. https://www.newscientist.com/article/2133963-theres-as-much-water-in-earths-mantle-as-in-all-the-oceans/

6. "Thus, ancestral anteriority can too easily be converted into anteriority *for us*." "By way of contrast, the posteriority of extinction indexes a physical annihilation which no amount of chronological tinkering can transform into a correlate 'for us,' because no matter how proximal or how distal the position allocated to it in space-time, it has already cancelled the sufficiency of the correlation." Ray Brassier, *Nihil Unbound: Enlightenment and Extinction* (Basingstoke: Palgrave Macmillan, 2010), 229.

I disagree with Brassier on this point. The past is what is more radically asymmetrical (not non-relational), because it cannot be affected by humans.

The future, however, is that which most definitely will retain traces of humans even in the final moments of the universe. The past is always related to the future as its immanent condition, but the future is not the condition of the past. The future as a hypothetical later present state is *not yet* and thus remains an immaterial and idealist speculation. In looking for a world without humans in the future Brassier finds the speculative mind without a world.

Brassier continues: "Accordingly, there can be no 'afterwards' of extinction, since it already corrodes the efficacy of the projection through which correlational synthesis would assimilate its reality to that of a phenomenon dependent upon conditions of manifestation. Extinction has a transcendental efficacy precisely insofar as it tokens an annihilation which is neither a possibility towards which actual existence could orient itself, nor a given datum from which future existence could proceed. It retroactively disables projection, just as it pre-emptively abolishes retention. In this regard, extinction unfolds in an 'anterior posteriority' which usurps the 'future anteriority' of human existence." Brassier, *Nihil Unbound*, 230.

I disagree here as well. Transcendental annihilation assumes absolute entropy as a law of the universe when it is actually an emergent feature in our region of the universe. Brassier tries to step outside our region of the cosmos but actually ends up universalizing entropy and solar death as Bataille does. But quantum gravity theories hold that the beginning of the universe was not a singularity and its death will not end in absolute stillness. The universe will begin and end through quantum fluctuations that are fundamentally indeterminate. Indeterminacy is more fundamental than vitalism or nihilism. Quantum indeterminacy is the source of the vital birth and rebirth but also of the particular history of death and decay in this universe.

In short, Brassier assumes a strictly classical, thermodynamic, and pre-quantum universe where there are no quantum fluctuations to alter the law of entropy. The heat death of the universe is not an ontological or necessary fact.

7. Alain Badiou has recently announced that "the rethinking of the univocity of ground is a necessary task for the world in which we are living today." Alain Badiou,

Deleuze: The Clamor of Being, trans. Louise Burchill (Minneapolis: University of Minnesota Press, 2000), 46. But he has at the same time abandoned "the philosophy of or as nature" to be a "contemporary impossibility." See Paul Patton, ed., "Deleuze, The Fold: Leibniz and the Baroque," in *Deleuze: A Critical Reader* (Oxford: Blackwell, 1997), 51–69; 64. For more discussion see Iain Grant, *Philosophies of Nature After Schelling* (London, New York: Continuum 2006), 199.

In contrast, I claim here that the earth gives us an ever-changing metastable ground that is both natural and historical. Badiou embraces the latter without the former, while Deleuze embraces the former without the latter. This is what *Theory of the Earth* hopes to overcome.

8. "The laws of nature," writes Deleuze, "govern the surface of the world," while a "transcendental or volcanic spatium" troubles its depths. Gilles Deleuze, *Difference and Repetition*, trans. Paul Patton (New York: Columbia University Press, 1994), 241.

9. Hazen, *The Story of Earth*, 97.

10. M.L. Wong, B.D. Charnay, P. Gao, Y.L. Yung, and M.J. Russell, "Nitrogen Oxides in Early Earth's Atmosphere as Electron Acceptors for Life's Emergence," *Astrobiology* 17, no. 10 (Oct 2017): 975–83. https://www.ncbi.nlm.nih.gov/m/pubmed/29023147/.

11. "Exactly how the stepped leader works is not understood." Marin A. Uman, *All About Lightning* (Mineola, NY: Dover, 2009), 74.

12. Karen Barad, "Transmaterialities: Trans/Matter/Realities and Queer Political Imaginings," *GLQ* 21, no. 2–3 (2015): 387–422; Kirby, *Quantum Anthropologies*.

Chapter 6: Centrifugal Atmospherics

1. Per Bak, *How Nature Works: The Science of Self-Organized Criticality* (New York: Springer, 2013).

2. See Addy Pross, *What Is Life? How Chemistry Becomes Biology* (Oxford: Oxford University Press, 2016), 58–81.

3. See Hazen, *The Story of Earth*, chapter 6; Peter Ward and Joe Kirschvink, *A New History of Life: The Radical New Discoveries about the Origins and Evolution of Life on Earth* (London: Bloomsbury Publishing Plc, 2016).

4. Hazen, *The Story of Earth*, 143–44.

5. Stuart A. Kauffman, *The Origins of Order: Self-Organization and Selection in Evolution* (Oxford, New York: Oxford University Press, 2015).

6. See Nail, *Marx in Motion*.

7. See image in Sean Carroll, *The Big Picture: On the Origins of Life, Meaning, and the Universe Itself* (New York: Dutton, 2016), 255.

8. On Rayleigh–Bénard Turbulence, see Eric D. Schneider and Dorion Sagan, *Into the Cool: Energy Flow, Thermodynamics, and Life* (Chicago: The University of Chicago Press, 2006).

Chapter 7: Archean Earth I: Pneumatology

1. Arthur Rimbaud, *Rimbaud: Complete Works, Selected Letters: A Bilingual*

Edition, trans. Wallace Fowlie and Seth A. Whidden (Chicago: University of Chicago Press, 2005), 187. My translation.Bottom of FormBottom of Form

2. Paul Valéry, *Cahiers*, vol. 1, ed. Judith Robinson (Paris: Gallimard, Bibliothèque de la Pléiade, 1973), C1, 990. English translation, Insook Webber.

3. See John Durham Peters, *The Marvelous Clouds. Toward a Philosophy of Elemental Media* (Chicago: The University of Chicago Press, 2015).

4. Ibid., 256.

5. See Geoffrey West, *Scale: The Universal Laws of Life, Growth, and Death in Organisms, Cities, and Companies* (New York: Penguin Books, Reprint edition, 2018).

6. For instance, Venus is born from the castrated falling seed of the sky god, Uranus, mixing in the ocean, and the sky woman falls through a hole in the sky to earth.

7. See Gleick, *Chaos*.

Chapter 8: Archean Earth II: Biogenesis

1. See Pross, *What Is Life?*; Eva Jablonka, Marion J. Lamb, and Anna Zeligowski, *Evolution in Four Dimensions: Genetic, Epigenetic, Behavioral, and Symbolic Variation in the History of Life* (Cambridge, MA: The MIT Press, 2014); Lynn Margulis and Dorion Sagan, *Microcosmos: Four Billion Years of Evolution from Our Microbial Ancestors* (Berkeley: University of California Press, 1998); Hazen, *The Story of Earth*.

2. Daniel J. Nicholson and John Dupré, eds., *Everything Flows: Towards a Processual Philosophy of Biology* (Oxford: Oxford University Press, 2018).

3. Georges Cuvier, *The Animal Kingdom: Arranged in Conformity with Its Organization*, trans. Henry M'Murtrie (New York: G. & C. & H. Carvill, 1833), 14.

4. T. H. E. Huxley, "Address to the British Association: Liverpool Meeting, 1870," *Nature* 2 (1870): 399–406; 402.

5. Charles Sherrington, *Man on His Nature* (Cambridge: Cambridge University Press, 1940), 83.

6. Carl R. Woese, "A New Biology for a New Century," *Microbiology and Molecular Biology Reviews* 68 (2004): 173–86; 176.

7. Alvaro Moreno and Matteo Mossio, *Biological Autonomy: A Philosophical and Theoretical Enquiry* (Dordrecht: Springer, 2015), 18.

8. See John Marks, "Molecular Biology in the Work of Deleuze and Guattari," *Paragraph* 29, no. 2 (Jul 2006): 81–97.

9. François Jacob, *The Logic of Living Systems: A History of Heredity*, trans. Betty E. Spillman (London: Allen Lane, 1974), 276.

10. Ibid., 292.

11. Ludwig von Bertalanffy, *Problems of Life: An Evaluation of Modern Biological and Scientific Thought* (New York: Harper & Brothers, 1952), 134.

12. Paul A. Weiss, "From Cell to Molecule," in *The Molecular Control of Cellular Activity*, ed. John M. Allen (Toronto: McGraw Hill), 1–72; 3.

13. Paul A. Weiss, "The Living System: Determinism Stratified," *Studium Generale* 22 (1969): 361–400; 369.

14. Richard C. Lewontin, "The Dream of the Human Genome," *The New York Review of Books* 39, no. 10 (1992): 31–40; 33.

15. Sarah Lewin, "New Equation Tallies Odds of Life Beginning," *Space.com*, 8 July 2016. https://www.space.com/33374-odds-of-life-emerging-new-equation.html.

16. "The hydrophobic force is interesting for our current discussion because it is not something a single molecule simply possesses, given its intrinsic properties." Stephan Guttinger, "A Process Ontology for Macromolecular Biology," in Nicholson and Dupré, eds., *Everything Flows*, 303–20; 318.

17. Elizabeth Howell, "Sun's UV Light Helped Spark Life," *Astrobiology Magazine*, 30 March 2017. https://www.astrobio.net/origin-and-evolution-of-life/suns-uv-light-helped-spark-life/.

18. Brian Mahy and Marc van Regenmortel, *Desk Encyclopedia of Animal and Bacterial Virology* (Burlington: Elsevier, 2009), 26; Luis P. Villarreal, *Viruses and the Evolution of Life* (Washington, DC: ASM Press, 2005). N. J. Dimmock, A. J. Easton, and K. N. Leppard, *Introduction to Modern Virology*, 6th edition (Malden, MA: Blackwell Publishing Ltd., 2007), 16.

19. E. V. Koonin and V. V. Dolja, "A Virocentric Perspective on the Evolution of Life," *Current Opinion in Virology* 3, no. 5 (Oct 2013): 546–57.

20. Ricard V. Solé and Santiago F. Elena, *Viruses As Complex Adaptive Systems* (Princeton: Princeton University Press, 2019).

21. Mary Jane West-Eberhard, *Developmental Plasticity and Evolution* (Oxford, New York: Oxford University Press, 2003), 192.

22. Jablonka, Lamb, and Zeligowski, *Evolution in Four Dimensions*.

23. Evelyn F. Keller and L. L. Winship, *Century of the Gene* (Cambridge: Harvard University Press, 2009), 63.

24. National Institutes of Health (NIH) News Release, "New Imaging Technique Overturns Longstanding Textbook Model of DNA Folding," U.S. Department of Health and Human Services, 27 July 2017. https://www.nih.gov/news-events/news-releases/new-imaging-technique-overturns-longstanding-textbook-model-dna-folding.

25. Carl Zimmer, "Everything You Thought You Knew about the Shape of DNA Is Wrong," *Stat News*, 31 March 2016, https://www.statnews.com/2016/03/31/dna-shape-double-helix-dekker/.

26. W. Wang, H. W. Hellinga, and L. S. Beese, "Structural Evidence for the Rare Tautomer Hypothesis of Spontaneous Mutagenesis," *Proceedings of the National Academy of Sciences* 108, no. 43 (Oct 2011), 17644–48.

27. McFadden and Al-Khalili, *Life on the Edge*, 226.

28. Ibid., 227.

29. See A. Datta and S. Jinks-Robertson, "Association of Increased Spontaneous Mutation Rates with High Levels of Transcription in Yeast," *Science* 268, no. 5217 (1995): 1616–19; Jürgen Bachl, Chris Carlson, Vanessa Gray-Schopfer, Mark Dessing, and Carina Olsson, "Increased Transcription Levels Induce Higher Mutation Rates in a Hypermutating Cell Line," *Journal of Immunology* 166, no. 8 (2001): 5051–57; P. Cui, F. Ding, Q. Lin, L. Zhang, A. Li, Z. Zhang, S. Hu, and J. Yu, "Distinct Contributions

of Replication and Transcription to Mutation Rate Variation of Human Genomes," *Genomics, Proteomics and Bioinformatics* 10, no. 1 (2012): 4–10; John Cairns, Julie Overbaugh, and Stephan Miller, "The Origin of Mutants," *Nature* 335 (1988): 142–45.

30. Johnjoe McFadden and Jim Al-Khalili, "A Quantum Mechanical Model of Adaptive Mutation," *Biosystems* 50, no. 3 (Jun 1999): 203–11.

31. Hazen, *The Story of Earth*.

32. The principal source of this region is what Vernadsky calls living matter: the collection of organisms and living bodies that are responsible for the creation of new compounds and that "exert a powerful permanent and continuous disturbing effect on the chemical stability of the surface of our planet." Cited in Emanuele Coccia, *The Life of Plants: A Metaphysics of Mixture* (Cambridge: Polity Press, 2019), 57–58. See also Vladimir Ivanovich Vernadsky, *The Biosphere*, eds. David B. Langmuir and Mark A. McMenamin (New York: Springer Science [1998] 2013).

Chapter 9: Tensional Vegetality

1. For example, Meillassoux, *After Finitude*, preface by Alain Badiou.

2. For example, Michael Marder, *The Philosopher's Plant: An Intellectual Herbarium, illustrations by Mathilde Roussel* (New York: Columbia University Press, 2015).

3. Immanuel Kant expressed doubt that biology would ever find a Newton to explain even a "single blade of grass." Immanuel Kant, *Critique of Judgment* [1790], Part 2: Critique of Teleological Judgment, trans. W. S. Pluhar (Indianapolis: Hackett Publishing Co., 1987), 282.

4. Erwin Schrödinger, *What Is Life? The Physical Aspect of the Living Cell; with Mind and Matter; & Autobiographical Sketches* (Cambridge: Cambridge University Press, 2017).

5. Peter McCourt, "Hormones and Developmental Timing Boundaries in Arabidopsis," abstract circulated by the University of Toronto Department of Botany, 18 March 2005. "A project of engaging philosophy on its classical terms and subjecting 'the plant' *to* those terms—terms of resemblance, difference as degrees from similarity of function, relevant functions and their relative value anchored by 'the human' *and* of hoping, as was the case with the animal, to find a common ground and a 'common logic between these two kingdoms' so that plants, now, too, can be *taken seriously*." Cited in Karen L. F. Houle, "Animal, Vegetable, Mineral: Ethics as Extension or Becoming? The Case of Becoming-Plant," *Journal for Critical Animal Studies* 9, no. 1–2, (2011): 95.

6. Michael Marder, *Grafts: Writings on Plants* (Minneapolis: University of Minnesota Press, 2016); Elaine Miller, *The Vegetative Soul: From Philosophy of Nature to Subjectivity in the Feminine* (Albany: State University of New York Press, 2002); Matthew Hall, *Plants as Persons: A Philosophical Botany* (Albany: State University of New York Press, 2011); Monica Gagliano, John C. Ryan, and Patrícia I. Vieira, eds., *The Language of Plants: Science, Philosophy, Literature* (Minneapolis: University of Minnesota Press, 2017); Craig Holdrege, *Thinking Like a Plant: A Living Science for Life* (Barrington, MA: Lindisfarne Books, 2013); Luce Irigaray and Michael Marder,

Through Vegetal Being: Two Philosophical Perspectives (New York: Columbia University Press, 2016); Michael Marder, *Plant-Thinking: A Philosophy of Vegetal Life* (New York: Columbia University Press, 2013); Marder, *The Philosopher's Plant*; Jeffrey T. Nealon, *Plant Theory: Biopower and Vegetable Life* (Stanford, CA: Stanford University Press, 2016).

7. See Nail, *Lucretius I* and Nail, *Being and Motion*.

8. Yutaka Sumino, Ken H. Nagai, Yuji Shitaka, Dan Tanaka, Kenichi Yoshikawa, Hugues Chaté, and Kazuhiro Oiwa, "Large-scale Vortex Lattice Emerging from Collectively Moving Microtubules," *Nature* 483 (March 2012): 448–52; R. E. Goldstein and J. W. van de Meent, "A Physical Perspective on Cytoplasmic Streaming," *Interface Focus* 5, no. 4 (2015): 2015003; W. F. Pickard, "Absorption by a Moving Spherical Organelle in a Heterogeneous Cytoplasm: Implications for the Role of Trafficking in a Symplast," *Journal of Theoretical Biology* 240, no. 2 (May 2006): 288–301; Yoshio Yotsuyanagi, "Recherches sur les phénomènes moteurs dans les fragments de protoplasme isolés II. Mouvements divers déterminés par la condition de milieu," *Cytologia* 18 (1953): 202–17; Tamás Vicsek and Anna Zafeiris, "Collective Motion," *Physics Report* 517, no. 3–4 (2012): 71–140.

9. William O. Hancock, "Cytoskeletal Organization: Whirling to the Beat," *Current Biology* 22, no. 12 (June 2012): R493-R495.

10. Schneider and Sagan, *Into the Cool*.

11. "Using live-cell imaging and computer simulations, we identify a flow pattern that produces vortices (eddies) on the upstream side of the septum. Nuclei can be immobilized in these microfluidic eddies, where they form multinucleate aggregates and accumulate foci of the HDA-2 histone deacetylase-associated factor, SPA-19. Pores experiencing flow degenerate in the absence of SPA-19, suggesting that eddy-trapped nuclei function to reinforce the septum. Together, our data show that eddies comprise a subcellular niche favoring nuclear differentiation and that subcompartments can be self-organized as a consequence of regimented cytoplasmic streaming." Laurent Pieuchot, Julian Lai, Rachel Ann Loh, Fong Yew Leong, Keng-Hwee Chiam, Jason Stajich, and Gregory Jedd, "Cellular Subcompartments through Cytoplasmic Streaming," *Developmental Cell* 34, no. 4 (Aug 2015): 410–20; 410.

12. S. Yoshiyama, M. Ishigami, A. Nakamura, and K. Kohama, "Calcium Wave for Cytoplasmic Streaming of Physarum Polycephalum," *Cell Biology International* 34, no. 1 (Dec 2009): 35–40.

13. On the formation of cellular subcompartments through eddies, see Pieuchot et al., "Cellular Subcompartments."

14. "It is not so much that the structures begin to move, but movements—for example in the assembly and self-organization of the cytoskeleton—begin to constitute structure." Hannah Landecker, "The Life of Movement: From Microcinematography to Live-Cell Imaging," *Journal of Visual Culture* 11, no. 3 (Dec 2012): 378–99; 393–94.

15. "As recently as two decades ago, college students were still being taught that organelles within eukaryotic cells probably 'pinched off' from the nucleus and subsequently evolved their separate specific functions. With the ability to analyze DNA,

however, it soon became clear that the pinch-off theory had a major failing: it couldn't explain the presence of DNA in certain organelles, nor the fact that this DNA was like that of bacterial genophores and different from that of the nucleus's chromosomes." Margulis and Sagan, *Microcosmos*, 120.

16. "At this late stage, the vortices showed a tendency to arrange their positions into a hexagonal lattice, although not perfectly. A given vortex consists of a sparse core and a dense peripheral annulus inside which microtubules move both clockwise and anticlockwise in small streams and slide past each other (Fig. 2 and Supplementary Movie 2). That is, the streams show nematic rather than polar order. Microtubules were never trapped in a vortex: they would circulate inside one vortex for some time before moving to a neighbouring one or travelling farther and starting to revolve around a more distant core. Large-scale vortex lattice emerging from collectively moving microtubules." Sumino et al., "Large-scale vortex lattice," 448.

17. Margulis and Sagan, *Microcosmos*, 110–15.

18. Ibid., 30.

19. Pieuchot et al., "Cellular Subcompartments."

20. Michiel T. M. Willemse, "Evolution of Plant Reproduction: From Fusion and Dispersal to Interaction and Communication," *Chinese Science Bulletin* 54, no. 14 (Jul 2009): 2390–403; 2397.

21. See Andrew J. Heidel, Oz Barazani, and Ian T. Baldwin, "Interaction Between Herbivore Defense and Microbial Signaling: Bacterial Quorum-Sensing Compounds Weaken JA-Mediated Herbivore Resistance in Nicotiana Attenuate," *Chemoecology* 20, no. 2 (Jun 2009): 149–54. See also R. Karban, I. T. Baldwin, K. J. Baxter, G. Laue, and G. W. Felton, "Communication Between Plants: Induced Resistance in Wild Tobacco Plants Following Clipping of Neighboring Sagebrush," *Oecologia* 125, no. 1 (Oct 2000): 66–71.

22. "We now know that plant chemical language, for example, is endowed with true semantic flexibility, so that new meanings may be assigned to old chemical words and used in novel interactions and new contexts." Monica Gagliano, "Breaking the Silence: Green Mudras and the Faculty of Language in Plants," in *The Language of Plants: Science, Philosophy, Literature*, eds. Monica Gagliano, John C. Ryan, and Patrícia I. Vieira (Minneapolis: University of Minnesota Press, 2017), 84–102; 92. See also J. K. Holopainen, "Multiple Functions of Inducible Plant Volatiles," *Trends in Plant Science* 9, no. 11 (Nov 2004): 529–33.

23. Paul W. Paré and James H. Tumlinson, "Plant Volatiles as a Defense against Insect Herbivores," *Plant Physiology* 121 (Oct 1999): 325–31; R. Karban et al., "Communication between Plants"; M. Heil and J. Ton, "Long-Distance Signalling in Plant Defence," *Trends in Plant Science* 13, no. 6 (Jun 2008): 264–72.

24. M. Dicke, A. A. Agrawal, and J. Bruin, "Plants Talk, but Are They Deaf?," *Trends in Plant Science* 8, no. 9 (Sep 2003): 403–5; 403.

25. M. A. Crespi and J. J. Casal, "Photoreceptor-Mediated Kin Recognition in Plants," *New Phytologist* 205, no. 1 (Jan 2015): 329–38.

26. Susan A. Dudley and Amanda L. File, "Kin Recognition in an Annual Plant,"

Biology Letters 3 (2007): 435–38; Guillermo P. Murphy and Susan A. Dudley, "Kin Recognition: Competition and Cooperation in *Impatiens* (Balsaminaceae)," *American Journal of Botany* 96, no. 11 (2009): 1990–96.

27. For discussion, see Alexander Kravchenko, "How Humberto Maturana's Biology of Cognition Can Revive the Language Sciences," *Constructivist Foundations* 6, no. 3 (Jul 2011): 352–62.

28. See František Baluška et al., "Introduction," in *Communication in Plants: Neuronal Aspects of Plant Life,* eds. František Baluška, Stefano Mancuso, and Dieter Volkmann (Berlin: Springer-Verlag, 2006).

29. Georg W. F. Hegel, A. V. Miller, and J. N. Findlay, *Hegel's Philosophy of Nature: Being Part Two of the Encyclopaedia of the Philosophical Sciences* [1830] (Oxford: Clarendon, 2007).

Chapter 10: Proterozoic Earth

1. Stefan Bengtson, Birger Rasmussen, Magnus Ivarsson, Janet Muhling, Curt Broman, Federica Marone, Marco Stampanoni, and Andrey Bekker, "Fungus-like Mycelial Fossils in 2.4-billion-year-old Vesicular Basalt," *Nature Ecology & Evolution* 1, no. 6, article no. 0141 (Apr 2017).

2. See Deleuze and Guattari, *A Thousand Plateaus.*

3. Barbara K. Kennedy, "First Land Plants and Fungi Changed Earth's Climate, Paving the Way for Explosive Evolution of Land Animals, New Gene Study Suggests," *Penn State: Eberly College of Science,* 9 Aug 2001, https://science.psu.edu/news-and-events/2001-news/Hedges8-2001.htm.

4. Gordon Grice, "Lichens: Fungi That Have Discovered Agriculture," *Discover Magazine,* 6 January 2010, http://discovermagazine.com/2009/nov/06-lichens-fungi-that-have-discovered-agriculture.

5. S. R. Gradstein and H. Kerp, "A Brief History of Plants on Earth," in *The Geologic Time Scale* 2012, eds. Felix M. Gradstein, James G. Ogg, Mark D. Schmitz, and Gabi M. Ogg (Amsterdam: Elsevier, 2012), 233–37. Increasing oxygenation levels from as far back as 700 million years ago suggest that photosynthesis may have already been going on. See Kennedy, "First Land Plants and Fungi."

6. Claire P. Humphreys, Peter J. Franks, Mark Rees, Martin I. Bidartondo, Jonathan R. Leake, and David J. Beerling, "Mutualistic Mycorrhiza-Like Symbiosis in the Most Ancient Group of Land Plants," *Nature Communications* 1, article no. 103 (2010).

7. Alice Klein, "Plants 'See' Underground by Channelling Light to Their Roots," *New Scientist,* 1 November 2016, https://www.newscientist.com/article/2111027-plants-see-underground-by-channelling-light-to-their-roots/.

8. Mei Mo, Ken Yokawa, Yinglang Wan, and František Baluška, "How and Why Do Root Apices Sense Light under the Soil Surface?," *Frontiers in Plant Science,* 26 October 2015, https://www.frontiersin.org/articles/10.3389/fpls.2015.00775/full.

9. "It is hardly an exaggeration to say that the tip of the radicle thus endowed, and having the power of directing the movements of the adjoining parts, acts like the brain of one of the lower animals; the brain being seated within the anterior end

of the body, receiving impressions from the sense-organs, and directing the several movements." Charles Darwin, *The Power of Movement in Plants* (New York: D. Appleton and Co., 1900), 576.

10. See František Baluška, Stefano Mancuso, Dieter Volkmann, and Peter W. Barlow, "The 'Root-Brain' Hypothesis of Charles and Francis Darwin: Revival after More than 125 Years," *Plant Signaling and Behavior* 4, no. 12 (Dec 2009): 1121–27.

11. See T. T. Kozlowski, *Seed Biology: Importance, Development, and Germination* (New York: Academic Press, 1972).

12. Paul Valéry, *The Collected Works, Vol. 2: Poems in the Rough*, ed. Jackson Mathews, trans. Hilary Corke (Princeton, NJ: Princeton University Press, 1969), 233–34.

13. Jeremy Rehm, "Dandelion Seeds Fly Using 'Impossible' Method Never Before Seen In Nature," *Nature Online*, 17 October 2018, https://www.nature.com/articles/d41586-018-07084-8.

Chapter 11: Elastic Animality

1. "Whilst, therefore, it is essential to consider the social construction of 'the animal' in contradistinction to the human, and to trace the processes through which animals are socially constructed, we cannot dissolve other species into their symbolic reference in human cultures." Erika Cudworth, *Social Lives with Other Animals: Tales of Sex, Death and Love* (Basingstoke: Palgrave Macmillan, 2011), 36.

2. Richard York and Stefano B. Longo, "Animals in the World: A Materialist Approach to Sociological Animal Studies," *Journal of Sociology* 53, no. 1 (Sept 2015): 32–46.

3. See Jacques Derrida, *The Animal That Therefore I Am* (New York: Fordham University Press, 2008).

4. This is why I reject the idea of "flat ontology." See Thomas Nail, "Kinopolitics: Borders in Motion," in *Posthuman Ecologies: Complexity and Process After Deleuze*, eds. Rosi Braidotti and Simone Bignall (London, New York: Rowman & Littlefield International Ltd., 2019), 183–203.

5. "Animality is like an unthinking, unthinkable mirror-twin of subjectivity. According to Lacan, looking into the mirror, the human being appropriates its own image as 'human' in a form originally exterior to itself. But what if it is the animal that exists outside of the mirror, where the human being has to recognize itself and at the same time cannot do so. Re-reading Lacan, Derrida specifies that the real enigma is to be found not in the human being gazing at its own mirror reflection, but rather in the animal that stares back at the human." Oksana Timofeeva and Slavoj Žižek, *The History of Animals: A Philosophy* (London, New York: Bloomsbury Academic, 2018), xv.

6. "Homo sapiens, then, is neither a clearly defined species nor a substance; it is, rather, a machine or device for producing the recognition of the human.... It is an optical machine constructed from a series of mirrors in which man, looking at himself, sees his own image always already deformed in the features of an ape. Homo

is a constitutively 'anthropomorphous' animal . . . , who must recognize himself in a non-man in order to be human." Giorgio Agamben, *The Open: Man and Animal*, trans. Kevin Attell (Stanford: Stanford University Press, 2013), 26.

7. Derrida, *The Animal That Therefore I Am*.

8. Jakob von Uexküll only gives worlds to animals, not plants. He is animal- and bio-centric. For Uexküll, "there is no space independent of subjects," and these subjects include nonhuman animals. Any human concept of a holistic, "all-encompassing world-space" is an anthropocentric "fiction" or "fable." Jakob von Uexküll, *A Foray into the Worlds of Animals and Humans: With A Theory of Meaning* (Minneapolis: University of Minnesota Press, 2010), 70. Merleau-Ponty is also biocentric because non-life has no field of action. He finds in the Umwelt a theory of "life as the opening of a field of action," an opening onto an "inter-animality" that connects beings within species and between different species. See Maurice Merleau-Ponty and Dominique Séglard, *Nature: Course Notes from the Collège De France*, trans. Robert Vallier (Evanston, Ill: Northwestern University Press, 2003), 170. See also Louise H. Westling, "Merleau-Ponty's Human-Animality Intertwining and the Animal Question," *Configurations* 18, no. 1–2 (Winter 2010): 161–80; and Derek Ryan, *Animal Theory: A Critical Introduction* (Edinburgh: Edinburgh University Press, 2015), 105.

9. "And yet, there remains something unsatisfying about the very manner in which Heidegger dismisses the possibility of animals relating to their environment 'as such.'" "In not considering that some relation to the 'as such' may be possible, he asserts that 'the animal is separated from man by an abyss'." Ryan, *Animal Theory*, 105, 106. See also Martin Heidegger, *The Fundamental Concepts of Metaphysics: World, Finitude, Solitude*, trans. William McNeill and Nicholas Walker (Bloomington and Indianapolis: Indiana University Press, 1995), 264. "[Merleau-Ponty] is clear that the animal—and like Heidegger it is often a singular, homogenised 'animal' that is under discussion in this lecture—engages 'a process of "giving shape" to the world.'" Ryan, *Animal Theory*, 59.

10. "In fact, this approach 'does not even privilege human bodies' because 'all bodies, including those of animals . . . evince certain capacities for agency. As a consequence, the human species, and the qualities of self-reflection, self-awareness, and rationality traditionally used to distinguish it from the rest of nature, may now,' according to Coole and Frost, 'seem little more than contingent and provisional forms or processes within a broader evolutionary or cosmic productivity' (20)." Ryan, *Animal Theory*, 74.

As an alternative, according to Ryan, "Barad therefore presents a non-anthropocentric conception of agency: 'refusing the anthropocentrisms of humanism and antihumanism, posthumanism marks the practice of accounting for the boundary-making practices by which the "human" and its others are differentially delineated and defined.' New materialism shifts the discussion away from subject/object, human/nonhuman dichotomies and towards messier entanglements of agency." Ryan, *Animal Theory*, 75. See also Karen Barad, *Meeting the Universe*

Halfway: Quantum Physics and the Entanglement of Matter and Meaning (Durham, NC: Duke University Press, 2007), 136.

11. Natalie Angier, "Bone, a Masterpiece of Elastic Strength," *New York Times*, 27 April 2009, https://www.nytimes.com/2009/04/28/science/28angi.html.

12. Barad, *Meeting the Universe Halfway*, 185.

13. See František Baluška and Michael Levin, "On Having No Head: Cognition throughout Biological Systems," *Frontiers in Psychology* 7, article no. 902 (2016).

14. "Complex signal processing and signal integration, including perception, memory and decision-making, should be extended from neural to aneural systems. Even though the concept of plant neurobiology has been controversially debated and the essential requirements for conscious as compared to unconscious cognition remain the topic of investigation and discussion an exploration of common principles underlying biological information processing in plants and animals leads to new insights and ideas. For instance, it has recently been suggested that both animals and plants are capable of active inference and anticipatory behavior. It is tempting to speculate that to achieve optimal accumulation of information and predictive coding, evolution would have developed strategies for exploiting the benefits of both classical as well as quantum information processing." Peter Jedlicka, "Revisiting the Quantum Brain Hypothesis: Toward Quantum (Neuro)biology?," *Frontiers in Molecular Neuroscience* 10, article no. 366 (2017). See also František Baluška and Stefano Mancuso, "Plant Neurobiology: From Stimulus Perception to Adaptive Behavior of Plants, Via Integrated Chemical and Electrical Signaling," *Plant Signaling and Behavior* 4, no. 6 (Jun 2009): 475–76; Peter Jedlicka, "Quantum Stochasticity and Neuronal Computations," *Nature Precedings* (2009), http://dx.doi.org/10.1038/npre.2009.3702.1; Anthony Trewavas, "Response to Alpi et al.: Plant Neurobiology—All Metaphors Have Value," *Trends in Plant Science* 12, no. 6 (2007): 231–33.

15. Margulis and Sagan, *Microcosmos*, 149.

16. Jedlicka, "Revisiting the Quantum Brain Hypothesis."

17. See Ahmed El Hady and Benjamin B. Matcha, "Mechanical Surface Waves Accompany Action Potential Propagation," *Nature Communications* 6, article no. 6697 (2015).

18. Mo Costandi, "Action Waves in the Brain," *The Guardian*, 1 May 2015, https://www.theguardian.com/science/neurophilosophy/2015/may/01/action-waves-in-the-brain.

19. See image in El Hady and Matcha, "Mechanical Surface Waves," 3.

20. Rodolfo R. Llinás, *I of the Vortex: From Neurons to Self* (Cambridge, Mass: MIT Press, 2008).

21. See P.W. Glimcher, "Indeterminacy in Brain and Behavior," *Annual Review of Psychology* 56, no. 1 (Feb 2005): 25–56; and Jedlicka, "Revisiting the Quantum Brain Hypothesis."

22. Glimcher, "Indeterminacy in Brain and Behavior."

23. See Gustavo Deco, Viktor K. Jirsa, Peter A. Robinson, Michael Breakspear, and Karl Friston, "The Dynamic Brain: From Spiking Neurons to Neural Masses and

Cortical Fields," *PLoS Computational Biology* 4, no. 8, article no. e1000092 (Aug 2008); Wolf Singer, "The Brain, a Complex Self-Organizing System," *European Review* 17, no. 2 (May 2009): 321–29; Emmanuelle Tognoli and J. A. Scott Kelso, "The Metastable Brain," *Neuron* 81, no. 1 (Jan 2014): 35–48.

24. Deco et al., "The Dynamic Brain."

25. See Glimcher, "Indeterminacy in Brain and Behavior"; D. J. Tolhurst, J. A. Movshon, and I. D. Thompson, "The Dependence of Response Amplitude and Variance of Cat Visual Cortical Neurones on Stimulus Contrast," *Experimental Brain Research* 41 (1981): 414–19; and D. J. Tolhurst, J. A. Movshon, and A. F. Dean, "The Statistical Reliability of Signals in Single Neurons in Cat and Monkey Visual Cortex," *Vision Research* 23, no. 8 (1983): 775–85.

26. Tolhurst, Movchon, and Thompson, "The Dependence of Response Amplitude"; Tolhurst, Movshon, and Dean, "The Statistical Reliability of Signals." Recently confirmed by Fred Rieke, David Warland, Rob de Ruyter van Steveninck, and William Bialek, *Spikes: Exploring the Neural Code* (Cambridge, MA: MIT Press, 1997); and Michael N. Shadlen and William T. Newsome, "The Variable Discharge of Cortical Neurons: Implications for Connectivity, Computation, and Information Coding," *JNeurosci* 18, no. 10 (1998): 3870–96.

27. See Glimcher, "Indeterminacy in Brain and Behavior"; Johnjoe McFadden, "The Conscious Electromagnetic Information (Cemi) Field Theory: The Hard Problem Made Easy?," *Journal of Consciousness Studies* 9, no. 8 (2002): 45–60.

28. Dante R. Chialvo, "Emergent Complex Neural Dynamics," *Nature Physics* 6 (2010): 744–50.

29. Jeffrey Satinover, *The Quantum Brain: The Search for Freedom and the Next Generation of Man* (New York, NY: John Wiley & Sons, Inc., 2001).

30. Haim Sompolinsky, "A Scientific Perspective on Human Choice," in *Judaism, Science, and Moral Responsibility*, eds. Yitzhak Berger and David Shatz (New York: Rowman & Littlefield, 2006), 13–44. "What cannot be ruled out is that tiny quantum fluctuations deep in the brain are amplified by deterministic chaos and will ultimately lead to behavioral choices." Christof Koch, "Free Will, Physics, Biology, and the Brain," in *Downward Causation and the Neurobiology of Free Will*, eds. Nancey Murphy, George F. R. Ellis, and Timothy O'Connor (Berlin: Springer-Verlag, 2009), 31–52; 40. See also McFadden, "The Conscious Electromagnetic Information (Cemi) Field Theory."

31. See Akira M. Lippit, *Electric Animal: Toward a Rhetoric of Wildlife* (Minneapolis: University of Minnesota Press, 2010).

Chapter 12: Phanerozoic Earth I: Kinomorphology

1. The "Cambrian explosion" of diversity cannot be explained by the mere increase of oxygen alone. See Douglas Fox, "What sparked the Cambrian explosion?," *Nature* 530 (Feb 2016): 268–70.

2. See Robert M. Hazen, Dominic Papineau, Wouter Bleeker, Robert T. Downs, John M. Ferry, Timothy J. McCoy, Dimitri A. Sverjensky, and Hexiong Yang, "Mineral Evolution," *American Mineralogist* 93 (2008): 1693–720.

3. "Just as evolution is largely ecology writ large, so the organism seems to be ecology writ small. Clearly work needs to be done in this area, but perhaps individual organisms can be understood as spatially and temporally condensed versions of ecological processes." Schneider and Sagan, *Into the Cool*, 256.

4. Fisher's "evolutionary fitness landscape" model is static. For a critique, see John Gribbin, *Deep Simplicity: Bringing Order to Chaos and Complexity* (New York: Random House, 2005).

5. "The relationship is non-linear because phenotype, or set of observable characteristics, is determined by a complex interplay between an organism's genes—tens of thousands of them, all influencing one another's behaviour—and its environment." Keith Bennett, "The Chaos Theory of Evolution," *New Scientist*, 13 October 2010, https://www.newscientist.com/article/mg20827821-000-the-chaos-theory-of-evolution/.

6. McFadden and Al-Khalili, *Life on the Edge*

7. Jablonka, Lamb, and Zeligowski, *Evolution in Four Dimensions*, chapter 4.

8. Luciana Parisi, *Abstract Sex: Philosophy, Bio-Technology and the Mutations of Desire* (London: Continuum, 2004).

9. Margulis and Sagan, *Microcosmos*.

10. "As argued by proponents of embodied cognition, intelligent behavior emerges from the interplay between an organism's nervous system, morphology, and environment [10]–[14]. In a given task environment certain morphologies can readily succeed with simple neural systems, while other morphologies require the discovery of more complex neural systems, or may prevent success altogether." Joshua E. Auerbach and Josh C. Bongard, "Environmental Influence on the Evolution of Morphological Complexity in Machines," *PLoS Computational Biology* 10, no. 1, article no. e1003399 (Jan 2014).

11. Jablonka, Lamb, and Zeligowski, *Evolution in Four Dimensions*, chapter 5.

12. Arhat Abzhanov, "The Old and New Faces of Morphology: The Legacy of D'Arcy Thompson's 'Theory of Transformations' and 'Laws of Growth,'" *Development* 144 (2017): 4284–97.

13. Briggs and Peat, *Turbulent Mirror*, 37–38.

14. See David R. Butler, *Zoogeomorphology: Animals As Geomorphic Agents* (New York: Cambridge University Press, 2007).

15. Per Bak, *How Nature Works: The Science of Self-Organized Criticality* (New York: Springer, 2013).

16. "Our views on natural selection owe much to Lotka and Wicken. Lotka held that natural selection works to increase both the mass of organic systems and the rates of circulation of matter through those systems." Schneider and Sagan, *Into the Cool*, 240. See also Alfred J. Lotka, *Elements of Mathematical Biology* (New York: Dover, 1956); and Jeffrey S. Wicken, *Evolution, Thermodynamics, and Information: Extending the Darwinian Program* (New York: Oxford University Press, 1987).

17. Gribbin, *Deep Simplicity*, chapter 6.

18. Flow produces pedetic mutation. Increased flow increases mutation. Indeed,

seen thermodynamically, variation is inevitable. The "Second Law," writes Wicken, "teleomatically promotes replication errors as ways of increasing configurational randomness. By this law, replication cannot be error-free." Wicken, *Evolution, Thermodynamics, and Information*, 89.

19. Free energy is the key to evolution. Diversity is produced through gradient reduction. The number of North American bird and plant species increases with evapotranspiration, which is the main activity of ecosystems. Indeed, judged by energy expenditure, transpiration is considerably more important than growth in climax ecosystems. Finding a close correlation between the amount of energy degraded and the number of species, Brown compliments the Yale group and "Hutchinson's emphasis on the fundamental role of energetics in evolutionary and community ecology. The acquisition and utilization of energy in accordance with the second law of thermodynamics remains the best place to start 'to understand the higher intricacies of any ecological system.'" James H. Brown, "Two Decades of Homage to Santa Rosalia: Towards a General Theory of Diversity," *American Zoologist* 21, no. 4 (1981): 877–88; 884. Any general theory of biological diversity, adds Brown, will have to deeply incorporate thermodynamics. See Schneider and Sagan, *Into the Cool*, 246–47. Island size, tree count, and temperature all increase species diversity. See also Francis G. Stehli and John W. Wells, "Diversity and Age Patterns in Hermatypic Corals," *Systematic Biology* 20, no. 2 (1971): 115–26.

20. "The animal's various limbs, organs, and tissues represent an increase in diversity similar to the growing biodiversity seen in a developing ecosystem. Then, as in an ecosystem, growth tapers off. An integrated, energy-efficient mature form appears. Adult organisms and mature ecosystems have achieved high levels of energy use and gradient reduction." Schneider and Sagan, *Into the Cool*, 208.

21. Zongjun Yin, Maoyan Zhu, Eric H. Davidson, David J. Bottjer, Fangchen Zhao, and Paul Tafforeau, "Sponge Grade Body Fossil with Cellular Resolution Dating 60 Myr before the Cambrian," *Proceedings of the National Academy of Sciences of the United States of America (PNAS)* 112, no. 12 (Mar 2015): E1453–60.

22. J. Alex Zumberge, Gordon D. Love, Paco Cárdenas, Erik A. Sperling, Sunithi Gunasekera, Megan Rohrssen, Emmanuelle Grosjean, John P. Grotzinger, and Roger E. Summons, "Demosponge Steroid Biomarker 26-Methylstigmastane Provides Evidence for Neoproterozoic Animals," *Nature Ecology and Evolution* 2 (2018): 1709–14.

23. B. F. Lang, C. O'Kelly, T. Nerad, M. W. Gray, and G. Burger, "The Closest Unicellular Relatives of Animals," *Current Biology* 12, no. 20 (Oct 2002): 1773–78.

24. See Thomas Cavalier-Smith, "Origin of Animal Multicellularity: Precursors, Causes, Consequences—The Choanoflagellate/Sponge Transition, Neurogenesis and the Cambrian Explosion," *Philosophical Transactions of the Royal Society B: Biological Sciences* 372, no. 1713 (Feb 2017): 20150476; Rosanna A. Alegado and Nicole King, "Bacterial Influences on Animal Origins," *Cold Spring Harbor Perspectives in Biology* 6, no. 11 (Nov 2014): a016162; and Manuel Maldonado, "Choanoflagellates, Choanocytes, and Animal Multicellularity," *Invertebrate Biology* 123, no. 1 (Mar 2004): 1–22.

25. Edward E. Ruppert, Richard S. Fox, and Robert D. Barnes, *Invertebrate Zoology: A Functional Evolutionary Approach*, 7th ed. (Belmont, CA: Thomson-Brooks/Cole, 2004), 83.

26. This is like what happens in the Sierpinski sponge.

27. "If the water be absolutely still, there is established between these afferent and efferent currents a re-entrant vortex, whose section is a circle in any radial plane through the osculum." G. P. Bidder, "The Relation of the Form of a Sponge to Its Currents," *Journal of Cell Science* (1923): 293–323; 296.

28. M. Nickel, "Kinetics and Rhythm of Body Contractions in the Sponge *Tethya wilhelma* (Porifera: Demospongiae)," *Journal of Experimental Biology* 207 (2004): 4515–24.

29. Ruppert et al., *Invertebrate Zoology*, 76–97.

30. Paulyn Cartwright, Susan L. Halgedahl, Jonathan R. Hendricks, Richard D. Jarrard, Antonio C. Marques, Allen G. Collins, and Bruce S. Lieberman, "Exceptionally Preserved Jellyfishes from the Middle Cambrian," *PLoS ONE* 2, no. 10, article no. e1121 (2007).

31. Ed Yong, "Why a Jellyfish Is the Ocean's Most Efficient Swimmer," *Nature*, 7 October 2013, https://www.nature.com/news/why-a-jellyfish-is-the-ocean-s-most-efficient-swimmer-1.13895.

32. Ruppert et al., *Invertebrate Zoology*, 111–24.

33. "It is possible the urbilaterian never had a brain, and that it later evolved many times independently. Or it could be that the ancestors of the acorn worm had a primitive brain and lost it—which suggests the costs of building brains sometimes outweigh the benefits. Either way, a central, brain-like structure was present in the ancestors of the vertebrates. These primitive, fish-like creatures probably resembled the living lancelet, a jawless filter-feeder." David Robson, "A Brief History of the Brain," *New Scientist*, 21 September 2011, https://www.newscientist.com/article/mg21128311-800-a-brief-history-of-the-brain/.

34. For more detail on the vortex motion of eel and the flatworm's twisting motion, see Figure 11.8 in George V. Lauder and Eric Tytell, "Hydrodynamics of Undulatory Propulsion," *Fish Physiology* 23 (Dec 2005): 425–68; 445.

35. See Rodolfo R. Llinás, *I of the Vortex: From Neurons to Self* (Cambridge, Mass: MIT Press, 2008).

36. This allows us to overcome the dualism posted by Hayles and others who biocentrically draw the line between cognitive and noncognitive activity at the level of "life." This only begs the question of how non-living matter produced life in the first place. No matter where one draws the line, ontological dualism creeps back in. Differences in the structure and circulation of affect do need to be attended to, but this can be done kinetically, without recourse to or need for metaphysical concepts like "cognition," "consciousness," or "thought." See N. Katherine Hayles, *Unthought: The Power of the Cognitive Nonconscious* (Chicago: University of Chicago Press, 2017).

37. H. M. Platt, Foreword to *The Phylogenetic Systematics of Freeliving Nematodes*, by Sievert Lorenzen (London: The Ray Society, 1994).

Chapter 13: Phanerozoic Earth II: Terrestrialization

1. See "List of Animals by Number of Neutrons," *Wikipedia*, accessed 23 August 2019, https://en.m.wikipedia.org/wiki/List_of_animals_by_number_of_neurons.

2. See West, *Scale*; and Adrian Bejan, *The Physics of Life: The Evolution of Everything* (New York: St Martin's Press, 2016).

3. Ruppert et al., *Invertebrate Zoology*, 518–22.

4. J. Dzik, "The Verdun Syndrome: Simultaneous Origin of Protective Armour and Infaunal Shelters at the Precambrian–Cambrian Transition,"*Geological Society of London: Special Publications* 286 (Jan 2007): 405–14.

5. Bernard L. Cohen, "Not Armour, but Biomechanics, Ecological Opportunity and Increased Fecundity as Keys to the Origin and Expansion of the Mineralized Benthic Metazoan Fauna," *Biological Journal of the Linnean Society* 85, no. 4 (Aug 2005): 483–90.

6. Carla Stecco, Warren Hammer, Andry Vleeming, and Raffaele De Caro, eds., "Connective Tissues," in *Functional Atlas of the Human Fascial System* (London: Churchill Livingstone Elsevier), 1–20.

Chapter 14: Kinocene Earth

1. See Nail, *Theory of the Object*.

2. See Nail, *Being and Motion*; and Nail, *Lucretius I*.

3. The American mathematician, physical chemist, statistician, and energy theorist Alfred James Lotka called this "the maximum power principle," or Lotka's principle, and it has been proposed as the fourth principle of energetics in open system thermodynamics, where an example of an open system is a biological cell. According to Howard T. Odum, "The maximum power principle can be stated: During self-organization, system designs develop and prevail that maximize power intake, energy transformation, and those uses that reinforce production and efficiency." H. T. Odum, "Self-Organization and Maximum Empower," in *Maximum Power: The Ideas and Applications of H. T. Odum*, ed. Charles A. S. Hall (Niwot, CO: University Press of Colorado, 1995), 311–30; 311.

My contribution to this theory is to have shown that this maximum power is expressed historically in four major kinetic patterns: centripetal, centrifugal, tensional, and elastic. The kinetic pattern is related to the rate of dissipation. "The concept of second law efficiency under maximum power . . . Neither the first or second law of thermodynamics include a measure of the rate at which energy transformations or processes occur. The concept of maximum power incorporates time into measures of energy transformations. It provides information about the rate at which one kind of energy is transformed into another as well as the efficiency of that transformation." Martha W. Gilliland, ed., *Energy Analysis: A New Public Policy Tool* (Boulder, CO: Westview Press, 1978), 101.

Lotka's work was important because he "provided the theory of natural selection as a maximum power organizer; under competitive conditions systems are selected which use their energies in various structural-developing actions so as to maximize their use of available energies. By this theory systems of cycles which drain less energy lose out in comparative development. However Leopold and Langbein have shown that streams in developing erosion profiles, meander systems, and tributary networks disperse their potential energies more slowly than if their channels were more direct. These two statements might be harmonized by an optimum efficiency maximum power principle (Odum and Pinkerton 1955), which indicates that energies which are converted too rapidly into heat are not made available to the systems own use because they are not fed back through storages into useful pumping, but instead do random stirring of the environment" H.T. Odum, "Energy Values of Water Sources," *Proceedings of the Nineteenth Southern Water Resources and Pollution Control Conference—April* 1970, 62. https://ufdc.ufl.edu/AA00004068/00001/2j.

The concept of maximum power can therefore be defined as the *maximum rate of useful energy transformation.* "... the maximum power principle ... states that systems which maximize their flow of energy survive in competition. In other words, rather than merely accepting the fact that more energy per unit of time is transformed in a process which operates at maximum power, this principle says that systems." Gilliland, *Energy Analysis*, 101–2. "The maximum power principle is a potential guide to understanding the patterns and processes of ecosystem development and sustainability. The principle predicts the selective persistence of ecosystem designs that capture a previously untapped energy source." T.T. Cai, C.L. Montague and J.S. Davis, "The maximum power principle: An empirical investigation," *Ecological Modelling* 190, no. 3–4 (2006): 317–35.

4. See Nail, *Theory of the Object*.
5. See image in Schneider and Sagan, *Into the Cool*, 266.
6. See Nail, *Being and Motion* and Nail, *Theory of the Object*.
7. Vaclav Smil, *Energy in Nature and Society: General Energetics of Complex Systems* (Cambridge, MA: MIT Press, 2008), 91.
8. Ibid., 29.
9. Ibid., 48.
10. Ibid., 54.
11. Ibid., 58.
12. Ibid., 53.
13. Ibid., 46.
14. Ibid.
15. Ibid., 39.
16. Schneider and Sagan, *Into the Cool*, 134.
17. Smil, *Energy in Nature and Society*, 39.
18. Ibid., 41.
19. Ibid., 46.
20. Ibid., chapter 10.
21. Ibid., 63.
22. Schneider and Sagan, *Into the Cool*, 256.

23. John P. DeLong, Jordan G. Okie, Melanie E. Moses, Richard M. Sibly, and James H. Brown, "Shifts in Metabolic Scaling, Production, and Efficiency across Major Evolutionary Transitions of Life," *PNAS* July 20, 2010 107 (29) 12941–45; https://doi.org/10.1073/pnas.1007783107.

24. A. I. Zotin, "Thermodynamic Aspects of Developmental Biology," *Monographs in Developmental Biology* 5 (1972): 1–59.

25. Schneider and Sagan, *Into the Cool*, 242.

26. Viviane Richter, "The Big Five Mass Extinctions," *Cosmos Magazine*, accessed 24 August 2019, https://cosmosmagazine.com/palaeontology/big-five-extinctions.

27. "Human gradient-reducing abilities pale in comparison to those of plants. . . . Plants are our great planetary gradient reducers—and, yes, they perceive the solar gradient, turning their leaves and flowers to follow the sun. Indeed if we accept the evidence for anthropogenic global warming, it is clear that humans, despite our minds, have depleted global gradient-reducing function by increasing temperatures near the surface. Human technics are amazing but literally globally dysfunctional." Sagan, *Cosmic Apprentice*, 215.

28. Schneider and Sagan, *Into the Cool*, 136.

29. See image in ibid., 222.

30. Christopher M. Gough, "Terrestrial Primary Production: Fuel for Life," *Nature Education Knowledge* 3, no. 10 (2011): 28.

31. J. C. Luvall and H. R. Holbo, "Measurements of Short-Term Thermal Responses of Coniferous Forest Canopies Using Thermal Scanner Data," *Remote Sensing of Environment* 27, no. 1 (1989): 1–10.

32. See E. D. Schneider and J. J. Kay, "Nature Abhors a Gradient," in *Proceedings of the 33rd Annual Meeting of the International Society for the Systems Sciences*, ed. Paul Ledington, vol. 3 (Edinburgh: International Society for the Systems Sciences, 1989), 19–23; and Schneider and Sagan, *Into the Cool*, 40.

33. See image in Schneider and Sagan, *Into the Cool*, 255.

34. See image in ibid., 193.

35. H. J. M. Bowen, *Trace Elements in Biochemistry* (London: Academic Press, 1966); Robert H. Whittaker and Gene E. Likens, "The Biosphere and Man," in *Primary Productivity of the Biosphere. Ecological Studies (Analysis and Synthesis)*, eds. H. Lieth and R. H. Whittaker, vol. 14 (New York: Springer, 1975): 305–28; Vaclav Smil, *Enriching the Earth: Fritz Haber, Carl Bosch, and the Transformation of World Food Production* (Cambridge, MA: MIT Press, 2001), cited in Smil, *Energy in Nature and Society*, 111.

36. See Paul Colinvaux and Cristina Eisenberg, *Why Big Fierce Animals Are Rare: An Ecologist's Perspective* (Princeton, NJ: Princeton University Press, [1978] 2018).

37. Vaclav Smil, *Energy and Civilization: A History* (Cambridge, MA: MIT Press, 2018), 19.

38. Smil, *Energy and Civilization*, 23.

39. Smil, *Energy in Nature and Society*, 134.

40. Ibid., 137.

Chapter 15: Kinocene Ethics

1. Sagan, *Cosmic Apprentice*, 235.

2. See Yinon M. Bar-On, Rob Phillips, and Ron Milo, "The Biomass Distribution on Earth," *Proceedings of the National Academy of Sciences* Jun 2018, 115 (25) 6506–11. For a nice summary and framing of human impact, see John Vidal, "The Rapid Decline of the Natural World Is A Crisis Even Bigger than Climate Change," *Huffington Post*, 03/15/2019. https://www.huffpost.com/entry/nature-destruction-climate-change-world-biodiversity_n_5c49e78ce4b06ba6d3bb2d44

3. Bar-On, Phillips, and Milo, "The Biomass Distribution on Earth".

4. See image in ibid.

5. J. M. Adams and H. Faure, "A New Estimate of Changing Carbon Storage on Land Since the Last Glacial Maximum, Based on Global Land Ecosystem Reconstruction," *Global and Planetary Change* 16–17 (1998): 3–24.

6. See R.A. Houghton, "Why Are Estimates of the Terrestrial Carbon Balance So Different?," *Global Change Biology* 9, no. 4 (Apr 2003): 500–509; and Sassan S. Saatchi, Nancy L. Harris, Sandra Brown, Michael Lefsky, Edward T. A. Mitchard, William Salas, Brian R. Zutta, Wolfgang Buermann, Simon L. Lewis, Stephen Hagen, Silvia Petrova, Lee White, Miles Silman, and Alexandra Morel, "Benchmark Map of Forest Carbon Stocks in Tropical Regions across Three Continents," *PNAS* 108, no. 24 (Jun 2011): 9899–904.

7. Benjamin D. Lieberman and Elizabeth Gordon, *Climate Change in Human History: Prehistory to the Present* (London and New York: Bloomsbury Academic, 2018); Sing C. Chew, *World Ecological Degradation: Accumulation, Urbanization, and Deforestation*, 3000 B.C.- 2000 A.D. (Walnut Creek: AltaMira Press, 2001); Christopher T. Fisher, J. Brett Hill, and Gary M Feinman, eds., *The Archaeology of Environmental Change: Socionatural Legacies of Degradation and Resilience* (Tucson: University of Arizona Press, 2009); Naomi Klein, *This Changes Everything: Capitalism vs. the Climate* (New York: Simon & Schuster, 2014); Marcel Mazoyer and Laurence Rodart, *A History of World Agriculture: From the Neolithic Age to the Current Crisis* (New York: Monthly Review Press, 2006); Bill McKibben, *Falter: Has the Human Game Begun to Play Itself Out?* (New York: Henry Holt and Co., 2019); Timothy Mitchell, *Carbon Democracy: Political Power in the Age of Oil* (London: Verso, 2011); Jason W. Moore, ed., *Anthropocene or Capitalocene? Nature, History, and the Crisis of Capitalism* (Oakland, CA: PM Press, 2016); Jason W. Moore, "The Capitalocene, Part I: On the Nature and Origins of Our Ecological Crisis," *The Journal of Peasant Studies* 44, no. 3 (May 2017): 594–630; Jason W. Moore, "The Capitalocene, Part II: Accumulation by Appropriation and the Centrality of Unpaid Work/Energy," *The Journal of Peasant Studies* 45, no. 2 (Feb 2018): 237–79; William Ruddiman, *Plows, Plagues, and Petroleum: How Humans Took Control of the Climate* (Princeton, NJ: Princeton University Press, 2005); and Michael Williams, *Deforesting the Earth: From Prehistory to Global Crisis, An Abridgment* (Chicago: University of Chicago Press, 2006).

8. Nail, *The Figure of the Migrant*; Nail, *Theory of the Border*.

9. For a full account of this long history, see Nail, *The Figure of the Migrant* and Nail, *Theory of the Border*.

10. Susan Casey, *The Wave: In Pursuit of the Rogues, Freaks, and Giants of the Ocean* (New York: Doubleday, 2010), 153.

11. Rafi Letzter, "Climate Change Could Make These Super-Common Clouds Extinct, Which Would Scorch the Planet," *Live Science*, February 25, 2019. https://www.livescience.com/64852-clouds-extinct-climate-change.html

12. Will Dunham, "Bolt from the Blue: Warming Climate May Fuel More Lightning," *Reuters,* 13 November 2014, https://www.reuters.com/article/us-science-light ning/bolt-from-the-blue-warming-climate-may-fuel-more-lightning-idUSKC NoIX2B020141113

13. Kathryn Yusoff, *A Billion Black Anthropocenes or None* (Minneapolis, MN : University of Minnesota Press 2018).

14. Craig Welch, "Half of All Species Are on the Move—And We're Feeling It," *National Geographic*, April 27, 2017. https://www.nationalgeographic.com/news/2017/04/climate-change-species-migration-disease/

15. Brad Plumer, "How More Carbon Dioxide Can Make Food Less Nutritious," *New York Times*, 23 May 2018, https://www.nytimes.com/2018/05/23/climate/rice-global-warming.html.

16. Perhaps the most notable proponent of this theory was Howard T. Odum, sometimes considered the father of ecosystems ecology.

17. See Smil, *Energy in Nature and Society*; and Vaclav Smil, *Cycles of Life: Civilization and the Biosphere* (New York: Scientific American Library, 2001).

18. According to Tennenbaum, Leontief's input-output method was adapted to embodied energy analysis by Hannon to describe ecosystem energy flows. Hannon's adaptation tabulated the total direct and indirect energy requirements (the energy intensity) for each output made by the system. The total amount of energies, direct and indirect, for the entire amount of production was called the embodied energy. See Stephen E. Tennenbaum, *Network Energy Expenditures for Subsystem Production* (Gainsville, FL: University of Florida, 1988); and Bruce Hannon, "The Structure of Ecosystems," *Journal of Theoretical Biology* 41, no. 3 (Oct 1973): 535–46.

19. Smil, *Energy in Nature and Society*.

20. See Smil, *Energy in Nature and Society* and Schneider and Sagan, *Into the Cool*, chapter 19.

21. See Gribbin, *Deep Simplicity*.

22. Adrian Bejan, *The Physics of Life: The Evolution of Everything* (New York: St Martin's Press, 2016).

23. McKibben, *Falter* and David Wallace-Wells, *The Uninhabitable Earth: Life After Warming* (New York: Tim Duggan Books, 2019).

24. Aisha Majid, "WHO Reveals 7 Million Die From Pollution Each Year in Latest Global Air Quality Figures," *The Telegraph*, 1 May 2018, https://www.telegraph

.co.uk/news/2018/05/01/estimates-7-million-die-pollution-year-reveals-latest-global/.

25. See Bataille, *The Accursed Share*; and Allan Stoekl, *Bataille's Peak: Energy, Religion, and Postsustainability* (Minneapolis: University of Minnesota Press, 2007). Stoekl agrees that Bataille is anthropocentric because he restricts the knowledge and ethics of expenditure to humans (religion). Animals are unaware of their limits, have no meaning, and no purposive act. They are just homogeneous movement. Stoekl's ethical solution to expenditure is the use of human muscle power and recycling. But additionally, we need to waste more as a planet by repopulating the biosphere. "To deny the ethical moment, the moment in which conservation and meaning are established only the better to affirm the destruction of expenditure, is to relegate that destruction to the simple, homogeneous movement of the animal, unaware of limit, meaning, and purposive act." Stoekl, *Bataille's Peak*, xvii.

For Stoekl and Bataille, the cosmos wastes but humans get to be the ones that know that they waste and thus choose ethically to waste and thus are religious—contra animals plants and the earth. I disagree.

26. "The second law tugs organisms to find ways to work together to stably reduce gradients, dissipating the energy that sustains them. The 'sustainable' part is crucial but often gets lost in the work of thermodynamic theorizers and those who would critique them as making an unreconstructed unthinking and politically dangerous contribution to neoliberalism, as objectionable in its way as is social Darwinism and neo-Darwinism's caricature of Darwinism." Sagan, *Cosmic Apprentice*, 214.

27. "The generalized sensitivities proposed here (whose values are reported in Table 1) demonstrate that the climate system becomes less efficient, more irreversible, and features higher entropy production as it becomes warmer." See image in Valerio Lucarini, Klaus Fraedrich, and Frank Lunkeit, "Thermodynamics of Climate Change: Generalized Sensitivities," *Atmospheric Chemistry and Physics* 10 (2010): 9729–37; 9736.

28. Blaise Pascal, *Pensées and Other Writings*, trans. Honor Levi (Oxford: Oxford University Press), 126.

29. Smil, *Energy in Nature and Society*, 386.

30. Ibid.

31. Lucretius, *On the Nature of Things: De Rerum Natura*, trans. Walter Englert, ed. Albert Keith Whitaker (Newburyport, MA: Focus Publishing, 2003), Book IV, line 833.

32. For a lovely critique of conservation, see Stoekl, *Bataille's Peak*.

33. Even new-materialist vitalism projects life onto inorganic matter. See Chris Gamble, Josh Hannan, and Thomas Nail, "What Is New Materialism?" *Angelaki: Journal of the Theoretical Humanities* 24, no. 6 (2019): 111–34. See also Nail, *Being and Motion*, chapter 3.

34. Lijing Cheng et al., "Record-Setting Ocean Warmth Continued in 2019," *Advances in Atmospheric Sciences* 37 (2020): 137–42.

35. C. Dyke, "Cities as Dissipative Structures," in *Entropy, Information, and*

Evolution: New Perspectives on Physical and Biological Evolution, eds. Bruce H. Weber, David J. Depew, and James D. Smith (Cambridge, MA: MIT Press, 1988), 355–67; 365.

36. Harold J. Morowitz, *The Kindly Dr. Guillotin: And Other Essays on Science and Life* (Washington, DC: Counterpoint, 1997), 121.

37. "[N]o other conception could be further from the correct interpretation of facts. Even if only the physical facet of the economic process is taken into consideration, this process is not circular, but unidirectional. As far as this facet alone is concerned, the economic process consists of a continuous transformation of low entropy into high entropy, that is, into irrevocable waste or, with a topical term, into pollution." Nicolas Georgescu-Roegen, *The Entropy Law and the Economic Process* (Cambridge, MA: Harvard University Press, 1971), 281.

38. For a book-length defense of this claim, see Nail, *Marx in Motion*.

39. For a nice, brief history, see Raj Patel and Jason W. Moore, *A History of the World in Seven Cheap Things: A Guide to Capitalism, Nature, and the Future of the Planet* (Oakland, CA: University of California Press, 2018).

40. J. E. Lovelock and M. Whitfield, "Life Span of the Biosphere," *Nature* 296 (1982): 561–63.

41. Smil, *Energy in Nature and Society*, 387.

42. Lorraine Chow, "The Climate Crisis May Be Taking a Toll on Your Mental Health," *Salon*, 22 May 2017, https://www.salon.com/2017/05/22/the-climate-crisis-may-be-taking-a-toll-on-your-mental-health_partner/.

43. Winona LaDuke, *All Our Relations: Native Struggles for Land and Life*, 2nd ed. (Chicago: Haymarket Books, 2016).

44. Vandana Shiva, *Who Really Feeds the World? The Failures of Agribusiness and the Promise of Agroecology* (Berkeley, CA: North Atlantic Books, 2016).

45. See Greta Gaard, "Ecofeminism Revisited: Rejecting Essentialism and Re-Placing Species in a Material Feminist Environmentalism," *Feminist Formations* 23, no. 2 (Summer 2011): 26–53; and Deane Curtin, "Recognizing Women's Environmental Expertise," in *Chinnagounder's Challenge: The Question of Ecological Citizenship* (Bloomington and Indianapolis, IN: Indiana University Press, 1999), 73–88.

46. See Karen Warren, ed., *Ecofeminism: Women, Culture, Nature* (Bloomington, IN: Indiana University Press, 1997); and Greta Gaard and Lori Gruen, "Ecofeminism: Toward Global Justice and Planetary Health," *Society and Nature* 2 (1993): 1–35.

47. "Women and Hunger Facts," *World Hunger: Hunger Notes*, accessed 24 August 2019, https://www.worldhunger.org/women-and-hunger-facts/.

48. Catriona Mortimer-Sandilands and Bruce Erickson, eds., *Queer Ecologies: Sex, Nature, Politics, Desire* (Bloomington, IN: Indiana University Press, 2010).

49. See Betsy Hartmann, *Reproductive Rights and Wrongs: The Global Politics of Population Control*, 3rd ed. (Chicago: Haymarket Books, 2016).

50. See Silvia Federici, *Caliban and the Witch: Women, the Body and Primitive Accumulation* (Brooklyn, NY: Autonomedia, 2014); Silvia Federici and Peter Linebaugh, *Re-Enchanting the World: Feminism and the Politics of the Commons* (Oakland, CA: PM Press, 2019); Floraine Clement, Wendy Harcourt, Deepa Joshi, and Chizu

Sato, "Feminist Political Ecologies of the Commons and Commoning," *International Journal of the Commons* 13, 1: 1–15; Peter Linebaugh, *The Magna Carta Manifesto: Liberties and Commons for All* (Berkeley, CA: University of California Press, 2009); David Bollier and Silke Helfrich, eds., *Patterns of Commoning* (Amherst, MA: The Commons Strategies Group, 2015).

Conclusion: The Future

1. See Nail, *The Figure of the Migrant*; Nail, *Theory of the Border*; Nail, *Being and Motion*; Nail, *Theory of the Image*; and Nail, *Theory of the Object*.

2. "My thoughts are to show me *where I stand*, but they are not to tell me where I am going to—I love the ignorance about the future and I do not want to perish in light of impatience and the anticipation of *augured* things." Friedrich Nietzsche, *The Gay Science: With a Prelude in Rhymes and an Appendix of Songs*, trans. Walter Kaufmann (New York: Vintage Books, 1974), 231.

Index

accretion, 66, 67, 68, 69, 76, 94; centripetal, 65, 75, 78, 79, 80, 91, 106
accumulation, 250, 256, 257, 258; in bone, 220; in capitalism, 263, 267; of cell nuclei, 149; central, 95; of energy, 52, 99, 148, 231, 234, 254; in evolution, 306n14; in flowers, 170, 171; of granite, 85; of heavy metals, 46; by humans, 2, 256, 257; by lightning, 86; luminous, 107; material, 231; of matter, 21, 33, 95; of microbes, 130, 200; of minerals, 83, 92, 97, 181; by nature, 231; of nitrogen oxides, 86; in nuclei, 301n11; patterns of, 118; of people, 249; of prebiotic materials, 116, 117; process of accretion, 75; of radiation, 207; of reproductive structures and cycles, 105; in stars, 52; of Theia's iron core, 78; vegetal, 238; of volatiles, 96; in water, 101
accumulation, centripetal, 52, 85, 95, 146, 171; of minerals, 78, 81, 87, 96
action potentials, 186, 188, 189
action waves, 186–187, 188

adaptation, 120, 128, 203, 210, 215, 218, 219, 223; in animal evolution, 194–195; cellular, 146
affective meshwork, 165–166
agency, 6, 126, 142, 170, 179, 283n39, 286n57, 307n10; of the earth, 2, 233, 242; geological, 2, 177, 182, 206, 247; human, 5
agriculture, 2, 158, 243, 245, 250, 265, 266
algae, 145, 147, 154, 158, 159, 167, 207; tubes, 155–156, 182
alterity, lunar, 78–81
amino acids, 46, 121, 122, 123, 125, 149, 183; as prebiotic material, 117, 119, 142
anemones, 201, 202
animal bodies, 198, 201, 204, 206, 215, 216, 222, 223; elasticity in, 181, 182, 183, 184, 190, 192, 194, 195; emergence of, 178; shape and function of, 196; skeletons of, 219
animal theory, 179–180
animality, 170, 177, 181, 193, 208, 250, 304n5, 305n8; elastic, 173, 182, 183,

319

186, 192, 217; history of, 180, 196, 204; theory of, 178, 179
animalization, 208, 218, 219
animals, wild, 243–244, 245, 250, 259, 265. *See also* mammals
Anthropocene, 7, 277n4, 279n15, 283n40, 287n61, 288n81, 293n6; definition of, 1–2, 254; and Kinocene, 14; science of, 287n64; unstable mobility of, 10, 20
anthropocentric energetics, 250
anthropocentrism, 19, 23, 253, 254, 259, 280n17, 288n75, 289n2; and agency, 305n10; destroying the earth, 260; in geophilosophy, 285n51; of Georges Bataille, 316n25
anti-form, 111–112
anus, 204, 206, 207, 209, 211
Archean Eon, 91, 100, 102, 104, 110, 124, 125, 233; amino acids production in, 122, 123; atmosphere in, 61, 91, 94, 106, 107, 109; centrifugal motion in, 94, 131; earth's movement in, 105, 106; emergence of clouds in, 111, 113; great die-off of, 254; ocean in, 101, 102; organisms in, 116, 130; oxygenation event of, 97, 153
archivalization, 69–70, 71, 72, 73
Aristophanes, 114
Aristotle, 24, 25, 137, 138, 229
arms, 212, 213, 215, 216
arthropods, 208, 215, 216–218, 219, 221
asteroids, 28, 46, 66, 75, 81, 85, 105, 280n15
atmospheric expenditure, 234–235, 239
atmospheric field, 91, 94. *See also* planetary fields
atmospheric geokinetics, 91–94
atmospheric pressure, 143, 163, 164, 235
atoms, 32, 33, 45, 48, 71, 100, 113, 292n2; accumulation of, 117; atomic elements, 30, 39, 70; atomism, 114, 290n14; circulations of, 72; in crystal formation, 63, 102; hydrogen, 81, 101, 295n4; patterns of, 69; pedetic motion of, 28; subatomic particles, 40, 128; weight of, 34
attraction, 52, 170–171, 172
attractor, periodic, 35, 36, 37
Australia, 84, 130
axons, 184, 186–187

Bak, Per, 196
balance, 3, 110, 146, 196, 203, 211, 234, 259; chemical, 135; electrically polarized, 186; of excitation and inhibition, 189; metabolic, 39; tensional, 53
Baragwanathia, 162
Bataille, Georges, 4, 253, 296n6, 316n25
Bertalanffy, Ludwig von, 119
Big Bang, 21, 22, 23, 38, 39, 45, 52, 259; in dualistic metaphysics, 24; in quantum gravity, 291n18
bilateralism, 203, 204, 205, 207, 209
biocentrism, 23, 44, 64, 84, 99, 228, 305n8, 310n36; and conservation, 15, 260; in evolutionary theory, 142; image of the earth in, 5, 282n33
biochemistry, 98, 102, 128, 129, 150, 160, 165
biodiversity, 15, 237, 243, 244, 247, 250, 265, 309n20
biogenesis, 116, 118, 123, 185
biomass, 238, 239, 240, 244, 247, 253, 262; destruction of, 243, 245, 250, 255, 256, 261
bio-mineralization, 181, 184, 200, 219
biosphere, 12, 27, 51, 97, 178, 234, 242, 244; absorbs solar radiation, 239; dark, 98; destruction of, 241, 243, 247, 257, 273; dissipative processes of, 15; dying, 256; feedback loop of, 191; functioning of, 279n15; human portion of, 251; repopulating, 316n25; survival of, 258, 259
birds, 170, 219, 221, 244, 273, 309n19

black holes, 52–53, 193, 231, 274
bodies. *See* animal bodies; celestial bodies; vegetal bodies
bonding, molecular, 70, 71, 101, 114, 123, 162, 236
bone, 8, 68, 93, 183, 220, 221, 222, 247; calcium in, 184, 221; elasticity of, 181, 219–220; remodeling, 220–221; role in adaptation, 223. *See also* skeleton
Brahe, Tycho, 47
brains, 71, 72, 93, 209, 215, 303–304n9, 307n30; costs of building, 310n33; elasticity of, 204–205; energy consumption of, 240; flatworms, 204–205; human, 114, 271; mollusks, 211, 212, 213; morphology of, 205; neuronal firing in, 113, 183; neuroplasticity, 220; rotifers, 207. *See also* neurons
bristles, 215, 217–218, 221
Buffon, Comte de, 69

calcium, 153, 181, 182, 186, 199, 216, 220, 221; metabolism of, 183; regulation of, 184
Cambrian period, 184, 201, 211, 212, 215, 237; explosion, 173, 177, 191, 197, 204, 219
capillaries, 148–149, 161–162, 163, 219
capitalism, 39, 253, 260, 261–263, 265, 267, 269; economics and, 245, 251, 261, 262; fossil fuel, 15, 245, 247, 251, 256, 258, 266; geoengineering and, 47; plants and, 249, 250
carbon dioxide, 83, 85, 106, 154, 234, 249, 250
Carboniferous period, 217
cartilage, 181, 183, 184, 220
celestial bodies, 42, 44, 48, 51, 53, 77, 80
cell membranes, 102, 103, 144, 145, 149, 181, 184, 186; elasticity of, 182, 185; and folding, 122, 141; for regulating flow, 142–143

cell walls, 140, 147, 149, 153, 155, 162, 185, 199; pressure of, 141, 142–143; rigidity of, 160, 167, 182, 183, 196
cells, animal, 182, 183, 185, 192
cells, vegetal, 143, 148, 182, 187
cellularity, 102, 144, 146, 148, 167, 180
cellularization, 140, 141, 144, 153, 154, 156
center of orientation, 103, 106
centrifugal motion, 52, 93, 99, 104, 107, 116, 130, 140; during Archean Eon, 131; and atmosphere, 13, 91, 94–95; balance with centripetal motion, 146; patterns, *14*
centripetal motion, 13, 51, 52, 65–66, 68, 73, 87, 94; patterns, *14*
cephalopods, 211–214
chemicals, 4, 127, 135, 186, 213, 217, 218, 236; as biomarkers, 197; catalyzation, 100, 108, 110; in DNA mutation, 124–125; earth's changes in, 24, 178, 300n32; electron bonding, 71; in metabolism, 101, 121, 122; mollusks detect, 211; plants release, 150, 170, 302n22; prebiotic, 117; reactions, 71, 102; stability, 300n32; transformation, 103; volatile, 150. *See also* biochemistry; electrochemistry
chlorophyll, 159, 182
choanoflagellates, 197, *197*, 199. *See also* flagella
circulation, 1, 19, 105, 122, 247, 261, 308n16, 310n36; centrifugal, 52, 53, 87, 98, 129, 130, 162; centripetal, 51, 82; elastic, 197–198; elliptical, 47; fields, 42, 43, 45–48, *48*, 50, 54–55; kinetic processes of, 96; planetary, 20, 44, 67, 76; vortical, 199–200
civilization, 228, 243, 255. *See also* culture
climate, 14, 76, 91–92, 264, 283n40, 287n72, 288n72, 316n27; climatology, 98; cycles, 38, 290n11;

forecasting, 281n21; history, 23, 28; lunar effects on, 46, 50; stability, 4, 11, 15
climate change, 2, 4, 243, 251, 262, 276, 286n53, 290n11; abruptness of, 280–281n20; earth's role in, 7, 46, 288n80; forces migration, 1, 249, 265
clouds, 96, 106, 108, 130, 131, 138, 165, 249; formation, 92, 111–115, 118; gas, 46, 47, 51; and geoaesthetics, 107; during Hadean Era, 85, 86; mineral, 77; nebular, 44, 67; and plants, 169, 178
clusters, 24, 45, 52, 53, 215
coevolution, 122, 123, 125, 129, 182, 204, 205, 208; of plants and animals, 173
collagen, 181, 192, 198, 199, 201, 205, 220, 222; muscular elasticity and, 183–184
collision, 66, 76, 77, 78, 85, 280n15; with Theia, 79, 81, 83
colonies, 147, 154, 157, 198
colonization (human), 245, 247, 250, 265, 266, 267
communication, 115, 130, 148, 170, 182, 183, 184, 190; cephalopod, 214; of the earth, 93; electrochemical, 185; kinetic structure of, 107; neurological, 187; of roots, 165, 186; theories of, 93; vegetal, 149–150
communism (kinetic/metabolic), 267–268
competition, 161, 166, 311n3, 312n3
composites, 34, 45, 50, 51, 69, 144, 148, 157; cosmic, 38; mineral, 66; mobile, 35; organisms, 158, 237; stone, 63, 70
composition, 27, 41, 51, 66, 110, 187, 199, 285n53
composting, 24, 25, 27, 258, 266, 267, 268, 274; ethics of, 263–264
conditions, material, 12, 43, 64, 69, 101, 108, 113, 269; for amino acids, 122; atmosphere, 106, 130; of the earth, 20, 22, 46, 54, 55, 62, 82; for energy expenditure, 234; of existence, 8; of geo-ontogenesis, 109; of humans, 6, 10; of life, 84, 129; minerality, 70; for new organisms, 177; phytality, 155; of sexuation and sexuality, 172
conjunctions, 35, 38–41, 40, 45, 47, 52–53
conservation, 2, 15, 21, 231, 254, 263, 267, 316n25; argument against, 258–260
constructivism, 9, 14, 92, 262, 269
contraction, 53, 180, 186, 192, 194, 196, 213, 259; centripetal, 77, 84; jelly bell, 204; medusa jelly, 201; muscle, 183, 184; sponge, 200. *See also* expansion
convection, 78, 80, 92, 96, 111, 153, 234, 236; cells, 102, 103, 107, 154; currents, 68, 102, 112; patterns, 85, 110
convection cycles, 97, 100, 102, 103, 118; of the atmosphere, 92, 94, 106, 140; of the earth, 80, 95, 99, 193
Cooksonia, 160
cool, the, 37, 231, 237, 243
Copernicus, 3, 26
coral, 201
cores, 65, 76, 78, 98, 290n11. *See also* earth's core
corporeal jelly, 198–199
cosmic matter, 19, 27, 54, 122, 151
cosmos, 151–152
Cretaceous period, 169
criticality, 94, 189, 195, 196
crust, earth's, 68, 76, 83, 96, 106, 234, 279n15, 295n4; formation of, 75; in Hadean Eon, 91; metastable, 77–78; mineral archives in, 113; molten state of, 79; seawater bends, 249
crystallization, 63, 77, 83, 86, 102, 108, 112, 196; of the earth's crust, 75, 79; gravitational, 80; of pedetic motions, 30; quartz, 71; of water, 113, 190
crystals, 63, 69, 70, 78, 82, 102, 103, 190; ice, 112, 113, 114; polysilicate, 71; zircon, 83

Index 323

culture, 12, 25, 43, 107, 265, 283n40, 289n81, 304n1; and deep history, 8; expends energy, 266; history of, 6, 7, 80, 139, 179; moon's role in, 81; vs. nature, 6, 62, 281n26, 283n38, 284n45; Western, 26, 44, 79, 254, 288n73
currents, 85, 101, 156, 187, 235, 236; air, 31, 140, 171, 235; convection, 68, 102, 112; ocean, 193, 235, 236, 280n15; water, 101, 140, 199, 218, 310n27
Cuvier, George, 117
cyanobacteria, 145, 153, 154, 158, 160, 200, 254
cycles, 34, 35–38, 259, 262, 263, 264, 292n37, 312n3; carbon, 244, 255, 261; chemical, 100, 178; closed, 11; cosmic, 5; elemental, 36, 40; extinction, 45; fluid, 100; folding, 99; human breath, 96; lunar, 91; metabolic, 101; orbital, 66; and period, 38; periodic, 40, 41, 42; planetary, 27, 39, 291n27; reproductive, 105; respiration, 99; rock, 39; solar, 110; vapor, 96; water, 82, 115, 235, 236, 249. *See also* convection cycles; kinetic cycles
cytoplasm, 125–126, 143–144, 145, 187, 301n11; strands, 141, 147, 149, 159, 199
cytoskeleton, 144, 149, 301n14

D double prime, 79, 86
dandelion, *172*
dark biosphere, 98
dark matter, 28, 33, 42, 44, 45, 47, 51, 190; galactic, 46, 53
Darwin, Charles, 137–138, 166
Darwinism, 119, 124, 195, 316n26
Dawkins, Richard, 119
debris, 46, 51, 66, 77, 247
decentralized tension, 156–157
deep earth, 92, 98, 100, 114, 125, 130, 138, 151; gases of, 83; minerals of, 81, 295n4

deep ecology, 92, 259, 260
deep history. *See* history, deep
deep space, 76, 92, 138
deep time, 92, 272–276
deep water, 82, 295n4
deforestation, 243, 244, 245, 262
deformation, 79, 80, 81, 111, 112, 186, 273, 274; in arthropods, 216, 217
degradation, 2, 233, 239, 262, 264, 267–268; of heat, 230, 235, 237. *See also* energy degradation
dendrites, 112, 113, 116, 184, 186, 190, 193, 231; of animal bodies, 206, *212*
dermal regeneration, 221–222
dermis. *See* skin
determination, 30, 31, 54, 69, 128, 286n58
dialectics of meteors, 75–78
diffusion, 114, 165, 205, 214, 237, 240, 271, 272
digestion, 165, 202, 205, 206, 207, 210
dispersal, 3, 273, 312n3
dissipation, 21, 238, 244, 257, 260, 262, 265, 311n3; vs. conservation, 231; cosmic, 230, 231, 234, 241, 272, 273; entropic, 19, 117; of matter, 21, 230, 274; metabolic, 243. *See also* energy dissipation; kinetic dissipation
dissipative patterns, 139, 179
dissipative processes, 15, 228
dissipative structures, 100, 103
dissipative systems, 99, 116, 232, 238, 243, 264
diversity, 182, 192, 196, 258, 264–267, 268, 309n19, 309n20; ecological, 239, 269. *See also* biodiversity
DNA/RNA, 18, 72, 142, 143, 145, 155, 156, 301–302n15; in biogenesis, 118–120; folding, 120–124, 127, 129, 195; multiplication of, 125–129; mutation, 125, 127, 146, 188, 198

earth processes, 5, 12, 22, 54, 228, 254

earth sciences, 8, 9, 23, 31, 98, 278n8, 279n15, 283n40
earthquakes, 5, 63, 79, 233, 234, 235, 249
earth's core, 65, 68, 74, 77, 95, 112, 131, 181; accumulated Theia's iron core, 67, 78, 79; liquid Hadean metals at, 64; moon pulls on, 87, 94
eccentricity, 7, 9, 27, 28, 42, 45, 46, 79
ecocide, 243, 244–247, 251, 253, 254, 263, 265
ecosystems, 111, 114, 130, 177, 233, 237, 250, 312n3; biologically diverse, 238; climax, 264, 309n19; developing, 309n20; dissipative, 265; energy flows, 315n18; jungle, 264
eddies, 36, 39, 100, 101, 118, 142, 219, 301n11; gravity, 52; liquid, 236; ocean water, 155; river, 85; self-organized, 301n11; spiral, 85, 193; turbulent, 102, 168; vortical, 144, 149, 193
Einstein, Albert, 23, 28
elastic motion, 53, 173, 180–182, 200, 201, 206, 210; and animals emergence, 13, 177; patterns, 14, 178
elasticity, 181, 182, 183, 201, 202, 207, 210–211; in brains, 204–205; nerve, 184–190, 203
electrochemistry, 187, 188, 189, 190; signaling, 156, 165, 184, 185, 186, 205
emergent orders, 48, 101, 105, 110, 116, 120, 123
energy, metabolic, 266
energy, potential, 229, 233, 234, 235
energy consumption, 194, 196, 204, 206, 218, 223, 247, 264; human, 2, 15, 245, 251, 258; for locomotion, 192, 201; of plants, 239
energy degradation, 177, 193, 194, 210, 215, 230, 238, 258; by compost cycles, 263, 309n19; planetary, 265; solar, 264; thermal, 205; vortices role in, 231, 236
energy dissipation, 100–101, 192, 194, 216, 222, 232, 235, 244; of atmosphere, 234, 236; humans and, 12, 256; patterns of motion and, 227, 228, 230. *See also* kinetic dissipation
energy expenditure, 223, 232, 233, 237, 243, 252, 262, 264; bodily elasticity and, 211, 219, 222; of humans, 15, 240, 247, 248, 251, 253; of plants, 245; predation and, 212; transpiration and, 309n19
energy flow, 94, 110, 117, 233, 237, 262, 315n18
energy gradients, 38, 103, 118, 154, 155, 177, 193; bodies locating, 192, 194; producing, 101, 105; reducing, 100, 144, 235, 239, 272
entropy, 65, 95, 100, 117, 237, 273, 316n27, 317n37; absolute, 296n6; cosmic, 20, 255, 286n57; cycles, 38; flow of matter and, 72, 261; heat and, 98, 151; maximizing, 253; pedetic expenditure vs., 229–231; process, 19; relations of, 286n57; reversing, 259; universe spreading is, 37, 47
environmentalism, 44, 92, 259, 266
epigenetics, 124, 125–126, 195, 216
equilibrium, 5, 21, 35, 100, 229, 230, 261, 262
erosion, 24, 69, 177, 182, 264, 280n15, 312n3
essences, 11, 26, 40, 41, 50, 124, 286n57
ethics, human, 5, 7, 92, 269
ethics of expenditure, 255, 256, 259, 267, 268, 269, 316n25
eukaryotes/prokaryotes, 116, 135, 142, 145, 154, 210, 301n15
evaporation, 92, 95, 96, 108, 149, 222, 238, 249; of global water cycle, 235; of the ocean, 97, 131; of volatiles, 107
evolution, 46, 125, 142, 193–203, 237, 258, 306n14; animal, 194–195, 201, 210; material, 193–194, 195, 214, 223, 236, 241, 268, 274; morphology in, 215, 216, 219. *See also* coevolution
exoskeletons, 216–217, 218, 219, 221

expansion, 180, 184, 186, 196, 213, 247, 258, 259; cell, 143; centrifugal, 91, 112; of the earth, 192; metabolic, 194; of motion, 53; of spacetime, 22; of sponges, 200; thermal, 264; of the universe, 19, 272. *See also* contraction
extension, 40, 163, 168, 186, 196, 215, 217; fallacy of, 139, 179
exteriorization, 65–67, 73, 76, 78
extinction, 228, 241, 245, 249, 250, 257, 261, 296n6; cycles, 38, 45; dinosaur, 28; first great, 238; human, 14, 259; Kinocene, 243–244, 255; mass, 242, 253, 256
extraction, 4, 127, 153, 159, 166, 257, 263, 265; of energy, 145, 234, 266; fossil fuel, 240; kinetic, 239; of resources, 267
extrusions, 69, 86, 105, 214, 215, 217, 218
eyes, 107, 209, 211–214

faint sun paradox, 84–85
feedback, 14, 84, 103, 215, 220, 237, 280n20; patterns, 7, 42, 45
feedback loops, 83, 161, 191, 212, 244, 249, 264; elasticity-energy expenditure, 194, 211; in evolution, 195, 241; kinetic, 210, 213; in nerve cells, 186, 187
fibers, 113, 180, 184, 198, 199, 205, 220, 222; elastic, 181, 183, 184
fields, 32, 35, 61, 72, 78, 138, 272, 275; atmospheric, 91, 94, 95; circulatory, 42, 43, 45–48, *48*, 50, 51–55; elastic, 53–55; geokinetic, 62, 135; mineral, 63, 65; Phanerozoic animal, 180; Proterozoic vegetal, 135, 140; quantum, 29, 33, 34, 39. *See also* planetary fields
filaments, 46, 144, 145, 146, 147, 156, 159, 182; actin, 183, 184; cytoskeletal, 149; fungi, 158; thalli, 157
fish, 98, 219, 221, 239, 244

flagella, 147, 197, 198, 200. *See also* choanoflagellates
flatworms, 204–205
flow of matter, 20, 21, 34, 39, 45, 78, 184, 215; atmospheric field is, 94; in brains, 209; continually changing, 35; cosmic, 12; entropy and, 72, 261; in Hadean Eon, 74; movement and, 28, 47, 65, 68, 112; pattern of motion of, 37; from the sun, 110; in theory of the earth, 19, 31; through the earth, 25
flower, 155, 164, 169–172, 313n27
flows, 2, 4, 238, 239, 283n40, 294n13, 308n18, 312n3; cosmic, 8, 12, 24; dendritic, 215, 242; electrochemical, 184, 188; of energy, 94, 109, 110, 117, 233, 237, 262, 315n18; entropic, 72, 98, 261; and flux, 22, 24, 74, 122; of gases, 47, 112; ice, 235; magma, 86; pattern, 209, 301n11; regulating, 142–143; sensory, 209. *See also* material flows
fluctuations, quantum, 33, 46, 189, 230, 296n6, 307n30; indeterminate, 21, 24, 34
fluid dynamics, 85, 86, 101, 136, 171
fluid reproduction, 100
fluidity, 79, 84, 294n13
flux, 2, 13, 15, 34, 37, 79, 96, 244; and flow, 22, 24, 74, 122; kinetic, 139; kinosemiotic, 187; of matter, 141; planetary heat, 235
fold of elements, 20, 32–35, 65, 196, 199, 200
forests, 1, 163, 239, 244, 249, 264, 265, 266
fossil fuel, 14, 228, 232, 241, 261, 265, 273, 289n82; capitalism, 15, 245, 250, 251, 256, 258, 266; classes that use, 247, 254, 257, 264
frequencies, 34, 35, 37, 110, 187, 288n72
fruiting bodies, 156, 157, *157*
fungi, 135, 148, 149, 150, 151, 177, 181, 263;

326   Index

biomass of, 244; as symbionts, 138; tension of, 139, 140, 156–158, 165, 182

Gaia, 5, 11, 19, 65, 84, 242, 279n11, 282n32; benevolence of, 260; in Greek myth, 45; required Theia, 81; theory of, 279n15, 282n33
galaxies, 29, 33, 42, 47, 48, 171, 193, 241; galactic clusters, 24, 45, 52, 53; spiral, 46
gases, 39, 44, 47, 51, 98, 113, 159, 177; affected by convection cycles, 97; atmospheric, 83, 84, 85, 92, 93, 194; atoms in, 28; clouds, 46, 47, 51, 111; complex orders of, 106; earth breathes in, 153; flows of, 112; galaxies formed from, 46; greenhouse, 1; liquids turn into, 63; in living bodies, 129; mineral crystallization, 77; pressurized, 164; as processes and patterns, 138; in soil, 24; volatile, 93, 105, 150
genome, 127, 129, 195
geoaesthetics, 107–108
geoanimality, 179
geocentrism, 23, 25, 47, 77, 80
geo-constructivism, 14
geokinetics, 11–12, 13, 20, 23, 63–65, 72, 177; acts, 170; analysis, 276; atmospheric, 91–94; core concepts of, 42; fields, 61, 62, 135; history of, 138; knots, 48; life, 182; material process, 26; mineral, 63–65; patterns, 180; phenomena of, 75, 106; subterranean, 275; tension, 140; theory of, 37, 228; waves, 239; zoology, 178–180
geolinguistics, 70–71
geological periods, 5, 7, 26, 227, 288n80
geological strata, 1, 51, 92, 182, 227, 247, 249
geology, 3, 7, 19, 25, 92, 98, 111, 293n6; hydrology immanent with, 82, 84; mechanical laws of, 4; of the present, 247, 250, 280n15
geometry, 47, 112, 113, 171, 190, 195, 283n39, 286n58; Platonic, 72
geomnemonics, 71–72
geomusicology, 110–111
geophilosophy, 7, 20, 78, 80, 285n51
geo-phyto-kinetics, 136–140
geosphere, 96
gift of death, 233, 258, 259, 263
global warming, 249, 313n27
gods, 23, 28–29, 41, 43, 80, 137, 272, 295n23; Greek, 76; Hades, 74; Uranus, 298n6
gradient reduction, 118, 233, 239, 241, 309n19, 309n20, 313n27
gradients, energy, 38, 100, 103, 105, 118, 154, 155, 177; animals locating, 192, 194; reducing, 144, 193, 235, 239, 272
granite, 84, 85
gravity, 44, 52, 149, 161, 162, 165, 189, 234; cephalopods detect, 213; pull of, 78, 79; quantum, 22, 23, 33, 230, 291n18, 296n6; of sun, 53
great chain of being, 11, 25, 64, 139, 267, 294n2
Great Red Spot of Jupiter, 37
greenhouse gases, 1, 84, 106, 264
growth, 103, 165, 166, 170, 182, 215, 249, 258; cell, 143; in ecosystems, 309n19, 309n20; geometric structures of, 195; plant, 164, 231, 238

Hadean earth, 75, 77, 78, 83, 84, 94, 105; mineral field of, 61, 63, 65; molten, 64, 69, 74, 81
Hadean Eon, 63, 65, 68, 69, 74, 75, 77
hair, 217, 218, 221
heads, 8, 192, 194, 203, 204, 205, 209; of cephalopods, 213; of rotifers, 207; of roundworms, 206
hearts, 183, 194
heat, 95–98, 102, 112, 130, 189, 235, 238, 312n3; death, 259, 296n6; degradation, 230, 235, 237;

dissipation, 234, 239, 249; earth's, 78, 82, 84; entropic, 98, 151; loss, 215, 263; mineral, 94, 193
Heisenberg, Werner, 28–29
helium, 39, 52
Heraclitus, 37
historical ontology, 7–10, 12, 25, 122, 273, 288n75
history, deep, 6, 10, 13, 15, 43, 55, 72, 241; animality and, 178, 179, 180; of the atmosphere, 106; being vs. becoming and, 108; brain evolution and, 205; climate and, 92; culture and, 8; and fallacy of extension, 139; figure vs. ground and, 107; geophilosophy and, 80; is the present, 9, 62; kinetic relationships and, 138; of material flows, 12; of matter, 137; minerality and, 63, 68, 75; moon and, 79; quantum cosmology part of, 34; theories of communication and, 93; of vegetality, 136
history, human, 2, 5, 6, 72, 139, 180, 245, 272; anthropocentrism and, 288n75; conditions of the present and, 13, 43, 55; cosmic patterns in, 54; from perspective of movement, 9, 11, 15, 62, 250; stable earth and, 242, 271
Holocene Epoch, 11, 23, 26, 43, 84, 92, 242, 244; Climatic Transition of, 279n15; glacial retreat, 5
homeostasis, 5, 7, 14
horizontal tension, 161, 163, 164
human thought, 7, 62, 109, 113, 280n17, 293n2
hurricanes, 5, 235, 236
Hutton, James, 3
Huxley, Thomas Henry, 117
hydras, 201, 202, 203
hydrogen, 39, 40, 46, 52, 82, 93, 96, 97; atoms, 81, 101, 295n4
hydrology, 2, 14, 82, 84
hydrosphere, 12, 27, 51, 96, 97, 98

hyphae, 156, 157, *157*, 158, 159, 165, 166, 182

ice, 40, 84, 111, 235, 249, 290n11; age, 5, 244, 290n11; crystals, 112–113, 114
identities, 35, 37, 40, 99, 138, 261
indeterminacy, 22, 29, 30, 31, 36, 37, 188, 230; cosmological, 41; nonhuman, 6. *See also* quantum indeterminacy
Indeterminate Epoch, 21, 230
indigenous people, 249, 250, 263, 265, 266
Inflationary Epoch, 22, 33
inner tension, 136, 146, 147, 167
insects, 150, 164, 170, 171, 216, 218, 244
instability, 7, 11, 20, 22, 38, 265. *See also* stability
intensity, 209, 210, 236, 239, 240, 261, 264, 315n18
intestine, 194, 206–207
iron, 46, 64, 76, 78, 79, 108, 130
iteration, 35, 62, 72, 99, 105, 116–117, 142, 215; of bones, 221; cosmic, 38, 272; of the earth, 34, 109; of energy, 21, 32; and folding, 121, 123; of human ideas, 80; of material elements, 48, 51; of metastable bodies, 47, 112, 118; of minerals, 100; of planetary cycles, 27; vortices and, 171

Jacob, François, 119
jellies, 198, 201, 202, *202*, 203, 204, 205, 211
junctions, 149, 182, 183
Jupiter, 27, 37, 46

Kant, Immanuel, 136–137
Kepler, Johannes, 47
Khaos, 45
kinemetrics, 12, 13
kinesthetics, 12, *13*
kinetic communism, 267, 268
kinetic conditions, 6, 21, 48, 54, 99, 142, 219, 263

kinetic cycles, 99, 101, 103, 105, 110, 115, 212, 234; multiplication and, 102
kinetic dissipation, 12, 103, 230, 242, 258
kinetic energy, 39, 217, 229, 232, 233, 235, 240, 272; degradation of, 215
kinetic expenditure, 228–229, 234, 239, 250, 253, 258, 264; ethics of, 256, 260, 269; optimal range of, 230, 241, 242, 254, 255; planetary, 214, 227, 247
kinetic history, 13, 276
kinetic morphology, 205, 258
kinetic patterns, 20, 84, 101, 190, 193, 273, 275, 276; atmosphere and, 107, 118; destruction of, 258; difference in, 136; of diffusion, 272; of earth, 51, 223, 244; elasticity and, 219; the environment, 155; four major, 51, 75, 106, 227, 311n3; gravity as, 52; human history as, 271; material-, 72; metastable, 78; minerality as, 63; of motion, 116, 146, 173, 234; of nature, 228; of organisms, 139, 142, 195, 213, 254; of singular movements, 179; temperature disrupts, 255; vegetal, 156, 164, 170
kinetic processes, 20, 24, 40, 140, 193, 221, 260, 287n68; of circulation, 96; of commoning, 267; DNA as, 122, 124; earth as, 54, 228; of fold cycles, 37; material, 34, 194, 229; minerals and, 69, 77; reproductive, 42; shift in physics to, 23; spacetime and, 68, 272; spiral, 78, 98
kinetic structures, 39, 79, 110, 117, 130, 190, 232, 275; of the atmosphere, 106; of communication, 107; of the earth, 154; of elasticity, 184, 186, 192, 214; of flow and fold, 41; of leaves, 170; of the past, 55; of sex, 171; stems as, 161; of turbulence, 28; of vegetal life, 250; of verticality, 157; of water, 142
kinetic theory, second, 29
kinetic theory of animality, 179, 180
kinetic theory of atmosphere, 98
kinetic theory of life, 118
kinetic theory of matter, 28
kinetic theory of the earth, 10–11, 12, 14, 20, 42, 267, 273
kinetic theory of the history of science, 275
kinetics, vortical, 144, 199
Kinocene, 14–15, 20, 227, 241, 242, 259, 265, 276; ethics of, 243, 260, 267, 269, 274; expenditure, 243, 255–256, 265; extinction, 243–244, 255
kinology, 13
kinomorphism, 190, 199, 202
kinomorphology, 215
kinopolitics, 13
kinospherology, 97–98
kinotopological neighborhoods, 51
knots, 48–49, 49, 50–51, 79, 122

land plants, 151, 159–160
language, human, 9, 70–71, 93
leaves, 155, 162–164, 169, 170, 186, 238, 313n27; nourishment, 151; pores, 149, 213; signaling, 150, 165
legs, 215, 216, 217, 218
Lemaître, Georges, 21
Lewontin, Richard, 119
lichen, 148, 155, 158–159, 160
life on land, 15, 135, 140, 157, 219; animal, 208, 222; insects, 216, 218; loss of, 243, 244, 247, 250; plant, 158, 159, 160
ligaments, 183, 222
light, 93, 108, 110, 111, 209, 210, 213, 214; absorption, 163, 169; attracting, 170; lunar, 107; moves through spacetime, 22; sensing, 165, 203; through a tensional network, 150; ultraviolet, 124, 125
lightning, 85–87, 138, 190, 231, 242, 256, 265; in Archean period, 91, 110; energy of, 235, 236; and global warming, 249; Hadean earth phenomenon, 75; quantity of, 234; solar rhythms and, 109

lignin, 160–161, 162
limbs, 194, 207, 208, 214–216, 222, 223, 309n20; in animal dendritic body, 212; of arthropods, 217, 218
links, 52, 53, 95, 144, 151, 158, 159, 292n2; cells, 164, 193; chain, 64, 139. *See also* tension, linked
lipids, 101, 117, 121, 141, 142
lithosphere, 12, 27
locomotion, 184, 192, 201, 202
Lorenz attractor, 35, *36*
Lucretius, 28, 258
lunar alterity, 78–81

magma, 63, 68, 69, 72, 77, 78, 83, 96; drive tectonic plates, 279n15; flows, 86; plumes, 79, 85
magnetosphere, 65, 78, 79, 80
mammals, 170, 183, 219, 222, 239, 240, 244, 250. *See also* animals, wild
mantle, 76, 77, 79, 81, 82, 96, 295n4, 296n4
Margulis, Lynn, 145
Mars, 47, 66, 77
Marx, Karl, 1, 262, 293n6
mass, 51, 52, 66, 198, 236, 239, 260, 308n16; of animal bodies, 200, 216, 237; of plants, 244; of the sun, 232. *See also* biomass; phytomass
material flows, 20, 27, 32, 33, 34, 66, 69, 150; history of, 12
material history, 54, 61, 94, 107, 109, 136
materialism, 25, 26, 29–31, 38, 75, 93, 293n2, 293n6; neo-, 288–289n81, 293n4, 305n10, 316n33
materialist theory, 29, 272–276, 289n81
materiality, 32, 64
materials, prebiotic, 101, 102, 116, 117, 118, 119–120, 137
matter, non-living, 137, 142, 255, 259, 293n2, 310n36
mechanism (theory), 4, 5, 26, 44, 71, 150, 188, 293n2; biology of, 118; kinetic history of the earth and, 13;

molecular biology and, 119
medusa, 201, 202, 203
membranes, 94, 110, 116, 117, 118, 161, 189, 207; lipid, 143. *See also* cell membranes
memory, 27, 63, 69, 72, 114, 130, 185, 306n14; archival, 102; human, 73; material, 272; mineral, 113, 136; molecular, 113
metamorphosis, 24, 86, 164, 191
metastable states, 35, 38, 94, 98, 99, 264
meteoroids, 46, 67, 68, 71, 75, 76, 80
microbes, 11, 97, 98, 111, 130, 138, 166, 177; symbionts, 200
migration, 1, 245, 247, 249, 250, 262, 263, 265
Milky Way, 28, 45, 48, 50, 244
Miller-Urey experiment, 101, 102, 121
mimesis, 48, 99, 100
mineral earth, 64, 115, 141, 158, 181, 234, 236, 249
minerality, 67, 70, 73, 75, 129, 136, 247–249, 283n39; centripetal, 65–66, 68, 69; geokinetics and, 63–65; of plants, 151
mineralization, 71, 72, 73, 79, 93, 181, 183, 196; centripetal, 81–82, 83, 84, 86, 87, 94, 95, 140; of collagen, 199; demineralization, 263; dendritic, 113; of the earth, 85; in geolinguistics, 70; in Hadean Eon, 74, 75; meteors, 76; ocean, 96; of spacetime, 65; water, 113. *See also* bio-mineralization
mineralogical expenditure, 233–234, 239
molecules, 30, 33, 39, 40, 44, 101, 113, 299n16; ATP, 236; biomolecules, 85, 103, 121, 123, 125, 141, 142, 145; DNA, 122; emergence of, 45; macromolecules, 124, 125; in vortices, 116, 117; water, 112, 118, 128, 142, 149, 154, 162
mollusks, 208, 210–211, 212

Monod, Jacques, 119
morphology, 162, 166, 192, 206, 215, 216, 219, 308n10; animal, 193, 194, 195, 196, 207, 223; kinetic, 205, 258; sponge, 199, 201
moss, 159, 168, 169
motion, 180, 209; pedetic, 28, 30, 259; planetary, 47, 227; terrestrial, 75, 106, 181; vegetal, 138, 148, 149, 151, 165, 251; vortical, 66, 101, 143, 144, 200; wave, 187, 198, 203, 204, 205. *See also* centrifugal motion; centripetal motion; elastic motion; locomotion; patterns of motion; tensional motion
mouths, 194, 198, 201, 204, 209, 210, 217, 218; elastic, 202, 207, 214; in mollusks, 211, 212; in worms, 206
movement, 2, 7, 10, 199, 292n2, 294n14, 301n14, 303n9; of animals, 316n25; bilateral, 209; cytoplasmic, 145; of dark matter, 53; dendritic, 216; earth's, 4, 5, 6, 9, 105, 106, 286n56, 291n19; eccentric, 28, 42; of flow of matter, 47, 65; of genetic matter, 131; human history and, 9, 11, 15, 62, 250; of light, 20, 22, 108; of matter, 29, 30, 44, 85, 148, 209, 215, 216; of microtubules, 302n16; of non-living matter, 255; orientation of, 10, 11, 12; pedetic, 29, 30, 86, 100, 188; periodic, 287n72; of respiration, 99, 103; symmetry, 203, 204; tectonic, 4; tensional, 53, 148, 151, 156, 162, 173; of the universe, 289n2; of volatiles, 131; of water, 200; of waves, 290n11
mucilage, 146–147, 149, 154, 155
multicellularity, 135, 147, 148, 155, 156
multiplication (biological), 102–103, 116–118, 124–129, 131, 163
muscle, 181, 183, 199, 200, 201, 202, 241, 316n25; animal, 222–223; fibers, 184, 205; folded layers of, 203

mushrooms, 157, 158
mutability, 182, 219
mutations, 39, 48, 131, 171, 215, 218–219, 283n39; coordinated, 203; of DNA, 127, 128–129, 146, 198; epigenetic, 125–126; not random, 120, 124, 194; pedetic, 308n18; rates, 125, 181, 196

natural selection, 116, 128, 236, 308n16, 311n3
naturalism, 256, 260, 268, 269
nematodes. *See* roundworms
nerves, 201, 205, 209, 220, 221, 222, 223, 240; dendritic patterns in, 194; elastic, 184–190, 203; in mollusks, 211, 213
nervous system, 184, 185, 192, 207, 214, 215, 217, 308n10; plasticity of, 187–190, 213
nests, 50–51
neurobiology, 185, 190, 306n14
neurologies, 113, 182, 187, 194, 203, 209, 212, 218; animal, 190, 192
neurons, 71, 186, 209, 210, 211, 214, 217, 221; in cephalopod arms, 212; firing of, 113, 183, 188, 189; plasticity of, 205; vibrations of, 187
neuroplasticity, 187–189, 195, 214, 215, 218, 220
Newton, 136–137
Nietzsche, Fredrich, 272
nitrogen, 86, 93, 103, 130, 182, 193, 264; in soil, 159, 177, 206; volcanos spew, 83, 96
noosphere, 51, 115
North Pacific Gyre, 37

ocean water, 143, 155, 181, 249
octopus, 211, 212, 213
orbits, 11, 28, 47, 77, 110, 249, 279n15, 292n37; celestial bodies,' 53; of debris, 66; earth's, 27, 46, 291n27; elliptical, 48, 66, 79; lunar, 50, 78, 79, 80, 94; molecular, 40; pedetic,

31; periodic, 37; planetary, 50, 51; spiral, 66
organelles, 135, 141, 142, 144, 145, 146, 159, 301–302n15
organs, sensory, 207, 208–210, 211, 213, 214, 303–304n9
orientation, 103–104, 111, 136, 203, 204, 205, 228; centered, 106, 107, 130; movement, 10, 11, 12
oscillation, 11, 35, 37, 53, 177, 181, 184, 187; bilateral, 204
osmosis, 142, 143, 144, 148, 149, 161, 165
osmotic tension, 165
ossein. *See* fibers
oxygenation, 91, 97, 153, 154, 181, 191, 192, 303n5
ozone, 85, 97, 131, 135, 154, 178, 249

particles, 35, 39, 52, 114, 124, 137, 145, 218; discrete, 229, 273; elemental folds and, 32–34; momentum of, 29; prebiotic, 119–120; protein, 127; subatomic, 40, 128
Pascal, Blaise, 257
patterns, 33, 103, 138, 179, 236, 250, 292n34, 311n3; atomic, 69; centrifugal, *14*, 94, 105, 106, 139; centripetal, *14*, 52, 74, 75; circulatory, 87; convection, 85, 110; cosmic, 54; dendritic, 113, 114, 147, 194, 209, 231, 272; of elastic motion, 178; electrochemical, 190; feedback, 7, 42, 45; flows, 209, 301n11; geokinetic, 180; metastable, 31, 93, 112, 118, 227, 261; tensional, 139, 141, 152, 154, 155. *See also* kinetic patterns; vortical patterns
patterns of motion, *13*, *14*, 54, 99, 127, 180, 231, 276; centrifugal, 94, 105, 106; centripetal, 74, 75, 116, 141; of convection cycles, 103; cosmic, 138; dendritic, 236; elastic, 192, 273; feedback, 42; geohistorical, 232; historical, 179, 241, 275; kinetic,
110, 146, 173, 234; of matter, 37; planetary, 177; self-organized, 229; shift in, 91; spiral, 272; tensional, 139, 152, 154, 155
pedesis, 31, 71, 94, 106, 123, 129, 232, 286n57; bioculture builds, 130; in bone remodeling, 221; in celestial fields, 46; and earth's core, 79; expenditure, 229–231; flows, 27–29, 51, 63, 65, 68, 112, 146; motion, 28, 30, 259; movement, 86, 100, 188; mutation, 308n18; sympedesis, 121; of weather, 114
pedosphere, 27, 28
periodicities, 110, 288n72
periods, 35–37, *38*, 292n37
petrogenesis, 67
Phanerozoic animal field, 61, 180
Phanerozoic Eon, 61, 155, 173, 177, 190, 191, 192, 208
phosphorus, 130, 159, 177, 182, 221
photons, 29, 39, 93, 97, 100
photosynthesis, 144, 145, 151, 154, 158, 161, 168, 303n5; filaments in, 147; of hyphae, 159; in kelp, 148; through stems, 160
phyllotaxis, 163
physiology, 98, 165, 240
phytality, 154–155
phytohistory, 137–140
phytomass, 244, 245, *246*, 247
Planck Epoch, 21
planetary expenditure, 243, 251, 254, 258, 265, 266, 268, 269; decline in, 245
planetary fields, 20, 45, 46, 47, 50, 54, 61, 65; dominant, 62, 63; geokinetic, 42, 135
planetary mobility, 3, 7, 243, 250
planetesimals, 66, 75, 76, 77
plantationocene, 250
plasticity, 190, 192, 205, 209, 212, 213, 220, 240; animal sense organs, 214; DNA, 126, 127; human, 264;

neuro-, 187–188, 195, 214, 215, 218, 220; osteo-, 220
pleats, 20, 32–35, 65, 196, 199, 200
pneumatological matter, 129–131
pollution, 98, 126, 251, 262
polymorphism, 202–203
pores, 70, 159, 162, 165, 200, 208, 217, 301n11; leaf, 149, 213; sensory, 209, 210, 214, 221; in skin, 222
porosity, 68, 142, 153, 163, 207, 214, 217, 222; of animal cells, 182, 209; of bones, 181, 220
predator/prey, 150, 199, 210, 211, 212, 214, 221, 237; feedback, 215; ratios, 195
process theory of earth, 11, 13, 20, 36, 38
processes, 22, 54, 82, 99, 120, 195, 241, 280n17; of accretion, 75; dissipative, 15; earth, 5, 12, 228, 254; entropic, 19, 117; gases as, 138; geological, 3, 7, 70, 85, 192; metabolic, 130, 220, 268; metastable, 29, 31, 69, 138, 264; terrestrial, 9; volcanic, 63, 96. *See also* kinetic processes
processes, material, 7, 43, 54, 75, 77, 93, 117, 267; earth as, 11, 19, 22; geokinetic, 26; memory is, 72; non-living, 253
prokaryotes/eukaryotes, 116, 135, 142, 145, 154, 210, 301n15
propulsion, 201, 203, 204
Proterozoic Eon, 141, 153, 154, 155, 173, 177, 191, 192; vegetal field of, 61, 135, 140, 152
protists, 138, 147, 155, 156, 167, 182, 197, 207; tension and, 139, 140, 173
Python, 65, 242, 260

quantum fields, 29, 33, 34, 39
quantum fluctuation, 33, 46, 189, 230, 296n6, 307n30; indeterminate, 21, 24, 34
quantum gravity, 22, 23, 33, 291n18, 296n6
quantum indeterminacy, 24, 30, 34, 36, 196n6, 259, 290n16, 296n6; emergence of cosmos from, 41. *See also* indeterminacy
Quark Epoch, 48

rabbit, *185*, 276
radiation, 53, 80, 95, 106, 107, 108, 207, 233; solar, 28, 64, 97, 110, 118, 238, 239, 279n15; ultraviolet, 97, 158
rain, 82, 87, 93, 114, 115, 138, 169, 235; in convection cycles, 95; in the Hadean Eon, 63; mineralogical, 83, 84, 85, 113; recycling process of, 239; rock, 77; of volatiles, 96, 131
randomness, 29, 30–31, 118, 121, 137, 229, 309n18
reciprocity, 47, 80, 170, 210, 228, 259, 263, 267; ethical, 274; life-death, 233, 243, 260–261
recycling, 94, 106, 221, 239, 263, 316n25
relational materialism, 29–31. *See also* materialism
replication, 99, 124, 128, 209n18
reproduction, 99, 116, 145, 161, 169, 206, 244, 258; asexual, 168, 203; and atmospheric field, 91, 95; autonomous, 125; cellular, 124, 131, 146, 147; center of orientation, 103–104, 106; and conjunction, 38; DNA in, 119, 122, 124, 125, 129, 142; fluid, 100–102; geosocial, 289n82; horizontal tension of, 161; membranes, 117; mineral, 100; multiplication, 102–103, 124; organs of, 170; resilin in, 217; seeds role in, 167, 168, 171; sexual, 99, 126, 167, 168, 169, 198, 203; of thalli, 148, 155, 156, 157; viruses role in, 125; women's, 266
reptiles, 170, 219, 221, 244
repulsion, 123, 162, 170, 171–173
resilin, 216, 217
respiration, 91, 93, 95–97, 105, 130, 177, 205, 240; biochemical, 102; biological, 96, 98; centrifugal, 106,

180; consumes volatiles, 129, 140; metabolic, 141, 239; movement of, 99, 103; planetary, 139, 153; of plants, 238; tubes, 218; vegetal, 191
rhythms, 36, 92, 96, 97, 109–110, 165, 187, 287–288n72
Rimbaud, Arthur, 78, 109
RNA. *See* DNA/RNA
roots, 149, 150, 151, 155, 161, 164–166, 238
rotifers, 204, 207
roundworms, 206–207, 211

satellites, 4, 51, 235, 247, 279n11
scales, 215, 217, 221
sea jellies. *See* jellies
seawater, 97, 143, 155, 181, 236, 249
sediment, 101, 111, 112, 114, 153, 200, 280n15; sedimentary layers, 97, 105, 247; sedimentary rocks, 39; sedimented cloud, 92
seeds, 155, 166–169, 171, 172–173, 265
sensation, 9, 186, 188, 210, 212, 214, 215, 218; in arthropods, 219; atmosphere's role in, 107, 108; bones' role in, 221; brain, 213; decentralization of, 203; mouth's role in, 211; noosphere, 115; and predation, 212; in skin, 208, 209
sessile polyp, 201, 202
setae, 217, 218. *See also* bristles
shells, 181, 184, 211, 212
Sherrington, Charles, 117
shield, magnetic, 64
signaling, 149, 150, 165, 170, 200, 205
signals, electrochemical, 156, 165, 184, 185, 186, 205
skeleton, 144, 149, 199, 205, 207, 220, 222, 301n14; cyto-, 144, 149, 301n14; exo-, 216–217, 218, 219, 221
skin, 161, 183, 206, 213, 215, 221, 223, 240; elasticity of, 184, 216, 218; pores in, 222; sensation in, 208, 209
slime molds, 144, 155
soil, 24, 25, 27, 32, 158, 161, 164, 178;

composting and, 263; erosion, 182; nitrogen in, 159, 177, 206; roots and, 165, 166
solar energy, 110, 148, 238, 264
solar expenditure, 232–233, 234, 239
solar rhythms, 92, 109–110
solar system, 23, 24, 27, 28, 44, 45, 51, 76; orbits in, 48, 50; self-degradation of, 267
solar winds, 42, 66, 97
sound, 71, 93, 108, 110, 111, 189, 209, 210
spacetime, 21, 23, 24, 32, 35, 51, 272, 290n14; in black holes, 52; dark matter and, 53; expansion of, 22; human ethics in, 92; mineralization of, 65, 75; petrogenesis of, 67–69; shaped universe, 33
spirals, 79, 85, 101, 108, 118, 144, 171, 292n2; of turbulence, 30; vortical dynamics of, 199
sponges, 197–200, *200*, 201, 202, 205, 207
spores, 157, 161, 166, 167, 168, 169
stability, 2, 23, 30, 96, 160, 182, 271, 281n29; chemical, 300n32; climate, 15; of the earth, 3; mobile, 34; planetary, 11; relative, 37, 196; unstable, 36. *See also* instability
stars, 32, 33, 44, 47, 53, 107, 178, 289n1; constellations, 276; dark matter pulls, 46; exploding, 45; fallen, 136, 151; kinetic materialism, 26; megastars, 52; order, 111; systems, 28
stasis, 2, 3–4, 5, 7, 11, 21, 34, 271; in Big Bang theory, 24. *See also* homeostasis
stems, 160–162
stomata, 160, 162, 163–164
storms, 33, 83, 101, 193, 258; systems, 85, 100, 193, 249; thunder, 31, 171, 235, 239
strands, 146, 147, 154, 156, 157, 182, 183, 198; cytoplasmic, 141, 144, 147, 149, 159, 199; RNA, 122, 123

strata, 69, 85, 131, 182, 192, 242, 279n15; geological, 1, 51, 92, 227, 247, 249
structures, 13, 45, 63, 71, 207, 215, 279n15, 301n14; dendritic, 190, *195*; dissipative, 100, 103; human, 6, 9; neurological, 187; periodic, 101, 110; reproductive, 105; rigid, 222; self-organized, 142; tensional, 144, 155, 157, 160, 161, 162, 164, 182; vascular, 185. *See also* kinetic structures
sunlight, 108, 125, 147, 148, 153, 154, 193, 232; in old-growth forests, 239; stems direct, 165; stored as carbohydrates, 168, 177; stored underground, 151; transformed into energy, 236
surface area, 194, 199, 216, 217, 218, 219, 220, 222; vegetal, 158, 161, 162, 163
symbionts, 125, 138, 144, 145, 146, 160, 200
symbiosis, 138, 140, 178, 182, 185, 195, 200, 263; in lichen, 158–159; mycorrhizal, 164, 165; and organelles, 145–146; in protists, 198
sympedesis, 121
sympoiesis, 12, 54, 121
symthanatosis, 181–182
synapses, 72, 188, 189

tails, 192, 197, 198, 203–204, 205, 206, 207, 216
taste, 170, 189, 209, 213
technology, 4, 46, 55, 251, 265, 275, 284n45, 294n16; capitalism and, 247; energy and, 229; is natural, 264; noosphere and, 115; studies of, 8; transportation, 1
temperature, 36, 38, 63, 70, 77, 84, 110, 218; at earth's core, 68, 78; fluctuations, 237; global, 4, 249, 250, 255, 264, 265, 313n27; of Greenland, 290n11; at ground level, 168; measurable, 22; minerals and, 295n4; of the ocean, 108; plants regulate, 244; of quantum fluctuations, 21; rising, 281n21; sensitivity to, 222; and species diversity, 309n19; thermosphere changed, 97
tendons, 181, 183, 184, 222
tension, 141–146, 147, 148, 149, 150, 164, 187; capillary, 162, 163; gravitational, 53; linked, 52, 155, 156, 166, 168, 171, 186, 192; terrestrial, 159–160; transcellular, 140, 151–152, 198; tubular, 155, 157; vascular, 161; vertical, 160; with the world, 167–169
tensional attraction, 170–171
tensional cellularity, 140, 141–145, 146–147, 148–153, 162, 180
tensional motion, 13, *14*, 52, 53, 140–141, 154, 161, 164; among living organisms, 135; in cells, 144, 146, 148; created organelles, 145; in vegetality, 139
tensional repulsion, 171–173
terrestrialization, 20, 21, 29, 34, 38
thallus/thalli, 146, 147–148, 154–160, 161, 164, 169, 185
Theia, 66–67, 77, 78, 79, 81, 83, 86
theory of matter, 25, 26, 28, 54
thermodynamics, 206, 251, 262, 267, 273, 296n6, 311n3, 316n26; in conduction, 154; energy flow and, 21; plant growth is, 238; in replication, 99; second law of, 309n19; variation and, 309n18
thermosphere, 97
tissues, 147, 151, 183, 194, 205, 216, 223, 309n20; bone, 221; elastic, 181, 201, 202; muscle, 184, 203; subcutaneous, *185*; vascular, 160; vegetal, 155
tongues, 210–211
topology, 86, 102, 111, 112, 118, 127, 181, 286n57; DNA/RNA, 123; of flowers, 170; hydrokinetic, 219; mineral, 130,

140; of nerve cells, 185; relational, 94; of the universe, 33, 231
tornados, 235, 236
touch, 209, 211, 213, 218, 222
transcellular strands, rigid, 198
transformation, 93, 168, 169, 199, 207, 208, 210, 273; chemical, 103; elastic, 198, 214, 220; energy, 39, 311–312n3; epigenetic, 195; evolutionary, 196; external, 124–125; geological, 24; internal, 126–129, 201; kinetic, 193, 203; material, 105, 187, 192, 193; mutual, 51, 87; planetary, 153; self-, 34, 108
transpiration, 149, 160, 161, 164, 165, 219, 238, 309n19; of plants, 168, 169
trees, 161, 166, 238, 243, 254, 257, 260, 309n19; dendritic patterns in, 113, 231; energy expenditure of, 240, 253, 262; water use of, 162, 164, 165
trophic level, 238, 239, 244, 250, 253
tubes, 68, 157, 160, 161, 182, 183, 214, 218; capillary, 162, 163; cellular, 149, 155, 156; elastic, 206–207
turbulence, 28, 29, 30, 38, 45, 71, 82, 118
turgor pressure, 158, 164, 165

uncertainty, 13, 28, 29, 38
uniformitarianism, 4, 14, 280n18
uniformity, 5, 23, 33
unity, 37, 45, 171, 206, 291n23

Valéry, Paul, 109, 171
vapor, 77, 85, 86, 93, 94, 95, 96, 112; water, 83, 239
vaporization, 77, 91, 92, 94, 95, 96, 151
vegetal bodies, 149, 151, 154, 155, 160, 168, 192, 196; thallic and vascular structure of, 185
vegetal expenditure, 236–237, 239
vegetal field, 135, 140. *See also* planetary fields
vegetality, 155, 160, 162, 164, 181, 186, 192; age of, 173, 177; tensional, 136, 137, 139, 140, 141, 149, 150, 151
vertebrates, 208, 213, 215, 221, 222, 239, 310n33; elasticity in, 214, 219–220, 223
verticality, 154, 157, 160, 161, 163, 182
vibrations, 11, 72, 100, 120, 129, 144, 210, 211; of the earth, 39; in language, 93; neuronal, 187; organs, 205; patterns of, 33; of sound, 110
viruses, 124, 125, 127, 129, 137, 195, 214
vision, 213
vitalism, 5, 12, 14, 64, 84, 137, 286n57, 296n6
volatiles, 78, 96, 97, 107, 130, 131, 150; earth's, 81, 93; respiration of, 105, 129, 140
volcanism, 64, 79, 81, 193, 235, 244, 249, 280n15; clouds are product of, 86; contributes aerosols, 234; deep-sea volcanic vents, 118, 140; as earth process, 5; as ecological issue, 92; effect on plants, 138; irregularities of, 109; patterns of, 68; processes of, 63; produced protective atmosphere, 84, 85; releases water, 296–297n4; respiration of, 83, 96; subsided by Archean Eon, 91, 106, 233; turbulence of, 71; underground effects of, 297n8; volcanic islands, 115; water a product of, 82
volume, 22, 40, 150, 166, 196, 199, 220, 235
vortex. *See* vortices
vortical patterns, 102, 103, 117, 140, 144, 146, 231, 236; of clouds, 115; cytoplasm in, 143, 145; smoke in, 106
vortices, 117, 147, 171, 207, 274, 292n2, 302n16, 310n27; of air, 168; circulation of, 199–201; clouds produce, 112; in cytoplasm, 143, 187; dandelion, *172*; as eddies, 301n11; electromagnetic, 193; fluid, 100, 198, 199; jellyfish, *202*; as

material kinetic forms, 193; plumes create, 102; role in degradation, 235, 236; sponge, *200*; transport, 184; turbulence has, 30; used in dissipation, 231; vortical life, 116–118; in water, 142, 205

water, ocean, 143, 155, 181, 249
water cycles, 34, 82, 115, 193, 235, 236, 249
water molecules, 101, 112, 118, 128, 142, 149
waves, 83, 101, 107, 154, 156, 169, 200, 222; action, 186–187, 188; cosmic, 39; diffractive, 118; electrical, 209; electromagnetic, 84; geokinetic, 239; gravitational, 33; kinetic, 40; motions of, 198, 203, 204, 205; movement of, 290n11; nerve, 222; pressure, 93, 110; radio, 43, 249; of rippling crust, 78
weather, 76, 97, 115, 135, 140, 244, 249, 290n11; formation, 114; meteoritical, 77
Weiss, Paul Alfred, 119
Western tradition, 3, 12, 25, 257, 271
whirlpools, 34, 37, 85, 100, 107, 117, 230, 292n2
Woese, Carl, 117
worms, 150, 184, 201, 204–205, 206–207, 209, 310n33

xylem, 160, 162, 163, 207

Lightning Source UK Ltd.
Milton Keynes UK
UKHW010157251121
394567UK00004B/262

9 781503 627550